TO COMPUTE NUMERICALLY

Little, Brown Computer Systems Series

Gerald M. Weinberg, *Editor*

Basso, David T., and Ronald D. Schwartz
Programming with FORTRAN/WATFOR/WATFIV

Chattergy, Rahul, and Udo W. Pooch
Top-down, Modular Programming in FORTRAN with WATFIV

Coats, R.B., and A. Parkin
Computer Models in the Social Sciences

Conway, Richard, and David Gries
An Introduction to Programming: A Structured Approach Using PL/1 and PL/C

Conway, Richard, and David Gries
Primer on Structured Programming: Using PL/1, PL/C, and PL/CT

Conway, Richard, David Gries, and E. Carl Zimmerman
A Primer on Pascal, Second Edition

Cripps, Martin
An Introduction to Computer Hardware

Easley, Grady M.
Primer for Small Systems Management

Finkenaur, Robert G.
COBOL for Students: A Programming Primer

Freedman, Daniel P., and Gerald M. Weinberg
Handbook of Walkthroughs, Inspections, and Technical Reviews: Evaluating Programs, Projects, and Products

Graybeal, Wayne, and Udo W. Pooch
Simulation: Principles and Methods

Greenfield, S.E.
The Architecture of Microcomputers

Greenwood, Frank
Profitable Small Business Computing

Healy, Martin, and David Hebditch
The Microcomputer in On-Line Systems: Small Computers in Terminal-Based Systems and Distributed Processing Networks

Lias, Edward J.
Future Mind: The Microcomputer — New Medium, New Mental Environment

Lines, M. Vardell, and Boeing Computer Services Company
Minicomputer Systems

Mills, Harlan D.
Software Productivity

Monro, Donald M.
Basic BASIC: An Introduction to Programming

Mosteller, William S.
Systems Programmer's Problem Solver

Nahigian, J. Victor, and William S. Hodges
Computer Games for Businesses, Schools, and Homes

Nahigian, J. Victor, and William S. Hodges
Computer Games for Business, School, and Home for TRS-80
Level II BASIC

Orwig, Gary W., and William S. Hodges
The Computer Tutor: Learning Activities for Homes and Schools

Parikh, Girish
Techniques of Program and System Maintenance

Parkin, Andrew
Data Processing Management

Parkin, Andrew
Systems Analysis

Pizer, Stephen M., with Victor L. Wallace
To Compute Numerically: Concepts and Strategies

Pooch, Udo W., William H. Greene, and Gary G. Moss
Telecommunications and Networking

Reingold, Edward M., and Wilfred J. Hansen
Data Structures

Savitch, Walter J.
Abstract Machines and Grammars

Shneiderman, Ben
Software Psychology: Human Factors in Computer and
Information Systems

Walker, Henry M.
Problems for Computer Solutions Using FORTRAN

Walker, Henry M.
Problems for Computer Solutions Using BASIC

Weinberg, Gerald M.
Rethinking Systems Analysis and Design

Weinberg, Gerald M.
Understanding the Professional Programmer

**Weinberg, Gerald M., Stephen E. Wright, Richard Kauffman, and
Martin A. Goetz**
High Level COBOL Programming

TO COMPUTE NUMERICALLY

Concepts and Strategies

Stephen M. Pizer

UNIVERSITY OF NORTH CAROLINA
AT CHAPEL HILL

WITH

Victor L. Wallace

UNIVERSITY OF KANSAS

LITTLE, BROWN AND COMPANY
Boston • Toronto

Library of Congress Cataloging in Publication Data

Pizer, Stephen M.
  To compute numerically.

  Includes bibliographies and index.
  1. Numerical analysis — Data processing.
I. Wallace, Victor L. II. Title.
QA297.P563 1983 519.4 82–14874
ISBN 0–316–70940–9

Library of Congress Catalog Card No. 82–14874

ISBN 0–316–70940–9

9 8 7 6 5 4 3 2 1

ALP

Published simultaneously in Canada
by Little, Brown & Company (Canada) Limited

Printed in the United States of America

Example of instability on pp. 12–13 adapted by permission of the publisher from G. E.
Forsythe, "Pitfalls in Computation, or Why a Math Book Isn't Enough," *American Mathematical Monthly*, 77(9): 949–951, November 1970.

To three inspiring teachers:

Eben T. Colby
George W. Morgan
Frederick P. Brooks, Jr.

— S.M.P.

To my wife, Mary

— V.L.W.

Foreword

There are many ways to learn about numerical computing, and some are much more pleasant than others. One of the least pleasant was the way I began. In 1953, I consumed a substantial fraction of my working hours inverting 10 by 10 matrices with the aid of a Friden, some paper, a pencil, and many erasers. Even though I was paid 90 cents an hour, I didn't find it a pleasant way to learn.

I added to my learning a few years later by writing programs to compute the orbits of artificial satellites. When I found that the satellites didn't go where our programs said they should go, the pressure to study numerical computing increased without bounds. This pressure didn't make the learning very pleasant, either.

By 1960, I had learned a lot about numerical computing — or so I thought — so I undertook to teach courses in the subject. The few textbooks that were available at that time were all written by mathematicians, which made the subject particularly unpleasant for both me and my students. We were computer scientists, or software engineers, and the mathematicians simply were not addressing our needs.

By the time Pizer and Wallace's manuscript fell into my hands, I had pretty much given up hope that learning about numerical computing could be pleasant. As a series editor, I receive about one numerical analysis manuscript each month, which is even less fun than using the Friden. Imagine my surprise, then, when I found myself actually *enjoying* Pizer and Wallace. And learning!

Now I suppose there are some instructors whose learning about numerical computing was as unpleasant as mine. And I suppose that some of them believe that the next generation has to suffer as much as we did. But I think there are many more who would just as soon minimize the suffering and maximize the learning — and that is why I decided to put *To Compute Numerically* in the Little, Brown Computer Systems Series.

Here is why I think learning with Pizer and Wallace should be pleasant and efficient:

1. Pizer and Wallace approach the subject as computer scientists, rather than as mathematicians. Although their mathematics is impeccable, it remains subservient to the true goal of the text: educating computer scientists, programmers, and program designers about what to expect and what to do when computing numerically. Undoubtedly, students of mathematics and engineering can also benefit immensely from this approach.

2. Pizer and Wallace have engineered the text and supporting materials for classroom use. They have prepared a solutions manual for teachers. As a result, *To Compute Numerically* is not only a pleasure to learn from, it is a pleasure to *teach* from. And, if teaching is pleasant, learning will be even more pleasant.

3. Not only is *To Compute Numerically* well engineered, it is well implemented. The authors have done their homework, and done it well. The reader and the teacher can concentrate on their own jobs, rather than completing the jobs left undone by the authors.

4. The authors write clearly. They choose clear and relevant examples. They have tested their materials with real students, and they have incorporated the results of that testing in revisions of the text. As a result, *To Compute Numerically* has the maturity of a fifth or sixth edition combined with the freshness and timeliness of a first edition.

As an old-timer, I suppose I should recommend that people start learning about numerical computing the same way I did. But after all these years, you'll probably have trouble finding a Friden in working condition. If so, I think the next best thing is to use Pizer and Wallace's *To Compute Numerically*.

Gerald M. Weinberg

Preface

Based on many years of teaching numerical computing as a core course in computer science, we asked ourselves "What is it that we really wish computer scientists to know about numerical computing?" Our answer was not "how to apply numerical methods." Although this knowledge is desirable for a computer scientist, it is not necessarily more desirable than for a physicist or an engineer who will use computers. Our answer was not "how to prove mathematical theorems or use other mathematical skills," although knowing how to do this is also beneficial. Our answer was that computer scientists need to understand how a computer is used numerically, what affects its performance of this task, and how this performance is evaluated, so that they can properly design, analyze, or model computer software or hardware to be used by programmers or users with numerical applications. Thus we set out to provide an introduction to numerical computing that emphasizes its basic concepts, approaches, and analysis methods. We noted that such an introduction seems attractive also for others who wish to develop more than a cookbook appreciation of numerical methods.

To produce such an introduction we had to discover what we believed to be the basic ideas of numerical computing and its performance evaluation. They are as follows:

1. When used numerically, computers approximate mathematical objects such as real numbers, vectors, functions, and operators by other objects more natural to the finite, discrete computer. The need to evaluate these approximations leads to concepts of accuracy and how it is measured. The need to produce good approximations leads to basic computing strategies.
2. Repetitive processes are especially well suited to computers. The fact that such repetition can cause large amounts of computer time to be used leads to concepts of efficiency, how it is measured, and how it is optimized.

Therefore the objective of this book is to present the basic strategies of numerical computing, the concepts of computer representation of mathematical objects, and the concepts of accuracy and efficiency and of ways of measuring them, for both simple and repetitive processes. We also want the student to obtain familiarity with some basic numerical methods which illustrate these ideas and which are useful in their own right.

If one concludes that this is what should be taught in a core course in numerical computing, one is immediately struck by the fact that the chapter headings of almost all books on numerical computing have to do not with these matters but with problems to be solved, for example, the solution of linear equations, approximation, and the solution of differential equations. From this, one can conclude that their objective is to teach solutions to these problems rather than the aforementioned concepts. We set out to develop an organization of the material in numerical computing for which the chapter headings were the basic concepts we felt needed to be taught. While doing this we were acutely aware that we had to avoid the trap of producing a course which was pure theory poorly supported by concrete application of the concepts. Both motivation for the concepts and illustration of their applicability depend on some understanding of the kind of problems for which numerical solution by computer is desired and of the kind of approaches that are used to solve these problems.

The result was an organization in which we began with a survey of problems to be solved numerically and of the simplest methods for their solution. Using these to motivate the basic concepts, we found that for each concept of accuracy, efficiency, or computing strategy which we discussed, improvements in many of these simple solution methods followed from a given concept. Thus, the methods of choice are seen to be examples of the application of the concepts which we were covering. This not only allowed us to focus on the concepts, but it made teaching the methods much more efficient since many methods were examples of the same concept. The result was that a course which had been the "ogre" of the department in its intellectual difficulty and time requirements became a normally challenging and normally time-consuming course covering the same material.

Every instructor of numerical methods faces the problem that there are far too many methods for solving problems of interest to cover all the methods of choice in a semester or even a year. So he or she must choose among methods, leaving out ones that are arguably as important as the ones covered. We believe that not only does our organization have advantages for a core course for computer scientists, but it also has advantages for the teaching of numerical methods in that once the basic concepts are understood many more methods are seen to be examples of the concepts than the ones covered in this book. The student can then be directed to the literature to read about these methods with understanding as he needs them. Thus, this book is desirable not only for computer scientists but also for people interested in the application of numerical computing. The approach taken does require a certain amount of mathematical sophistication, so the book is not appropriate for sophomore level numerical methods courses, but it is appropriate for senior level courses for people with the amount of mathematical training common in programs in the mathematical sciences, the natural sciences, and engineering.

In particular, to use this book the student should have two or three semesters of calculus and a semester of linear algebra as well as knowledge of programming in a higher level language. In calculus the student should be familiar with functions, limits, continuity, differentiation, integration; the Rolle, Taylor, and mean value theorems and their multivariable forms; and point sets, sequences, and series and the related notions of limit and limit point, convergence, and the Taylor and binomial series. In linear algebra the student should be familiar with vectors and matrices and their determinant, transpose, and inverse, as well as their relation to the representation and solution of systems of linear equations; vector spaces, basis, dimension, inner product, and orthogonality; and linear transformations including their definition, matrix representation, rank, and null space. It is also useful but not necessary for the student to be familiar with vector norms. Eigentheory is not required for the use of this book.

This book can also be used in courses where students have previously taken a course in numerical methods. In this case Chapter 3 can be treated as a review. A previous course in numerical methods will strengthen the student's appreciation of problems in numerical computing and ease this still somewhat challenging course. In fact, this book can be used as a leisurely two-term introduction to numerical computing by covering Chapters 1–3 in the first term and Chapters 4–10 in the second. In this case one might use an introductory book on numerical methods to supplement the first-term readings, and in the first term one might cover some of the aforementioned mathematical prerequisites to Chapters 4–10. For mathematically prepared students who will take only one course in numerical computing, however, a course covering this whole book is entirely tractable and, we believe, is preferable to one in numerical methods.

All the material in this book has been taught in a one-semester course at the University of North Carolina at Chapel Hill and at the University of Kansas. The course organization has been class tested by five instructors in eleven different offerings of the course, which is required at the University of North Carolina for senior undergraduates majoring in computer science or applied mathematics and for graduate students in computer science. It has been taken as an elective by undergraduate mathematics majors and by graduate students in many different sciences. Teaching the course with the organization of this book has resulted in striking improvement of the understanding of basic concepts, by all students as measured on course examinations and by computer science graduate students as measured by comprehensive examinations in later years, as compared to students who had taken courses organized in the classical way.

All programs in this book are presented in Pascal, because of the cleanness and popularity of this programming language and the good program structure it encourages. With a warning that FORTRAN programming can be dangerous to a program's health, they are also presented in FORTRAN in Appendix C. These programs are intended as a complete specification of the algorithms being discussed. Thus, although they are directly usable and tested subprograms, they were written with clarity more than use in mind and thus may not in all cases be fully general or optimally efficient and do not include every bell and whistle of mathematical software, such as input validation.

It is not intended that every student using the text be fluent in Pascal, and familiarity with FORTRAN, ALGOL, BASIC, PL/I, or any similar programming language should allow the students to read the Pascal programs, since only a basic subset is used. Nevertheless for the aid of those unfamiliar with Pascal, a summary of the program structures of Pascal that are used in the programs in the book is provided in Appendix B. We suggest that it is beneficial for the student to write a few programs while covering Chapters 1–3, but these can be in any programming language with which the student is familiar.

A few further comments on organization will help the reader of this book. Equations are numbered by chapter, with the numbers of the most important equations in boldface. Appendix A summarizes the notational conventions and usages in the book. Finally, the organization of the book into chapters is discussed in the last section of Chapter 1 (Section 1.5), after the basic ideas to be covered are presented.

We wish to acknowledge gratefully the work of Miss Kathy Yount, Mrs. Kristine Brown, and Ms. Marcia Hunt for the typing of the difficult manuscript; of Dr. Gyula Magó, Dr. Richard Hetherington, Dr. Martin Hanna, Dr. John Hedstrom, and Dr. William Meyers for class testing the manuscript and suggesting improvements; of Ms. Sandra Bloomberg for preparing and testing the programs; of Mr. Michael Pique for helping to produce the figures; of Mr. Jon Cohen, Ms. Rose Motley, Dr. Lee Nackman, and Dr. Gerald Weinberg for useful advice; of Ms. Susan Warne and Mr. Bert Zelman for their careful and cooperative editing; and of the many UNC students in COMP 151 who found errors, gave constructive criticism, and put up with using the manuscript as a text. Special thanks go to Lyn Pizer for long hours of proofreading and her continuous support.

Stephen M. Pizer
Victor L. Wallace

Contents

Table of Numerical Methods

Function Evaluation

General, including polynomial evaluation: *pages* 1, 19–22, 27–29, 32–34, 139, 148–149, 174–178, 183–184, 188–191, 201–202, 204, 211–225, 231–233; *problems* 2.1–2.4, 6.2–6.4, 6.9, 6.12, 7.5, 7.8, 10.3
Least-squares approximation
  General: *pages* 51–58, 135–137, 169, 313; *problems* 3.8–3.16, 4.10–4.12
  Fourier approximation: *pages* 34, 55–59, 90–91, 98–99, 312–313, 318; *problems* 3.13–3.15, 3.17, 10.2
  Polynomial least-squares approximation: *pages* 52–53, 55; *problem* 12.2
Polynomial exact-matching
  General: *pages* 149–155, 168–169, 181, 312, 318; *problems* 3.2, 3.20, 5.2, 5.15, 7.6, 7.27
  Lagrange approximation: *pages* 29–32, 35, 41, 92; *problems* 3.3–3.4, 5.2
  Newton divided-difference method: *pages* 35–41, 58, 93–95, 159–160, 181–182, 288; *problems* 3.3–3.4, 6.7, 6.16
  Equal-interval methods
    General (Gauss forward/backward approximation): *pages* 41–45, 153, 155–159, 178–181, 244–247; *problems* 3.5, 3.7, 3.17, 5.3–5.8
    Bessel approximation: *pages* 244–247; *problems* 7.21–7.24
    Stirling approximation: *pages* 244–247; *problems* 7.21–7.25
Rational approximation: *pages* 34, 312; *problem* 3.1
Splines
  General: *pages* 46, 313, 318

Numerical Integration

Numerical Differentiation

Solution of Differential Equations

Modified-Euler predictor: *pages* 22–23, 69–70, 161–165, 272, 276–278
Partial differential equations: *pages* 13, 65, 318
Runge-Kutta method: *pages* 236–239, 315; *problems* 7.14–7.15, 7.19
Stiff differential equations: *pages* 284, 315, 318
Systems of differential equations: *page* 318

Solution of Nonlinear Equations

General: *pages* 1–2, 9, 11–12, 19–20, 58, 65, 71–72, 82, 139, 264, 288;
    problems 3.20, 9.7
Aitken's δ^2-acceleration: *pages* 247–249, 299, 315–316; *problems* 7.26,
    9.9
Bisection method: *pages* 72–73, 104
Hybrid methods: *pages* 81–82, 108–109, 315
Method of constant slope: *page* 298; *problem* 9.2
Method of false position: *pages* 73–76, 82, 105, 108–109, 264,
    315–316; *problems* 3.28, 7.26, 9.3, 10.6
Newton's method: *pages* 3–4, 13–14, 76–81, 89, 106, 127–128, 137,
    264, 288, 294, 297–299, 302–303, 315, 318; *problems* 3.23,
    3.26–3.28, 9.1, 9.6, 9.14, 9.16
Picard iteration: *pages* 79–80, 89, 247, 249, 288–299; *problem* 9.15
Secant method: *pages* 80–82, 89, 107–108, 297, 300–303, 315–316,
    318; *problems* 3.23, 3.28, 9.10, 9.11

Solution of Linear Equations

General: *pages* 1, 15, 20, 29, 71, 82–83, 184–186, 199, 202, 288;
    problems 6.8, 6.13–6.14
Conjugate gradients method: *page* 317
Gaussian elimination: *pages* 9–11, 15, 83–85, 88, 109–110, 191–199,
    204–205, 233, 264, 288, 316–317; *problems* 3.29–3.31,
    6.10–6.11, 6.13
Gauss-Seidel method: *pages* 87–88, 113–114, 204–205, 233, 316–317;
    problem 3.29
Jacobi method: *pages* 85–89, 111–112, 288, 316
Kaczmarz projection method: *pages* 316–317

TO COMPUTE NUMERICALLY

CHAPTER ONE

Computing Numerically

1.1 Numerical Applications

Numerical computing involves solving problems which produce numerical answers given numerical inputs. Such problems arise whenever mathematics is used to model the real world. Examples of numerical problems are

1. the evaluation of functions at given arguments, for instance, the evaluation of sin(0.241);
2. the evaluation of the derivative of a function at a given argument, a definite integral of a function, or some other operator applied to a function;
3. the determination of maxima and minima of a function;
4. the solution of equations, linear or nonlinear, that is, the determination of values for unknowns which satisfy the equations, often called the roots;
5. the solution of differential equations, that is, the determination of the value, at a specified argument, of a function which is given implicitly by a differential equation.

Many of these problems can be solved analytically. For example, the quadratic equation

$$ax^2 + bx + c = 0 \qquad [1]$$

is solved by the formula

$$x = \frac{-b \pm \sqrt{b^2 - 4ac}}{2a} \qquad [2]$$

Numerical methods to solve these problems consist of evaluating the solution formula. This evaluation will require some computation but normally not much.

On the other hand, many integrable functions are not analytically integrable, many nonlinear equations have no closed form solution, etc. For these, numerical methods based on approximation can be developed. However, they produce only roughly correct results and are often much more time consuming than methods involving evaluation of analytically developed formulas. Therefore, when an analytic solution is available for any part of a problem, it is to be preferred on the grounds of accuracy and also usually on the grounds of efficiency.

1.2   Numerical Methods: Properties and Strategies

1.2.1   APPROXIMATION, FINITENESS, AND DISCRETENESS

In almost all cases numerical methods involve approximation. For example, integrating a function which is not analytically integrable is accomplished by approximating it by one which is integrable and integrating the approximation. Similarly, the roots of an equation with no closed form solution are found approximately by finding successively better approximations whose roots are determined analytically until values are found for the unknowns which satisfy the original equation to an adequate tolerance.

A second source of inaccuracy in numerical methods is the finiteness and discreteness of the digital computer. It is possible to represent only a finite number of real numbers with complete accuracy in a computer because of the finite number of digits which are allowed in a number in the computer. It is not possible to evaluate an infinite series but only a finite number of terms of that series. In fact, any iteration must be terminated after a finite number of steps. It is not possible to specify a function at all arguments on a continuous interval but rather only at a discrete number of points in the interval. Movement in infinitesimal steps along a function is not possible.

The need for approximation to allow solution of problems combined with the finiteness and discreteness of a digital computer make numerical methods inherently inaccurate. Much of this text will be concerned with developing methods which minimize the inaccuracy of the result of numerical computing and with analyzing methods so as to predict the amount of inaccuracy in any result.

1.2.2   COMMON COMPUTING STRATEGIES

When we wish to apply an operator, such as integration or root finding, to a function to which it is not analytically applicable, the standard strategy is to approximate the function by one to which the operator is analytically applicable and then to apply the operator to the approximation. This strategy, illustrated twice above, is called *approximate and operate*.

The most common approximating function is a straight line. The reason is that approximation by lines is straightforward and many of the operators we wish to apply, including differentiation, integration, and root finding, are easily applied analytically to lines. For example, to integrate the function illustrated in Figure 1–1 from $x = a$ to $x = b$, we can approximate the function by the straight line through the points $(a, f(a))$ and $(b, f(b))$ and then calculate the integral of the straight line approximation, that is, the area of the trapezoid, which is given by $[(b - a)/2] [f(a) + f(b)]$. As another example, to solve the equation $f(x) = 0$, that is, to find the value of x for which the graph of f crosses the abscissa (the root), as illustrated in Figure 1–2, we may estimate the root from the graph, approximate the function by a line tangent to it at the root estimate, and produce an improved root estimate by finding where the line crosses the axis — a problem which can be solved analytically.

Because functional approximation gives only approximate answers, it is often desirable to improve these answers still further by making another approximation which is better than the first because it occurs nearer the correct answer. With the more accurate answer thus produced, we can reapproximate the function and apply the operator of interest,

FIGURE 1–1
Integration of $\int_a^b f(x)\,dx$ by Straight-Line Approximation

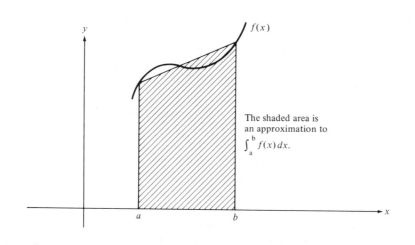

The shaded area is an approximation to $\int_a^b f(x)\,dx.$

FIGURE 1–2
Solution of $f(x) = 0$ by Straight-Line Approximation

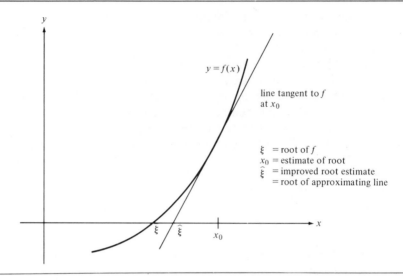

producing an even more accurate answer. Thus we produce a sequence of answers x_0, x_1, x_2, ... which we intend to converge to the correct answer. This approach of repeatedly applying some procedure, given an estimate of a desired value, to produce a better estimate is called *iteration*. Though in theory the iterative step can be repeated infinitely, actually the iteration must be stopped at some point to produce a still approximate answer.

Another form of repetitive application of a procedure is also common in numerical computing. A common example is in the solution of differential equations. If we have a differential equation which gives the value of a function $y(x)$ at a single argument x_0 and the derivative of $y(x)$ at any point (x, y) in the plane, we may be interested in the value of y at some argument $x_{goal} \neq x_0$ (see Figure 1–3). In calculus we would move from $y(x_0)$ in the direction of the derivative $y'(x)$ for an infinitesimal distance in x, from this new point we would move in the direction of the derivative at the new point for an additional infinitesimal distance in x, etc., perhaps tracing out the function illustrated in Figure 1–3. But in numerical computing we cannot move for an infinitesimal distance in x; we must move in the direction of the derivative for a finite distance in x. If we move too far, however, we will poorly approximate the function $y(x)$ which was the solution to the differential equation. This will not only produce an inaccurate answer for the step in question but will also result in using an inaccurate derivative value in the next step, since the function giving the derivative depends on y, that is, the result of the previous step. To keep these inaccuracies small, moving from x_0 to x_{goal} we take many small steps, x_0 to x_1, x_1 to x_2, etc., as illustrated in Figure 1–3. Each step involves the same procedure, moving in the direction of the derivative for a fixed finite interval in x. This procedure is repeated until we arrive at the argument x_{goal}.

The procedure repetition just discussed differs from the repetition discussed with iteration in that after each step of the repetition with the differential equation there is a different correct answer, $y(x_i)$. Though each step gives an answer which is used to produce the next and ultimately to produce the final answer, each answer is not an approximation to the final answer but to some intermediate answer. This form of repetition we will call *recurrent,* as opposed to the iterative form of repetition in which the answer at the end of any given step is an estimate of the final answer and has no correct answer other than the limit of the sequence.

Recurrent and iterative repetition are very common in numerical computing. In this book each is analyzed in detail in its own chapter. Here, however, we note that when any procedure is repeated many times to achieve a final answer, as in either of these forms, the computation can be very time consuming, even if a single application of the procedure is fast. Thus the repetitive application of procedures leads us to a concern for the time efficiency of numerical methods.

FIGURE 1–3
Recurrent Approximate Solution to a Differential Equation

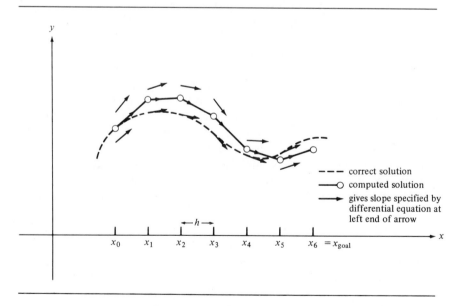

1.3 Grounds for Choice of Numerical Methods

We have already discussed the two major criteria by which numerical methods are chosen: accuracy and time efficiency. Two other criteria are storage efficiency and ease of programming.

Accuracy is measured by the error in a result, which is defined as the difference between the computed result and the desired result. If x is the desired result, x^* is the computed result, and ε_x is the error in x^*, then

$$\varepsilon_x \overset{\Delta}{=} x^* - x \qquad [3]$$

This error is often called the absolute error, to distinguish it from the relative error, defined as ε_x/x. Relative error is often more useful than absolute error since an absolute error of 10^{-5} in an answer of 1000.51, which corresponds to a relative error of about 10^{-8}, is much less important than the same absolute error in an answer of 2.01036×10^{-5}, which corresponds to a relative error of about 50%. The accuracy which we can achieve will depend on the quality of our input data as well as the quality of the numerical method. In many cases we will be trying to obtain relative errors in the range 10^{-5} to 10^{-10}, though in many cases we will be talking about relative errors on the order of 1%. Relative errors of more than 10% will usually be considered large, though what is a large error depends on the problem.

Time efficiency is measured by the amount of computer time taken to execute the numerical method. Although many methods will require only a few milliseconds or less on a modern computer, we will encounter complex problems and large problems which require minutes and, if done improperly, days, years, or centuries.

With the increase in the size of computers, minimization of the amount of storage required by a method has become less of a concern than it was in the past. However, large problems often tax the memories of even modern large computers, and methods which minimize storage are sometimes of concern. It is often the case that storage and time can be traded. For example, a value can be recomputed instead of being stored. Thus a weighted average (arithmetic or geometric) of time efficiency and storage efficiency might be the most appropriate measure of overall efficiency. In this book we will make little attempt to optimize storage efficiency, and when we refer to efficiency we will mean time efficiency.

The time of the programmer is clearly also an important issue. However, since programs for many numerical methods, once written, will be used again and again, modestly large programming time is of negligible concern compared to time efficiency for execution and accuracy. More important is that the program be well structured so that it can be easily understood and modified. In this book we will attempt to provide such programs, but we will expend little attention on the ease of programming of algorithms.

The overall requirements of method quality and the relative importance of the various criteria for quality vary with how the method will be used. Clearly, with a method that will be used once with one input, ease of programming may have considerable importance, time and space efficiency may have little importance, and the accuracy should be optimized only for the particular input in question. In contrast, a subroutine that will be used again and again will need to have its accuracy optimized across the whole range of inputs for which it might be used, and both time and space efficiency would have great importance, but programming time would not be important. In general, we will see that it is desirable before designing a numerical method to specify the conditions of its use. This will include not only the amount of use and milieu of its use but also what will change between one use and another. For example, we must know the variables that are likely to change significantly from use to use and the ones that are not, the requirements for accuracy and efficiency and whether these are likely to change from use to use, the range of inputs possible, and the range of input errors which can be expected.

1.4 Calculating and Its Failures

It might be assumed that with modern computers, with their speed, large storage, and relatively great precision of floating-point numbers, concern for accuracy and efficiency might be misplaced. One might expect that adequate accuracy and efficiency can almost always be obtained with little attention. Unfortunately this is not the case. In the following we will give a few examples of the kinds of problems that can arise.

1.4.1 FLOATING-POINT ARITHMETIC

Because of the need for finite representation, digital computers can not even do arithmetic accurately. We must understand how they do arithmetic. We will be concerned with floating-point arithmetic, as for almost all the calculations in which we will be interested this is the most appropriate modality, for it provides a far wider range of values than does fixed-point (integer) representation.

A floating-point number consists of two parts, the fraction and the exponent. Associated with the floating-point number is a base, B, commonly 2, 8, 10, or 16. The fraction† has a fixed number, n, of base-B

---

† In a base-B fraction the kth digit after the radix point is the multiplier of B^{-k} and the digits are chosen from 0, 1, ..., $B-1$ (if $B = 16$, the digits representing 10 through 15 are written as the letters A through F).

digits, and the magnitude of the fraction is strictly less than 1. Usually the fraction is normalized so that a nonzero fraction has magnitude greater than or equal to B^{-1}, that is, has a nonzero digit in the first place to the right of the radix point. The number represented by the fraction f and the exponent exp is $f \times B^{exp}$. For example, if $B = 10$ and $n = 6$, $f = -0.163589$ and $exp = 3$ represents the number $-0.163589 \times 10^3 = -163.589$.

Addition of floating-point numbers occurs as follows. A zero is appended at the low order end of each operand, making it $n + 1$ digits long. This extra digit is called the guard digit; its importance will be discussed in Chapter 5. The operand with the smaller exponent is shifted right the number of positions which is the difference between the larger exponent and its exponent. In the shifting, all digits beyond the $n + 1$th are lost. The shifted operand is added to the other operand, producing a result with $n + 2$ digits: one digit, possibly 0, to the left of the radix point, and $n + 1$ digits to the right. The exponent of the result is that of the operand with the larger exponent. If the result is nonzero, it is normalized by shifting the fraction right or left so that the digit to the left of the radix point is zero and that to the right of the radix point is nonzero, and modifying the exponent accordingly. The fraction is then truncated to n digits. As an example, take the case with $B = 16$ and $n = 6$ of the addition of $0.10003A \times 16^2$ and -0.789012×16^{-2}, where the fractions are given in hexadecimal. A zero is appended to the right of each fraction, and the second operand, having the smaller exponent, is shifted four places to the right and truncated to 7 hexadecimal places, producing

$$0.10003A0 \times 16^2 \quad \text{and} \quad -0.0000789 \times 16^2$$

Adding these two hexadecimal fractions produces $0.0FFFC17 \times 16^2$. Normalization of this fraction is accomplished by shifting left one position and subtracting 1 from the exponent, giving the final result of $0.FFFC17 \times 16^1$.

Another example in the same computer is the addition of $0.3209C1 \times 16^3$ and $0.FE135B \times 16^4$. Here the former operand has the smaller exponent; so after appending zeros and shifting, the problem becomes

$$0.03\,209C1 \times 16^4$$
$$+\ 0.FE135B0 \times 16^4$$

the addition of which produces $1.0133F71 \times 16^4$. Normalization and truncation of this result produces $0.10133F \times 16^5$.

Subtraction is done by addition with the reversal of the sign of one of the operands. Multiplication of floating-point numbers is quite straightforward. The fraction parts are multiplied together to full precision ($2n$ digits), and the exponents of the operands are added. The double precision product is then normalized and truncated to n digits.

Division is also straightforward. If both the divisor and the dividend are normalized and nonzero, the magnitude of the quotient of the fractions will be greater than $0.1/1 = 0.1$ and less than $1/0.1 = 10$. Therefore, $n + 1$ digits of a nonzero quotient, beginning with the place to the left of the radix point, will have at least n significant digits. Thus, division of the fractions to these $n + 1$ digits and subtraction of the exponent of the divisor from that of the dividend produces an $n+1$-digit result which can be normalized and truncated to produce the n-digit answer with exponent.

Note that all arithmetic operations can generate error by truncation of results beyond the nth digit to the right of the radix point of the fraction. In an expression consisting of more than one arithmetic operation, the fraction in the result of each operation is truncated to n digits before succeeding operations are begun.

1.4.2 SENSITIVITY

Small input or arithmetic truncation errors can produce large output errors. In this case the intervening arithmetic operations are said to *propagate* error badly and the whole procedure is said to be *sensitive*. As an example, consider the following problem assuming a computer with decimal floating-point arithmetic with a 3-digit fraction. Though this fraction size is small, it is chosen so that the student can easily check all the arithmetic. The effects demonstrated can also be demonstrated with a fraction of more normal size.

Assume we wish to find the root of smaller magnitude of the quadratic passing through the points (1, 102), (1.5, 153.25), and (2, 205). This quadratic is

$$y = x^2 + 100x + 1 \qquad [4]$$

whose root of smaller magnitude is approximately equal to -0.0100. We will assume that we do not know this equation or its root.

The straightforward approach which occurs to us is as follows. We know

$$y(x) = a_2 x^2 + a_1 x + a_0 \qquad [5]$$

If we plug into this equation each of the three given (x, y) values, we will produce three linear equations in the three unknowns a_0, a_1, and a_2. We will solve these equations and then use the quadratic formula

$$x = \frac{-a_1 + \sqrt{a_1^2 - 4a_2 a_0}}{2a_2} \qquad [6]$$

to find the desired root.

We will first put the three equations in matrix form. The result is

$$\begin{bmatrix} 1 & 1 & 1 \\ 2.25 & 1.5 & 1 \\ 4 & 2 & 1 \end{bmatrix} \begin{bmatrix} a_2 \\ a_1 \\ a_0 \end{bmatrix} = \begin{bmatrix} 102 \\ 153 \\ 205 \end{bmatrix} \tag{7}$$

Note that because the computer can only represent three digits 153.25 has been represented as 153, generating error in the fourth significant digit.

We will solve the foregoing matrix equation by a method called Gaussian elimination. In this method we multiply the second and third equations by appropriate constants to make their first coefficients the same as the first equation. Thus, we wish to multiply the second row by $1/2.25$ and the third equation by $1/4 = 0.250$. The multiplier $1/2.25$ is calculated as 0.444 with a representation error in the fourth decimal place. In the multiplications, truncations in multiplying 0.444 times 153 and 0.250 times 205 produce errors in the fourth significant digit. The result is

$$\begin{bmatrix} 1 & 1 & 1 \\ 1 & 0.666 & 0.444 \\ 1 & 0.5 & 0.25 \end{bmatrix} \begin{bmatrix} a_2 \\ a_1 \\ a_0 \end{bmatrix} = \begin{bmatrix} 102 \\ 67.9 \\ 51.2 \end{bmatrix} \tag{8}$$

After this multiplication, the second and third equations are subtracted from the first equation to produce two equations in the second two unknowns, a_1 and a_0:

$$\begin{bmatrix} 0.334 & 0.556 \\ 0.5 & 0.75 \end{bmatrix} \begin{bmatrix} a_1 \\ a_0 \end{bmatrix} = \begin{bmatrix} 34.1 \\ 50.8 \end{bmatrix} \tag{9}$$

In this case no errors are made in the subtraction.

With these two equations we now multiply the second by an appropriate constant, $0.334/0.5$, to make the second equation have the same first coefficient as the first equation. The multiplier $0.334/0.5$ is calculated as 0.668 with no error. In the multiplication of 0.668×0.75 and 0.668×50.8, errors due to truncation are made in the fourth significant digit, and the result is

$$\begin{bmatrix} 0.334 & 0.556 \\ 0.334 & 0.501 \end{bmatrix} \begin{bmatrix} a_1 \\ a_0 \end{bmatrix} = \begin{bmatrix} 34.1 \\ 33.9 \end{bmatrix} \tag{10}$$

Subtracting these two equations produces

$$0.055a_0 = 0.2 \tag{11}$$

where no error was introduced in the subtraction.

From equation 11 we compute a_0 as $0.2/0.055 = 3.63$, where a truncation in the fourth significant digit has been made in this division.

Note that the correct answer for a_0 is 1, so we have 263% error in this coefficient, even though relatively few errors were introduced by truncation in the arithmetic operations and all were less than 1%. It is such sensitivity which we must avoid and which is one of the reasons for care in designing numerical methods.

With the value of $a_0 = 3.63$, we can solve for a_1 using the first equation in equation 10. From this equation we produce $0.334a_1 = 34.1 - 0.556(3.63)$. In our computer $0.556(3.63) = 2.01$, with an error in the fourth significant digit, and subtracting 2.01 from 34.1 produces 32.0, again with an error in the fourth significant digit due to truncation occurring in the subtraction process. Thus $a_1 = 32.0/0.334 = 95.8$, with an error introduced in the division in the fourth significant digit. Since a_1 should be 100, the error in a_1 is 4.2%, not as dramatic as that in a_0, yet larger than the introduced errors of less than 1%.

The value of a_2 is calculated from the first equation of matrix equation 7, using the already computed values of $a_0 = 3.63$ and $a_1 = 95.8$. The result is $a_2 = 102 - 95.8 - 3.63$. Assuming that the operations occur from left to right, we compute $102 - 95.8 = 6.2$, introducing no error, and $6.2 - 3.63 = 2.57$, again introducing no error. Note that if we had by chance written our program so that we subtracted 3.63 from 102 and 95.8 from the result, there would have been an error introduced in subtracting $102 - 3.63 = 98.4$ and the final result would have been 2.60 rather than 2.57. Even without this error, however, a_2, which should have the value 1, has 157% error in it.

According to our computation the quadratic equation whose smaller root we want is

$$2.57x^2 + 95.8x + 3.63 \qquad [12]$$

Its correct root is -0.0379, which is off from the correct value of -0.0100 by 279%. But let us do the calculation to compute this root, assuming that even the square root can be done accurately to three significant digits. Plugging the coefficients into the quadratic formula (equation 6) produces the root as

$$\frac{-95.8 + \sqrt{(95.8)^2 - 4(2.57)(3.63)}}{2(2.57)} = \frac{-95.8 + \sqrt{9170 - 37.0}}{5.14}$$

$$= \frac{-95.8 + \sqrt{9130}}{5.14}$$

$$= \frac{-95.8 + 95.5}{5.14}$$

$$= \frac{-0.3}{5.14}$$

$$= -0.0583$$

Errors due to truncation were made in the fourth significant digit in the multiplications of 4 by 2.57, the result by 3.63, and 95.8 by itself, in the subtraction of 37.0 from 9170, in the square root, and in the division of -0.3 by 5.14. The result, -0.0583, has an error of 54% from the correct root, -0.0379, of the incorrect quadratic equation which we solved and an error of 483% from the correct result, -0.0100, the root of the correct equation.

Had we solved the correct quadratic, $x^2 + 100x + 1$, the root would have been calculated as

$$\frac{-100 + \sqrt{(100)^2 - 4(1)(1)}}{2(1)}$$

which after the multiplication and subtraction under the radical are carried out, each to three digits accuracy, would produce $(-100 + \sqrt{10000})/2 = 0$. Even this result has 100% error.

We must understand why such large errors were produced when only a few small errors were introduced in the process of computation. We must then develop methods which avoid such problems. So as not to prolong the suspense for many chapters, let us note that in each case where a large error was made it was produced when two numbers of approximately the same value, each correct to three significant digits, were subtracted from one another, producing, if you will, zeros in high-order places. For example, in the solution of equation 12, assuming the equation is correct as given, 95.8 with no error was subtracted from 95.5 with error in the fourth significant digit to produce -00.3. Thus, after normalization, errors which were in the fourth significant digit became errors in the second significant digit. When these numbers were then used in further calculations, the errors were propagated into the high-order decimal digits, resulting in hundreds-of-percent error.

It is clear that we will have to avoid the subtraction of approximately equal numbers. We will see that there are other sensitive operations which must be avoided.

1.4.3 INSTABILITY

Another problem which sometimes arises is called instability. This problem arises in recurrent repetition of a procedure. The problem is that a small error made at any step of the procedure grows very quickly as the procedure is repeated. Forsythe (1970) gives the following example. Consider the problem of determining the temperature, as a function of time, of a rod of length 1 which is initially at temperature 0 and beginning at time 0 is kept at temperature 0 at one end and at temperature 1 at the other end. It is clear that the temperature at the center should begin at 0 and with time should monotonically approach 1/2. The problem is solved numerically by dividing the length of the rod into a number of

discrete intervals and time into a number of discrete intervals and recurrently marching along the grid in solution of the differential equation
— the heat equation — which describes the process of conduction of
heat. The derivatives in the heat equation are approximated by quotients
of simple differences. With time intervals of 0.01 and space intervals of
0.1, we find that on a decimal floating-point computer with an 8-digit
fraction the computed solution after 0.15 time units is a temperature of
132276; after 0.99 time units it is 1.02×10^{44}; and after 1 time unit it is
-2.96×10^{44}. These clearly ridiculous results, oscillating from one time
interval to the next with wild excursions, are due to the behavior called
instability. The surprising result is that when the time interval is halved
to 0.005 the behavior alluded to does not occur when using precisely the
same algorithm: at time 1 the computed temperature at the center of the
rod is 0.49997. It is clear that we must learn to choose our numerical
methods and discretization intervals so that instability does not occur;
therefore we must understand under what circumstances such instability
occurs.

1.4.4   DIVERGENCE

Iterative methods are designed to take an estimate of a value and repeatedly apply an improvement procedure until the estimate is appropriately accurate. The difficulty is that the improvement procedure sometimes is not satisfactory and the sequence of results fails to converge.
In some cases the sequence diverges to $\pm\infty$; in other cases the sequence
oscillates or wanders about, never converging. For example, consider
the problem of finding the root marked ξ of the function graphed in
Figure 1–4. A common method, sketched in Section 1.2.2, is Newton's
method, which involves fitting a tangent line to the function at the most
recent estimate and having the next estimate be the root of that tangent
line. This method is usually very successful. Note how well it works
with the initial estimate x_0 illustrated in Figure 1–4a. But consider what
happens if the initial estimate x_0 is as illustrated in Figure 1–4b. Then
the sequence of root estimates, the zeros of the tangent lines in Figure
1–4b, will diverge to $+\infty$. Alternatively, consider the case when x_0 is
as illustrated in Figure 1–4c. In this case the sequence of iterates oscillates about the argument where the valley occurs.

It is clear that if we are to use iterative methods we have to be able
to determine in what situations they converge and choose for any given
problem a method that will produce convergence.

1.4.5   INEFFICIENCY

A common source of inefficiency is repetitive methods where very many
repetitions are required. For example, consider the solution of a differential equation with an initial value y_0 at x_0, where we must compute

FIGURE 1–4

Convergence and Lack of Convergence with Newton's Method. *Key*: ξ = desired root; x_0 = initial root estimate; x_1, x_2, ... = successive root estimates.

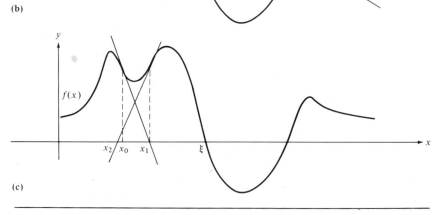

$y(x_{goal})$. If stability considerations require a very tiny step in x to be taken for each repetition of the procedure, inordinately many steps may be required to arrive at x_{goal}. Similarly, with iterative methods, there are convergent methods that converge so slowly that they require hundreds of iterations to achieve desired accuracy even from a reasonable initial estimate.

Another source of inefficiency is a combinatorial algorithm for the solution of a relatively large problem. For example, consider the problem of solving 26 linear equations in 26 unknowns by Cramer's rule, a method many learned in high school, which involves the calculation of 27 determinants. The method that we learned to evaluate determinants was evaluation by minors. It can be shown that this method for solving these linear equations will require approximately $1.7(27!) \approx 2 \times 10^{28}$ multiplications. On a computer which could do 10 multiplications per microsecond, these multiplications alone would require 100 billion centuries.

In comparison consider the Gaussian elimination algorithm which we used above. It can be shown that this method will require under 7000 multiplications, quite an improvement. Attention to efficiency in many problems of even moderate magnitude is sometimes well justified.

1.5 Summary and Complements

We have seen that errors are generated in numerical computing by functional approximation and by the finiteness and discreteness of the digital computer. We have seen that if we are not careful the small errors thus produced in a computation can result in very large errors in our final result. This is especially true because many methods involve repeating a procedure many times, providing many computations to generate these errors and propagate errors from previous computations.

The number of computations made are not only of concern because of the errors that they generate and propagate but also because of the amount of computer time that they can use. It is therefore of concern to design and use numerical methods that are both accurate and efficient. To do this, we must understand where the inaccuracies and inefficiencies arise and how to analyze numerical methods in terms of the accuracy of their results and the efficiency of their execution.

We will see that though there exist innumerable numerical methods, each useful for a special class of problems, a relatively small number of computing strategies are used to develop numerical methods and a relatively small number of analysis techniques are necessary to deal with the collection of numerical methods. Therefore, this book is organized to develop these computing strategies and analysis techniques. In this presentation the student will learn the common numerical methods because they are used for motivation of these strategies and techniques and illustration of the application of these strategies and techniques. But, more important, he or she will learn to understand, develop, and analyze any numerical method.

This book is organized by approaches to development and analysis of numerical methods. In Chapter 2 we tackle the basic concepts of efficiency. In Chapter 3 we consider many of the common numerical problems and a few of the basic methods for their solution, in the process gaining an understanding of most of the basic computing strategies. In Chapter 4 we generalize from these problems and the methods thus developed, and in Chapters 5 through 9 we develop analysis techniques and computing strategies that follow from these generalizations. Chapters 5 through 7 are devoted to the analysis of accuracy — first of error generation by simple operators, then of error propagation by simple operators, and finally of error generation and propagation by compound operators. Chapters 8 and 9 are devoted to the analysis of both the accuracy and efficiency of repeated operators. Finally, Chapter 10 summarizes the analysis techniques and based on such analysis makes conclusions on numerical methods of choice.

REFERENCES

Forsythe, G. E., "Pitfalls in Computation, or Why a Math Book Isn't Enough." *Amer. Math. Monthly,* **77**(9):931–955, November 1970.

PROBLEMS

1.1 The following exact numbers are given in a decimal floating-point computer with a two-digit normalized fraction:

$$a = 9.9 \qquad b = 9.8 \qquad c = 0.098$$

Perform the following operations, assuming a single guard digit is maintained during each arithmetic operation but all results including intermediate results are truncated to two digits. Assume all expressions are evaluated left to right. Give the generated absolute and relative error in each case. (Use ε_x/x as the definition of relative error.)

(a) $a + b$ (d) $a - c$ (g) $a \cdot b$
(b) $a - b$ (e) $a - b - c$ (h) $a \cdot c$
(c) $a + c$ (f) $a - (b + c)$ (i) a/b

CHAPTER TWO

Efficiency in Numerical Computing

Besides accuracy, which will be discussed in Chapters 5 through 9, numerical algorithms are evaluated on the bases of time required for the execution of the algorithm, storage required, and ease of programming. We will say little about programming and storage efficiency. Rather, we will concentrate on time efficiency and accuracy. In this chapter we will develop measures of efficiency and methods to optimize efficiency.

2.1 Measures of Efficiency

The efficiency of a numerical method is strictly speaking measured by the amount of computer time required to execute the algorithm. However, this measure is dependent on the particular computer involved, and a more general measure is required. Although parallel computers are coming into use, the largest amount of numerical computing is still done on sequential Von Neumann computers, and we will assume that a sequential computer is being used throughout this book. Since in numerical methods for sequential computers the floating-point arithmetic operations usually dominate all other operations, for example, those for initializing, incrementing, and testing index variables, it is customary to measure efficiency by the number of floating-point operations of each type required by the method. Since in most computers addition and subtraction take approximately equal time and also multiplication and division take approximately equal time, it is common to count operations in two categories: additions and subtractions (A/S), and multiplications and divisions (M/D). In minicomputers and older computers addition and subtraction are done more quickly than multiplication and division, but in modern large computers these operations all require approximately the same amount of time, so we will often measure efficiency simply by the total number of floating-point arithmetic operations required.

It is often the case that a numerical method is dominated by evaluations of functions. For example, in solving a nonlinear equation $f(x) = 0$, many methods operate by making a guess at the root, x_0, and applying some operator to x_0 to produce an improved estimate x_1. This improved estimate is then used as input to the improvement operator, producing x_2, and so on. For most methods of this form, in each step of the iteration the work is dominated by an evaluation of the function f. In other repetitive methods the work is dominated by the evaluation of common functions such as sin and log. When the work is dominated by function evaluations, it is common to evaluate efficiency simply by counting the number of them required by the method. Note that each function evaluation requires not only the execution of a subroutine but also the overhead of calling and returning from a subroutine.

In order to evaluate the efficiency of a method, we must have a prescription for the method, that is, an algorithm, not just an equation. For example, consider the evaluation of $1/(1 + x^4) + 1/(1 - x^4)$. The most straightforward algorithm involves multiplying x times x times x times x, which requires three multiplications; adding 1 and dividing the result into 1, which requires two more operations; multiplying x times x times x times x, which requires three multiplications; subtracting the result from 1 and dividing the result into 1, which requires two more operations; and adding the two fractions, requiring a final operation. This method requires 11 operations. Clearly the three operations of computing x^4 for the second time can be saved by storing the originally computed value of x^4, with the result that only eight operations are required. A little cleverness allows us to see that x^4 can be computed by multiplying x times x and then multiplying the result by itself, requiring two operations instead of three, so the overall result requires only seven operations. We can do even better by doing a little algebra to note that the original expression can be rewritten as $2/(1 - x^8)$; x^8 requires only the multiplication of x^4 by itself, that is, requires three operations when starting with x, so in this form the value desired can be computed in five operations. It is thus seen that one cannot answer how much time is required to calculate any given expression but only how much time is required to execute a particular algorithm or how many operations are the minimal number required to evaluate the given expression. In Section 2.2 we will further discuss methods for optimizing the efficiency of the evaluation of an expression.

We will see that many numerical methods operate by subdividing the problem into a number of steps, each the same, and repeating that step, often after a certain amount of setup. Thus, for example, methods for the solution of nonlinear equations are obtained by iterating the improvement operator discussed briefly above, and methods for solving initial value differential equations to compute $y(x_{goal})$ given $y(x_0)$ often operate by subdividing $x_{goal} - x_0$ into n parts, $[x_0, x_1]$, $[x_1, x_2]$, ..., $[x_{n-1}, x_n]$, where $x_n = x_{goal}$, applying an operator to compute $y(x_1)$ from $y(x_0)$, reapplying the operator to compute $y(x_2)$ from $y(x_1)$, etc., until $y(x_n)$ is computed from $y(x_{n-1})$. One other example is methods involving a series, where a value is computed by successively evaluating higher-order terms and adding them to the previous result. Often the evaluation of a term to be added involves a fixed number of arithmetic operations given the term previously added or components of it. For example, e^x for small x can be computed by evaluating a truncated form of its Taylor series, and the nth order term, $x^n/n!$, can be computed as x/n times the $n-1$th order term.

In methods involving the repetitive application of an operator after setup, the total time required by the method is given by

$$t = t_{setup} + (\text{number of steps}) \times t_{step} \qquad [1]$$

The number of operations for setup, t_{setup}, and the number of operations per step, t_{step}, depend on precisely what algorithm is used for setup and for each step. If these are nonrepetitive, t_{setup} and t_{step} can be determined by counting, as illustrated above. If setup or each step involves repetition, these can be analyzed using equation 1, with this process repeated until a nonrepetitive operator is arrived at. The calculation of the number of steps required for repetitive methods is not always straightforward, and it will be put off until other chapters, especially Chapters 8 and 9. Here let us note that the number of steps required will depend not only on the algorithm used but also often on the input parameters of the problem being solved.

Most algorithms are applicable to problems of various sizes. Their full specification involves a parameter which describes the size of the problem, and the amount of time required by the method depends on this parameter. For example, in repetitive methods, the number of repetitions is such a parameter, and in methods involving n-vectors, that is, arrays of n real numbers, or matrices, the amount of time required depends on the size of the vectors or matrices. Thus, we may wish to evaluate the efficiency of a method to solve linear equations, but its efficiency must be measured with the number of equations and unknowns as a parameter. Similarly, we may have an algorithm for evaluating a polynomial, but the degree of the polynomial and thus the number of coefficients must be a parameter. Let the parameter in question be denoted by n.

We wish to know what function of n describes the number of operations required (or the number of A/S and the number of M/D) for a problem of size n. Since efficiency is of greatest concern when n is large, we often ignore the behavior of the algorithm for small n and ask about its behavior for large n. For large n a few operations are negligible compared to the total number of operations required, so we often wish to know simply how fast the number of operations grows as n increases. We thus define the order notation: we say $f(n)$ is of order $g(n)$ and write

$$f(n) = O(g(n)) \tag{2}$$

if $f(n)$ grows no faster than $g(n)$ as n gets large. More precisely $f(n) = O(g(n))$ if there exist positive constants M and n_0 such that for $n \geq n_0$,

$$|f(n)| \leq M|g(n)| \tag{3}$$

or, more succinctly, if

$$\lim_{n \to \infty} \sup \left| \frac{f(n)}{g(n)} \right| < \infty \tag{4}$$

Thus,

$$2n^4 + 30n^3 - n^2 + 1 = O(n^4) \tag{5}$$
$$2n^4 + 30n^3 - n^2 + 1 = O(2n^4) \tag{6}$$

and

$$2n^4 + 30n^3 - n^2 + 1 = O(n^5) \tag{7}$$

but

$$2n^4 + 30n^3 - n^2 + 1 \neq O(n^3 \log n) \tag{8}$$

Most often we will use O in the sense described by equation 5, that is, giving the high-order term of the function describing the amount of work required in terms of n, but strictly Knuth (1976) has specified that the notation $\Theta(g(n))$ should be used for this purpose.

Thus we will say one method is an $O(n^3)$ method and another is an $O(n^2)$ method, and we will prefer the latter. Similarly, we will prefer an $O(\log(n))$ method to an $O(n)$ method. Note that two methods both $O(n^2)$ may be very different in efficiency, in that the factors multiplying n^2 in the two methods may be very different. Also, two $O(n^2)$ methods with the same factor may differ strikingly because one of them has a much larger factor multiplying n than the other. Finally, though knowing that two methods have different high-order terms allows us to distinguish between them for large n, it may be that the one that is better for large n is not as good for small n. For example, a method that requires $100n^2$ operations is preferable to one that requires n^3 operations for large n, but for $n < 100$ the method requiring n^3 operations is preferable.

As a concrete example consider the evaluation of the polynomial

$$p(x) = \sum_{j=0}^{n} a_j x^j \tag{9}$$

at an evaluation argument $x = x_0$. The most obvious algorithm, evaluating each term separately, requires j multiplications to evaluate the jth term, so it requires $n(n + 1)/2$ multiplications and n additions. Thus, this method requires $O(n^2)$ operations.

One general strategy for improving the efficiency of the computation is to reuse already computed values. In the evaluation of polynomials this can be done by evaluating the terms in order of increasing j and computing $x^{j+1} = x \cdot x^j$, where x^j has been computed at the previous step. With this improvement the evaluation requires $2n - 1$ multiplications and n additions. Thus we have a method which requires $O(n)$ operations, fewer than the $O(n^2)$ operations required by the previous method.

An even more efficient algorithm, and in fact the most efficient way to evaluate polynomials on a sequential computer, is obtained by writing the polynomial in the nested form:

$$p(x_0) = (\cdots ((a_n x_0 + a_{n-1})x_0 + a_{n-2})x_0 + \cdots)x_0 + a_0 \qquad [10]$$

Thus the algorithm which should normally be used to evaluate a polynomial is

```
PX  := A[N];
FOR I  := N − 1 DOWNTO 0
    PX := PX * X0 + A[I]
```

This algorithm, sometimes called *synthetic division* (or *Horner's method*), requires only n multiplications and n additions. Like the method discussed in the previous paragraph, this method requires $O(n)$ operations, but the constant in front of the n is 2 instead of 3.

2.2 Tactical Improvement of Efficiency

Strategic optimization of efficiency basically involves designing numerical methods that require a small number of repetitions, that is, finding iterative methods which converge quickly or recurrent methods where the number of steps into which the problem is divided is small. Here we will discuss the tactical improvement of efficiency that is obtainable by careful programming of a given numerical method. Basically this optimization involves organizing the computation so as to allow the storage and reuse of intermediate results.

First, anything inside a loop that can be computed once and for all should be put into the setup before the loop. For example, one method for solving a differential equation involves the repeated step

$$y_{i+1} = y_{i-1} + 2hf(x_i, y_i) \qquad [11]$$

The interval width h is known at the beginning of the process, and therefore the multiplication of 2 by h should be computed in the setup before the loop so that each step requires only one multiplication and one addition beyond the evaluation of the function f.

In cases of nested loops some computations may be removable not into the setup for the outer loop but outside of some of the inner loops. The less that is done in inner loops, the more efficient will be the computation.

The next category of calculation and reuse of values involves using at one step what was computed in a previous step. Of particular importance is the reuse of values obtained by function calls. For example, one method for the solution of differential equations involves recurring the pair of steps

$$y_{i+1}^{(p)} = y_{i-1} + 2hf(x_i, y_i) \qquad\qquad [12]$$

$$y_{i+1} = y_i + \frac{h}{2}\left(f(x_i, y_i) + f(x_{i+1}, y_{i+1}^{(p)})\right)$$

The value of $f(x_i, y_i)$ needed in computing y_{i+1} should not be reevaluated, but rather the value computed in calculating $y_{i+1}^{(p)}$ should be used. Similarly, if $\sin(x) + 1/\sin^2(x)$ is to be calculated, $\sin(x)$ should be evaluated only once and the value should be multiplied by itself to produce the denominator of the second term in the expression.

Another method for avoiding computation is achieved by factoring of expressions, which produces nesting of the type discussed with polynomial evaluation. In particular, polynomials should normally be evaluated using the nested form. This rule should be applied not only for polynomials in a simple variable but also for polynomials in arbitrary functions. For example, to evaluate $(x^2 + 2xy + y^2)e^{2x}e^{2y} + 2(x + y)\,e^x e^y + 5$, the expression should be rewritten as $[(x + y)e^{x+y}]^2 + 2[(x + y)e^{x+y}] + 5$, a polynomial in $(x + y)e^{x+y}$. The function $g(x, y) = (x + y)e^{x+y}$ should be evaluated once and for all, and then the polynomial in g, namely, $g^2 + 2g + 5$, should be evaluated using the nested form $(g + 2)g + 5$.

Other algebraic manipulation can often save arithmetic operations. For example, the manipulation shown in Section 2.1 of $1/(1 - x^4) + 1/(1 + x^4)$ into $2/(1 - x^8)$ was shown to save two additions and a division in exchange for one multiplication.

The last method we will mention for saving computation is recognizing symmetries. In the evaluation of an even function such as $f(x) = x^2 \cos(x) + 3$ at the integers between -10 and $+10$ one should evaluate only at the nonnegative integers and use the already computed value of $f(x)$ at a positive integer for the value at the corresponding negative integer. Similarly, odd functions allow such a reuse of values. It is not possible to specify all symmetries that can be recognized or in fact all ways subexpressions can be reused, but the programmer should be on the lookout for such cases.

2.3 Summary and Complements

It might be supposed that with the speed of modern computers concern for efficiency is not needed. This is not true for a number of reasons:

First, some of the most straightforward algorithms are so inefficient that even for small problems fast computers require a very long time. In this case improved algorithms are necessary, although this falls into the category of strategic rather than tactical improvement. The evaluation of a determinant by expansion by minors is an example of such an inefficient algorithm.

Second, it is often the case that numerical methods must be used to solve very large problems. Modern science often produces linear equations in hundreds of unknowns or experimental results consisting of thousands of numbers or models involving hundreds of simultaneous differential equations. The limitation on the complexity of the scientific problems which can be solved is often the efficiency of the computation involved in solving these problems.

Third, many numerical methods are written in the form of subroutines which will be used many times by many users. Though the time saved in one application of the subroutine may be negligible, over millions of applications the timesaving will not be negligible at all.

At the tactical level, improvement in efficiency can be very important when it is in a deeply nested inner loop. Admittedly, there are times when the effort to find a small improvement is not worth the computer time saved. Furthermore, such an improvement is often not worth the decrease in program readability that it may cause. Finally, some of the improvements can be left to optimizing compilers and if done explicitly may even thwart the compiler's optimization. However, in this book we are trying to develop the habit of producing efficient programs, though from time to time we will comment that a particular improvement is not necessarily desirable for one or more of the reasons indicated above.

REFERENCES

Knuth, D. E., "Big Omicron and Big Omega and Big Theta." *SIGACT News,* **8**(2):18–24, ACM, 1976.

PROBLEMS

2.1 Give an algorithm to efficiently evaluate and tabulate the function

$$f(x) = \left(\frac{1}{1 + 4x^2} - 1 + \frac{100}{101} \frac{|x|}{5} \right) \text{sign}(x)$$

at the argument points $-4(1)5$, that is, -4 through 5 in steps of 1, where

$$\text{sign}(x) = \begin{cases} 1, & x \geq 0 \\ -1, & x < 0 \end{cases}$$

Use this algorithm to produce tabular data which will be used on problems in Chapter 3. Determine (analytically) the number of multiply/divides (M/D's) and add/subtracts (A/S's) required when $x > 0$.

2.2 Let $h(x) = c_1 x \sin^3(x) - x^2 \cos^2(x) + c_2 x^3 \sin(x) + x^2 - c_3 x^4$, and let

$$S_N(x) = \sum_{\substack{i=-N \\ i \neq 0}}^{N} h(x - ia)$$

(a) Give an algorithm to evaluate $S_N(x)$ efficiently at any x, given a, c_1, c_2, c_3, and N. Assume N is large.
(b) Assume that evaluating $\sin(x)$ requires 25 arithmetic operations and that evaluating $\cos(x)$ requires 25 arithmetic operations. Give in terms of N the number of arithmetic operations required to evaluate $S_N(x)$ at any x, given a, c_1, c_2, c_3, and N.
(c) Assume that a recursive algorithm for evaluating a new function $g_N(x)$ at any x, given a, c_1, c_2, c_3, and $N =$ integer power of 2, operates as follows:

if $N = 1$ then return (x)

else return $\left[g_{N/2}\left(x + \frac{a}{2} \right) + g_{N/2}\left(x - \frac{a}{2} \right) + S_N(x) \right]$

Assuming that your algorithm (from part a) is used to compute $S_N(x)$, give, as $O(f(N))$ for some f, the number of arithmetic operations required to evaluate $g_N(x)$.

2.3 Assume that we wish to compute

$$f(x, y) = x^2 y^2 (\log(x) + \log(y)) + 3xy - 2$$

Assume we have a subroutine to compute $\log(z)$ for any given z with 50 arithmetic operations. How would you compute $f(x, y)$ for a given x and y, and how many arithmetic operations would it take?

2.4   Assume that you need to compute

$$v = 2y^2x^8 - 4yx^8 + 7x^8 + 6y^2x^4 - 12yx^4 + 21x^4 - 2y^2 + 4y - 7$$

Give an algorithm which computes v in the fewest number of arithmetic operations, given values for x and y.

CHAPTER THREE

Some Familiar Problems: Algorithms

For most of this book we will study numerical problems as examples that illustrate some basic concepts of accuracy, efficiency, or algorithm design. To provide motivation for these concepts we will present here a survey of the kinds of numerical problems that are commonly encountered and some simple approaches to the numerical solution of these problems. In the process we will also practice analyzing the efficiency of the methods that we will produce.

3.1 Function Evaluation

Probably the most common problem requiring numerical solution is the evaluation of a function at a given argument value, called the *evaluation argument*. The function can be given analytically, such as $\sin(x^2 + 3)$ or $5x^3 - x + 4.2$, or it can be given as a table of measured or computed values for each of a set of argument values (the *tabular arguments*); see Figure 3–1.

For each way of giving the function, it can sometimes be evaluated directly, that is, with only the use of the simple arithmetic functions of addition, subtraction, multiplication, and division and with complete accuracy, assuming these functions can be accurately evaluated. For functions given analytically, the only ones that can be evaluated directly are the polynomials and the rational functions, which are quotients of polynomials. For tabulated functions direct evaluation is possible only if the evaluation argument is one of the tabular arguments or if the tabulated function is known to be a polynomial of appropriately low degree or a

FIGURE 3–1

A Tabulated Function: $f(x) = \sin(x)$

Tabular argument x_i	Tabulated value $f(x_i)$
0	0
$\pi/6$	$1/2$
$\pi/3$	$\sqrt{3}/2$
$\pi/2$	1

quotient of such polynomials. In all other cases the function can only be approximately evaluated. Let us discuss direct evaluation first.

3.1.1 DIRECT FUNCTION EVALUATION: POLYNOMIALS

Evaluation of a polynomial has been discussed in Section 2.1. We have seen that the best way to evaluate a polynomial is using the synthetic division algorithm, that is, using the nested form of the polynomial. We also saw there that this requires n multiplications and n additions to evaluate

$$p(x) = \sum_{j=0}^{n} a_j x^j \qquad [1]$$

at an evaluation argument $x = x_0$.

Rational functions are evaluated by evaluating the numerator and denominator polynomials by synthetic division and computing the quotient of the results.

To evaluate a tabulated function at an evaluation argument equal to one of the tabular arguments is very straightforward: one simply reads the appropriate tabulated value. If the evaluation argument is not a tabular argument, direct evaluation is possible in general only if the tabulated function is a polynomial of degree less than the number of tabulated values or a quotient of two polynomials, one monic, the sum of whose degrees is less than the number of tabulated values. In this case values specifying the polynomial(s), such as the coefficients, can be determined and the resulting polynomial or rational function evaluated by synthetic division or some other algorithm. We will prove this result for polynomials and leave the proof for rational functions to an exercise (Problem 3.1).

The nth† degree polynomial $p(x) = \sum_{j=0}^{n} a_j x^j$ can be determined from

---

† We will routinely in this book use n or $n + 1$ to count the coefficients of an approximating function and N or $N + 1$ to count the number of tabular points used in the approximation. In this section the polynomial coefficients will be numbered from 0 to n and the tabular points from 0 to N or, because often $N = n$, from 0 to n.

$\{(x_i, y_i) \mid i = 0, 1, 2, ..., n; \; y_i = p(x_i)\}$ by solving the $n + 1$ linear equations in the $n + 1$ variables a_j:

$$y_i = \sum_{j=0}^{n} a_j x_i^j \qquad (0 \leq i \leq n) \qquad \qquad [2]$$

We will show below a more efficient method of finding these coefficients and in doing so will demonstrate the existence of the polynomial. This polynomial can be shown to be unique by noting that if there are two such nth degree polynomials they agree at $n + 1$ points, so their difference is an nth degree polynomial with $n + 1$ zeros. By the fundamental theorem of algebra, it follows that this difference polynomial must be the zero polynomial, i.e., the two original polynomials are the same.

Given that any $N + 1$ points with distinct x_i determine a unique Nth degree polynomial, we can justify our statement above that if $N + 1$ points on an nth degree polynomial are given with $N \geq n$ we can determine that polynomial. The result follows from the fact that the nth degree polynomial passing through the $N + 1$ points is an Nth degree polynomial whose coefficients a_j for $n < j \leq N$ are zero. Since there is only one polynomial of degree N through the points, that one must be the nth degree polynomial we desire.

All the above depends on our being able to construct the exact-matching polynomial. We have said that it must satisfy the constraints

$$y_i = \sum_{j=0}^{n} a_j x_i^j \qquad (i = 0, 1, ..., n) \qquad \qquad [3]$$

The basis of the construction of the polynomial is to write it in the form

$$p(x) = \sum_{j=0}^{n} y_j L_j(x) \qquad \qquad [4]$$

where each $L_j(x)$ is a polynomial of degree n. Equation 3 will be satisfied if $L_j(x)$ is 0 at every tabular argument except x_j and is 1 at x_j. If each $L_j(x)$ has the above properties, $p(x_i)$ will have the value y_i, because the fact that $L_j(x_i) = 0$ for $j \neq i$ will cause the y_j for $j \neq i$ to be multiplied by 0 and the fact that $L_i(x_i) = 1$ will cause y_i to be multiplied by 1. Thus, we have reduced our problem to finding a set of polynomials $L_j(x)$, called the *Lagrange polynomials*, such that

$$\text{for all } j \qquad L_j(x_i) = \begin{cases} 0 & (i \neq j) \\ 1 & (i = j) \end{cases} \qquad \qquad [5]$$

Since $L_j(x)$ is a polynomial, to have a zero at $x = x_i$, it must have a factor $(x - x_i)$. Thus, we have

$$L_j(x) = q_j(x) \prod_{\substack{i=0 \\ i \neq j}}^{n} (x - x_i) \qquad [6]$$

where $q_j(x)$ is some polynomial. But $L_j(x)$ is a polynomial of degree n or less, so $q_j(x)$ must be a constant. The constant is determined by the one constraint remaining to be satisfied, namely,

$$L_j(x_j) = 1 \qquad [7]$$

producing

$$1 = q_j \prod_{\substack{i=0 \\ i \neq j}}^{n} (x_j - x_i) \qquad [8]$$

from which follows

$$L_j(x) = \frac{\displaystyle\prod_{\substack{i=0 \\ i \neq j}}^{n} (x - x_i)}{\displaystyle\prod_{\substack{i=0 \\ i \neq j}}^{n} (x_j - x_i)} \qquad [9]$$

We have now constructed the polynomial $p(x)$ with the required properties.

Let us illustrate the use of Lagrange polynomials to find the first-degree polynomial, a straight line, determined by the two points (x_0, y_0) and (x_1, y_1). The Lagrange polynomials are

$$L_0(x) = \frac{x - x_1}{x_0 - x_1} \qquad [10]$$

and

$$L_1(x) = \frac{x - x_0}{x_1 - x_0} \qquad [11]$$

Thus, the approximating line is

$$p(x) = y_0 \frac{x - x_1}{x_0 - x_1} + y_1 \frac{x - x_0}{x_1 - x_0} \qquad [12]$$

which can be written

$$p(x) = y_0 + \frac{x - x_0}{x_1 - x_0} (y_1 - y_0) \qquad [13]$$

in which the polynomial value is written as the value at the beginning of the interval plus the product of the tabular y interval and the fraction of the full tabular x interval taken by $x - x_0$.

Before moving on to function evaluation using approximation, let us determine the efficiency of this method of polynomial evaluation where the *n*th degree polynomial is represented tabularly. For a given evaluation argument, the evaluation of each $L_j(x)$ requires $2(n - 1)$ multiplications, 1 division, and $2n$ subtractions. Thus the whole evaluation requires $2n(n + 1)$ M/D's and $n(2n + 3)$ A/S's. A considerable reduction can be made by some clever algebra. The reduction depends upon noting that the polynomial $p(x) = 1$ fits exactly through a table which has $y_i = 1$ for all x_i, with $0 \leq i \leq n$. Since there is only one polynomial of degree *n* or less passing through these points,

$$1 = \sum_{j=0}^{n} 1 \cdot L_j(x) = \sum_{j=0}^{n} \left(\frac{\prod_{\substack{i=0 \\ i \neq j}}^{n} (x - x_i)}{\prod_{\substack{i=0 \\ i \neq j}}^{n} (x_j - x_i)} \right) \quad [14]$$

Let $p(x)$ be the Lagrange polynomial through the data points (x_i, y_i) of some given problem. Then we can divide $p(x)$ by 1 without changing its value, producing

$$p(x) = p(x)/1 = \frac{\sum_{j=0}^{n} y_j L_j(x)}{\sum_{j=0}^{n} L_j(x)} \quad [15]$$

If we divide both the numerator and the denominator of equation 15 by $\psi(x) \triangleq \prod_{i=0}^{n} (x - x_i)$ and let

$$A_j = \frac{1}{\prod_{\substack{i=0 \\ i \neq j}}^{n} (x_j - x_i)} \quad [16]$$

we produce

$$p(x) = \frac{\sum_{j=0}^{n} y_j A_j / (x - x_j)}{\sum_{j=0}^{n} A_j / (x - x_j)} \quad [17]$$

Using this so-called barycentric form of Lagrange interpolation (see Program 3–1)†, computing each A_j requires *n* multiplications and divisions, so the whole computation requires $(n + 2)(n + 1) + 1$ M/D's, approximately half that required by the direct Lagrange method. Furthermore, if we are computing the approximation for many different *x*

---

† Programs for Chapter 3 start on p. 92.

values, we can compute the A_j once and for all; then each evaluation will require only $2(n + 1) + 1$ multiplications and divisions. Of course, the method can not be used when the evaluation argument is the same as a tabular argument to within the number of digits in the computer representation since a division by zero would be caused in both the numerator and denominator.

3.1.2 FUNCTION EVALUATION BY APPROXIMATION

If a function can not be evaluated directly, it must be evaluated by an approximation. Approximations are functions which can be evaluated directly and are chosen as being appropriately close to the function which we desire to evaluate. Approximation is necessary either when the function is known only by its values at a discrete set of argument points or when it is analytically defined but can not be efficiently evaluated at arbitrary argument values in a direct way. A function is given only at a set of tabular arguments if it comes from some real-world process and is known only by measuring its value at these argument points or when the data values are a result of a previous computation.

An example of an analytically defined function that can not be evaluated directly in an efficient manner is

$$f(x) = \int_0^x \sin\left(\frac{1}{u^2 + 2}\right) du$$

In this case we use a two-part strategy: (1) evaluate the function in some inefficient way at a set of data points to whatever accuracy is required; and (2) for routine evaluation, use techniques of approximation to calculate the function from the discrete number of data points tabulated.

For analytically defined functions approximation can often be carried out by evaluating a truncated infinite series, but this is usually an inefficient process so it is only used to compute tabular values, leaving us with evaluation using a table as described above in this section. Thus, for our purposes, the *approximation* problem can be stated as follows: Given a discrete set of argument values and a corresponding set of function values, each with some error due to computation and/or measurement, find a function defined on a desired real interval such that this function is an appropriately close approximation to the function that produced the data points (the *underlying function*). With respect to operations on the function such as differentiation, integration, and transforms, the strategy is to apply these operations to the approximating function after the approximation has been carried out.

In this book, we will direct our attention to functions of one variable. But the concepts that we will develop can be extended to functions of many variables.

FIGURE 3–2

Approximating Functions Fitted to Data Points

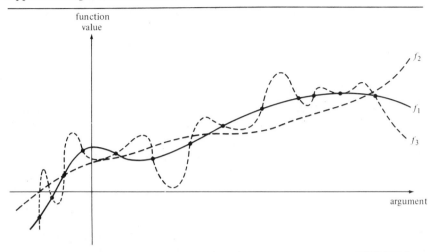

What do we mean by the "best" approximation? We must exclude the answer that the approximating function should be close to the function underlying the tabulated data because we either do not know this underlying function or can not evaluate it for the purpose of measuring its closeness to the approximating function. Two properties suggest themselves. First, we usually assume the underlying function is *smooth*, that is, has *low average curvature*, so the smoother the approximation is, the better it is.† Second, we have measured the underlying function at the tabular argument points, so we can say that the closer the approximating function is to the data at those points, the better it is. Thus, for example, the function f_1 in Figure 3–2 is a better approximation to the data than f_2 because it is closer to the data values at the tabular arguments, and f_1 is a better approximation than f_3 because it is smoother.

The aforementioned properties must be translated into a precise criterion for goodness of approximation. After this is done, one must specify the set of functions over which the criterion is to be applied to find a "best" approximating function. Most often the form of the function is constrained so that the best approximating function, according to the criterion we have established, is unique. The smoothness requirement is often met by restricting the set of functions to "appropriately smooth" functions.

Let us illustrate the above by listing some possible function forms and some possible criteria. Probably the most commonly used function

---

† In this book we use the word "smooth" to mean something stronger than the common mathematical usage of "appropriately differentiable."

form is the polynomial of degree n for some fixed integer n. Another common form is the Fourier expansion: a linear combination of functions of the form $\sin(2k\pi x/N)$ or $\cos(2k\pi x/N)$, for a fixed N and with k ranging over integers between 0 and some integer less than or equal to N. Another common form for approximating functions is the rational function, defined as a quotient of two polynomials each of degree less than some fixed value n.

An infinite number of function forms can be used; the one that should be chosen depends, first, on physical knowledge of the kind of function that produced the data and, second, on numerical matters dealt with in this book. We can illustrate the concepts we need using only approximating function forms which are a linear combination of basis functions, each having no variable other than the independent variable of the function. For example, the polynomials mentioned above are a linear combination of the functions x^j for $0 \le j \le n$; similarly the Fourier series are linear combinations of the sinusoids indicated above. The rational functions are not of the form indicated. The student should not infer from our choice to treat only linear combinations of functions that there are not cases where an approximating form such as the rational functions can make important contributions. Rather, because of limited space, a choice to deal with only the somewhat simpler form has been made.

The criterion of fit of our approximating function specifies what we mean by the "closeness" with which the approximating function fits the data points. The simplest criterion for approximation is that the approximating function must pass through all of the data points. This criterion, which we will call *exact matching*, is commonly used, though there are many cases where it is not the best criterion. Other criteria depend upon minimizing some "average" of the differences, $\hat{f}(x_i) - y_i$, over all data points (x_i, y_i), where $\hat{f}$ is the approximating function. In general, one chooses the exact matching criterion if the data points have little or no error, for then the underlying function is known to pass through or very near them, and one chooses an "average" minimizing criterion, which requires the approximating function only to go near the data points, if they have significant error. In this section we will first cover exact-matching approximation with polynomials and then we will cover the ideas of approximation using other criteria or other functions.

Polynomial Exact-Matching Approximation

Simple Exact-Matching Constraints.   The most common approximating functions are the polynomials. The most common criterion for approximation is exact matching. In both cases, the popularity is due to the mathematical tractability of the analysis and is not due to nor indicative of the quality of the approximation thus produced; polynomial exact matching is used far too often when it should not be. We will develop these methods first, pointing out their advantages, but we will be careful to point out their disadvantages as well so that the reader will not be inclined to overuse them.

We have already shown, in Section 3.1.1, how to find the unique polynomial passing through a given set of tabular points using the barycentric form of the Lagrange algorithm. This is our first approximation algorithm. But the set of points to choose is not immediately clear.

Given a table of data, we may decide to use only a subset of the tabular points to determine our approximation. When we analyze the error in polynomial exact-matching approximation, we will see that the error at any evaluation argument depends upon the number of points we use, the choice of points, and derivatives of the functions being approximated. It will often be the case that too many as well as too few points will result in an approximation of low accuracy.

If one knows beforehand which data points are to be used in an approximation, the barycentric form of Lagrange approximation provides a reasonable method for computation. In most cases, however, one does not know how many points will provide either the best approximation or an approximation of desired accuracy, because the function f being approximated is either not analytically known or too complicated to use to compute and evaluate derivatives of high order. We need a formulation that produces approximations using, successively, one data point, two data points, and so on, and that allows us to determine the optimum number of points during the computation. The Lagrange method is not satisfactory for this purpose because if we have used it to compute the approximation through $N + 1$ points, a great deal of new work is required to evaluate the polynomial with a single point added. Each old Lagrange polynomial must be multiplied by a factor in both the numerator and the denominator, and a new Lagrange polynomial term corresponding to the new data point must be computed. An alternative approach, the *Newton divided-difference formulation*, avoids this difficulty. The approach is based on attempting to write a series like the Taylor series that will permit us to increase the degree of the approximating polynomial by simply adding a term to the preceding approximation, where that term depends on a derivative of the function being approximated. If we have only a table of arguments and function values, we do not have available values for the derivative, but these can be approximated using finite differences of the data values.

We define the first *divided-difference* of f at the points x_0 and x_1 as

$$f[x_0, x_1] \triangleq \frac{f(x_1) - f(x_0)}{x_1 - x_0} \qquad [18]$$

Note that this divided-difference is an approximation to the derivative; in fact if f and f' are continuous between x_0 and x_1, the first divided-difference must be equal to the derivative at some point in the interval $[x_0, x_1]$.

We define higher divided-differences by a straightforward extension of equation 18:

$$f[x_0, x_1, x_2, ..., x_k] \triangleq \frac{f[x_1, x_2, ..., x_k] - f[x_0, x_1, x_2, ..., x_{k-1}]}{x_k - x_0} \quad \text{[19]}$$

We generate an approximation formula at an arbitrary value x by noting that

$$f[x, x_0] = \frac{f(x_0) - f(x)}{x_0 - x} \quad \text{[20]}$$

so

$$f(x) = f(x_0) + (x - x_0) f[x, x_0] \quad \text{[21]}$$

We can not compute $f[x, x_0]$ because doing so requires that we know $f(x)$, but we can continue the process started with equations 20 and 21 by writing

$$f[x, x_0, x_1] = \frac{f[x_0, x_1] - f[x, x_0]}{x_1 - x} \quad \text{[22]}$$

so

$$f[x, x_0] = f[x_0, x_1] + (x - x_1) f[x, x_0, x_1] \quad \text{[23]}$$

and therefore

$$f(x) = f(x_0) + (x - x_0) f[x_0, x_1] + (x - x_0)(x - x_1) f[x, x_0, x_1] \quad \text{[24]}$$

Applying the same technique again and again, we can show

$$\begin{aligned} f(x) = f(x_0) &+ (x - x_0) f[x_0, x_1] + (x - x_0)(x - x_1) f[x_0, x_1, x_2] \\ &+ \cdots + (x - x_0)(x - x_1) \cdots (x - x_{n-1}) f[x_0, x_1, ..., x_n] \quad \text{[25]} \\ &+ (x - x_0)(x - x_1) \cdots (x - x_n) f[x, x_0, x_1, ..., x_n] \end{aligned}$$

THEOREM.   Consider the polynomial $p(x)$, which is equal to the sum of all but the last term of equation 25. Then $p(x_i) = f(x_i)$ for $0 \le i \le n$.

Proof.   The proof proceeds by induction on the terms of equation 25. As a basis step, the first term of equation 25, $f(x_0)$, agrees with f at x_0. Assume that the polynomial made up of the first $m + 1$ terms of equation 25 agrees with f at x_i, for $0 \le i \le m$. Since the $(m + 2)$th term has a factor $x - x_i$ for $0 \le i \le m$, the first $m + 2$ terms agree with f at x_i, for $0 \le i \le m$. We have only to show that the first $m + 2$ terms of equation 25 agree with f at x_{m+1}. This claim is proved using the fact that the divided-difference is a symmetric function, that is, the order of the argument points does not affect the value; the proof of this fact is left to the reader.

We note that the identity given by equation 25 applies for $n = m$ and $x = x_{m+1}$:

$$
\begin{aligned}
f(x_{m+1}) = {}& f(x_0) + (x_{m+1} - x_0)f[x_0, x_1] \\
& + (x_{m+1} - x_0)(x_{m+1} - x_1)f[x_0, x_1, x_2] + \cdots \\
& + (x_{m+1} - x_0)(x_{m+1} - x_1) \cdots (x_{m+1} - x_{m-1}) \qquad [26] \\
& \times f[x_0, x_1, \ldots, x_m] \\
& + (x_{m+1} - x_0)(x_{m+1} - x_1) \cdots (x_{m+1} - x_m) \\
& \times f[x_{m+1}, x_0, x_1, \ldots, x_m]
\end{aligned}
$$

But, by the symmetry of the divided-difference,

$$
f[x_{m+1}, x_0, x_1, \ldots, x_m] = f[x_0, x_1, \ldots, x_{m+1}] \qquad [27]
$$

Therefore, equation 26 is the same as the first $m + 2$ terms of equation 25 evaluated at x_{m+1}; the first $m + 2$ terms of equation 25 equal $f(x_{m+1})$.

Q.E.D.

We have shown that the nth-degree polynomial consisting of all but the last term of equation 25 agrees with f at x_i, with $0 \leqslant i \leqslant n$. Since there is only one nth-degree polynomial with this property, this polynomial must be the same as that produced by Lagrange approximation using those points.

Computing polynomial exact-matching approximations using the Newton divided-difference method proceeds as follows:

1. $\hat{f}(x) := f(x_0)$
2. factor $:= 1$
3. for $i := 1$ by 1 until stopping criterion is met:
   a. factor $:=$ factor $\times (x - x_{i-1})$
   b. $\hat{f}(x) := \hat{f}(x) +$ factor $\times f[x_0, x_1, \ldots, x_i]$

The method assumes that the divided-differences are available or are computed as needed. (See also Program 3–2.)

The order in which tabular points are added into the approximation, that is, which tabular argument is called x_0, which x_1, etc., must be determined. After any number of points have been used, the resulting polynomial, we have seen, is the exact-matching polynomial through these tabular points — a unique polynomial and thus independent of the order in which the points have been added in. Thus, if the number $N + 1$ to be finally used is known, the order does not matter. But we have designed this method on the assumption that N is not known but rather that points will continue to be added until a decision is taken to stop. Thus points should be added in an order such that whenever the decision to stop is made the resulting approximation to the value $f(x)$ is the best for that number of points.

We will see that error analysis supports our intuition that the approximation error is normally least for $N + 1$ tabular points if the tabular arguments used are the $N + 1$ closest to the evaluation argument, x. Thus, tabular points should be added into the approximation in order of the closeness of the tabular argument values to the evaluation argument. To characterize this order more fully, assume a divided-difference table of the form of Figure 3–3, in which the x_i are assumed to be in numerical order and their subscripts do not correspond directly to the subscripts in the divided-difference formula, but there exists a correspondence between the tabular arguments and the x_i of the divided-difference formula as discussed below.

Then adding the points in this order produces a notion of a *path* through the divided-difference table (see Figure 3–4). This path begins at a function value (0th difference) in the first column of the divided-difference table and moves from that 0th difference diagonally to an adjacent first difference; then diagonally again, but not necessarily in the same direction, to an adjacent second difference; and so on, stopping at the last difference used. The data values used in the approximation corresponding to any given path ending can be determined by drawing a triangle from the path ending to the 0th-difference column along the

FIGURE 3–3
A Divided-Difference Table

x_0	$f(x_0) = y_0$		
		$f[x_0, x_1]$	
x_1	$f(x_1) = y_1$		$f[x_0, x_1, x_2]$
		$f[x_1, x_2]$	
x_2	$f(x_2) = y_2$		$f[x_1, x_2, x_3]$
		$f[x_2, x_3]$	
x_3	$f(x_3) = y_3$		$f[x_2, x_3, x_4]$
		$f[x_3, x_4]$	
x_4	$f(x_4) = y_4$		
$\vdots$	$\vdots$	$\vdots$	$\vdots$
x_{k-2}	$f(x_{k-2}) = y_{k-2}$		
		$f[x_{k-2}, x_{k-1}]$	
x_{k-1}	$f(x_{k-1}) = y_{k-1}$		$f[x_{k-2}, x_{k-1}, x_k]$
		$f[x_{k-1}, x_k]$	
x_k	$f(x_k) = y_k$		

FIGURE 3–4
Building a Path in a Difference Table

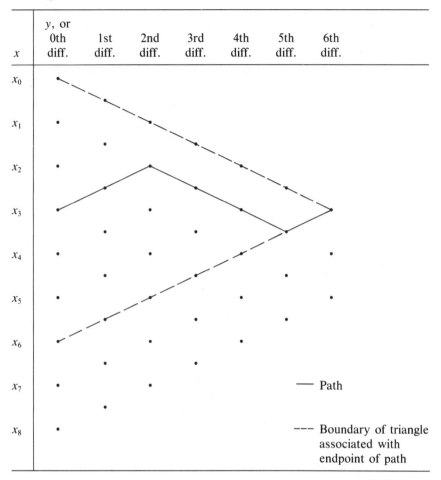

x	y, or 0th diff.	1st diff.	2nd diff.	3rd diff.	4th diff.	5th diff.	6th diff.
x_0							
x_1							
x_2							
x_3							
x_4							
x_5							
x_6							
x_7							
x_8							

—— Path

––– Boundary of triangle associated with endpoint of path

diagonals of the table (look again at Figure 3–4). The y_i values so enclosed and their corresponding x_i values are the data points used. For any temporary path ending, the difference in the next column to which the path should go is chosen from the two differences adjacent to the present path ending. This difference is the one for which the additional argument value enclosed by the new triangle is closest to the evaluation argument.

Thus, to be in the best position to stop at any place along the path, for any evaluation argument x, we choose x_0 as the tabular argument closest to x and begin our path at $f(x_0)$. Then we move to the next column in the direction specified by the rule above, calling the newly involved tabular argument value x_1. We build our path to the second-difference column, calling the newly involved tabular argument point x_2, and so on.

These points — the points x_0, x_1, and so on, in the approximating formula given by equation 25 — depend on the path chosen, which in turn depends on the evaluation argument. The divided-difference table is ordered so that the tabular arguments increase in order, but for any given x the subscripts corresponding to tabular values used in the formula increase in order of the distance of the corresponding tabular arguments from x.

For example, consider approximating $f(2.5)$ to the third difference using Figure 3–5. Then

$$
\begin{aligned}
\hat{f}(2.5) &= f(2) + (2.5 - 2)f[2, 3] + (2.5 - 2)(2.5 - 3)f[2, 3, 1.5] \\
&\quad + (2.5 - 2)(2.5 - 3)(2.5 - 1.5)f[2, 3, 1.5, 1] \qquad [28] \\
&= 83 + 0.5(-16) + (0.5)(-0.5)(-1.33) \\
&\quad + (0.5)(-0.5)(1)(0.33) = 75.25^-
\end{aligned}
$$

Program 3–2 (p. 93) summarizes the Newton divided-difference algorithm we have developed. In it the stopping criterion is that the magnitudes of successive terms increase or the magnitude of a term is less than the allowed error tolerance. The basis for this criterion is discussed in Chapters 5 and 6.

We see from Figures 3–3 and 3–4 and equation 19 that to compute an approximation for a path ending on the nth difference requires the use of only n differences of order 1 or higher but that computing these requires the computation of all the differences enclosed by the triangle drawn back through the differences from the nth difference used. That is, the computation of $n(n + 1)/2$ differences of order 1 or greater is required.

Let us evaluate the efficiency of the Newton divided-difference al-

FIGURE 3–5
Using a Divided-Difference Table to Approximate $f(2.5)$

x	y			
-1	107			
		$-11/2 = -5.5$		
1	96		$-6.5/2.5 = -2.6$	
		$-6/0.5 = -12$		$0.6/3 = 0.2$
1.5	90		$-2/1 = -2$	
		$-7/0.5 = -14$		$0.67/2 = 0.33$
2	83		$-2/1.5 = -1.33$	
		$-16/1 = -16$		$0.33/3.5 = 0.095$
3.0	67		$-3/3 = -1$	
		$-38/2 = -19$		
5	29			—— Path

gorithm. Assuming that the divided-differences are available, each step requires 2 M/D's and 2 A/S's so an approximation through $n + 1$ points requires $2n$ M/D's and $2n$ A/S's. If the differences have to be calculated, an additional 2 A/S's and 1 M/D are required per divided-difference. Since $n(n + 1)/2$ differences must be calculated, the total algorithm requires $n(n + 5)/2$ M/D's and $n(n + 3)$ A/S's.

If n is known, a few operations can be saved with the Newton method by using a fully nested form (cf. the nested polynomial evaluation algorithm):

1. $\hat{f}(x) := f[x_0, x_1, \ldots, x_n]$
2. for $i := n - 1$ to 0 by -1
   $\hat{f}(x) := (x - x_i) \times \hat{f}(x) + f[x_0, x_1, \ldots, x_i]$

This algorithm saves n M/D's.

Comparing the Lagrange and Newton algorithms for a single evaluation argument, we find that the Lagrange method requires approximately $n^2/2$ more M/D's but approximately $n^2/2$ fewer A/S's (assuming that multiplications by -1 to produce $x_j - x_i$ from $x_i - x_j$ are ignored). Since M/D's are generally more expensive than A/S's, the Newton divided-difference method is preferred for a single evaluation even if n is known.

For many evaluation arguments that use the same data points, both algorithms can do certain computations once rather than for each evaluation. In the case of the barycentric Lagrange method, the values A_j and $y_j A_j$ can be computed once so that each evaluation requires additionally only $2n + 3$ M/D's and $3n + 1$ A/S's. In the same situation with the Newton divided-difference method, the differences can be computed once, so that each evaluation requires additionally only $2n$ M/D's and $2n$ A/S's, that is, 3 fewer M/D's and $n + 1$ fewer A/S's. Thus even in this situation the Newton method is preferable.

Until now, we have seen no situation where the Lagrange method is preferred to the Newton method. Such a situation does arise, though unusually. If we are given a number of functions, say m, all tabulated at the same tabular arguments and must evaluate each at the same evaluation argument, the Lagrange method requires only $m(n + 1)$ M/D's and mn A/S's whereas the Newton method, which must recompute the divided-differences for each function, requires $mn(n + 5)/2$ M/D's and $mn(n + 3)$ A/S's.

In the special but common case in which the tabular arguments are equally spaced, that is, $x_{i+1} - x_i = h$ for all i (where the tabular arguments are subscripted in numerical order), because the denominators of all divided-differences of the same order will be the same we might expect to be able to save operations in the Newton method. We note that

$$f[x_i, x_{i+1}] = \frac{f(x_{i+1}) - f(x_i)}{x_{i+1} - x_i} = \frac{f(x_{i+1}) - f(x_i)}{h} \qquad [29]$$

By defining an undivided equal-interval difference

$$\delta f(x_{i+1/2}) \overset{\Delta}{=} f(x_{i+1}) - f(x_i) \qquad [30]$$

we can rewrite equation 29 as

$$f[x_i, x_{i+1}] = \frac{\delta f(x_{i+1/2})}{h} \qquad [31]$$

Now we see that

$$\begin{aligned} f[x_i, x_{i+1}, x_{i+2}] &= \frac{f[x_{i+1}, x_{i+2}] - f[x_i, x_{i+1}]}{x_{i+2} - x_i} \\ &= \frac{\frac{1}{h}\delta f(x_{i+3/2}) - \frac{1}{h}\delta f(x_{i+1/2})}{2h} \\ &= \frac{\delta(\delta f(x_{i+1}))}{2h^2} \overset{\Delta}{=} \frac{\delta^2 f(x_{i+1})}{2h^2} \end{aligned} \qquad [32]$$

and in general

$$f[x_i, x_{i+1}, x_{i+2}, \ldots, x_{i+m}] = \frac{\delta^m f(x_{i+m/2})}{m! h^m} \qquad [33]$$

We note that

$$x_i = x_0 + ih \qquad [34]$$

that is, x_i is i intervals of width h from x_0. We can also think of the evaluation argument in terms of the number s (not necessarily an integer) of intervals from x_0:

$$x = x_0 + sh \qquad [35]$$

With equal intervals, if the closest tabular argument to the evaluation argument is called x_0, then the next closest will be x_1 or x_{-1} and the next closest will be the other of those two. Choosing additional tabular arguments in order of their closeness to the evaluation argument x will produce either the sequence $x_0, x_1, x_{-1}, x_2, x_{-2}, x_3, x_{-3}, \ldots$ or the sequence $x_0, x_{-1}, x_1, x_{-2}, x_2, x_{-3}, x_3, \ldots$. Both cases correspond to a zigzag path through the difference table (see Figure 3–6), with the first sequence corresponding to a path which begins in a downward diagonal direction

FIGURE 3–6
Paths in an Equal-Interval Difference Table

x_{-3}	y_{-3}					
		$\delta y_{-5/2}$				
x_{-2}	y_{-2}		$\delta^2 y_{-2}$			
		$\delta y_{-3/2}$		$\delta^3 y_{-3/2}$		
x_{-1}	y_{-1}		$\delta^2 y_{-1}$		$\delta^4 y_{-1}$	
		$\delta y_{-1/2}$		$\delta^3 y_{-1/2}$		$\delta^5 y_{-1/2}$
x_0	y_0		$\delta^2 y_0$		$\delta^4 y_0$	
		$\delta y_{1/2}$		$\delta^3 y_{1/2}$		$\delta^5 y_{1/2}$
x_1	y_1		$\delta^2 y_1$		$\delta^4 y_1$	
		$\delta y_{3/2}$		$\delta^3 y_{3/2}$		
x_2	y_2		$\delta^2 y_2$			
		$\delta y_{5/2}$				
x_3	y_3					

Solid path is for $x_{-1} < x < x_0$ and $|x - x_0| < |x - x_{-1}|$
Broken path is for $x_0 < x < x_1$ and $|x - x_0| < |x - x_1|$

and the second corresponding to a path which begins in an upward diagonal direction.

The Newton divided-difference formulas corresponding to these two paths are respectively

$$\begin{aligned}\hat{f}(x) = f(x_0) &+ (x - x_0)f[x_0, x_1] + (x - x_0)(x - x_1)f[x_0, x_1, x_{-1}] \\ &+ (x - x_0)(x - x_1)(x - x_{-1})f[x_0, x_1, x_{-1}, x_2] \\ &+ (x - x_0)(x - x_1)(x - x_{-1})(x - x_2) \\ &\times f[x_0, x_1, x_{-1}, x_2, x_{-2}] + \cdots\end{aligned} \quad [36]$$

and

$$\begin{aligned}\hat{f}(x) = f(x_0) &+ (x - x_0)f[x_0, x_{-1}] + (x - x_0)(x - x_{-1})f[x_0, x_{-1}, x_1] \\ &+ (x - x_0)(x - x_{-1})(x - x_1)f[x_0, x_{-1}, x_1, x_{-2}] \\ &+ (x - x_0)(x - x_{-1})(x - x_1)(x - x_{-2}) \\ &\times f[x_0, x_{-1}, x_1, x_{-2}, x_2] + \cdots\end{aligned} \quad [37]$$

With the notation of equations 33 and 35, equation 36 can be written as

$$\begin{aligned}\hat{f}(x) = f(x_0) &+ sh\frac{\delta f(x_{1/2})}{h} + sh(s - 1)\,h\frac{\delta^2 f(x_0)}{2h^2} \\ &+ sh(s - 1)h(s + 1)h\frac{\delta^3 f(x_{1/2})}{3!h^3} \\ &+ sh(s - 1)h(s + 1)h(s - 2)h\frac{\delta^4 f(x_0)}{4!h^4} + \cdots\end{aligned} \quad [38]$$

which simplifies to

$$\hat{f}(x) = f(x_0) + s\delta f(x_{1/2}) + \frac{s(s-1)\,\delta^2 f(x_0)}{2}$$
$$+ \frac{s(s-1)(s+1)\,\delta^3 f(x_{1/2})}{3!} \tag{39}$$
$$+ \frac{s(s-1)(s+1)(s-2)\,\delta^4 f(x_0)}{4!} + \cdots$$

Similarly equation 37 simplifies to

$$\hat{f}(x) = f(x_0) + s\delta f(x_{-1/2}) + \frac{s(s+1)\,\delta^2 f(x_0)}{2}$$
$$+ \frac{s(s+1)(s-1)\,\delta^3 f(x_{-1/2})}{3!} \tag{40}$$
$$+ \frac{s(s+1)(s-1)(s+2)\delta^4 f(x_0)}{4!} + \cdots$$

We hope that we have saved operations by the cancellation of the powers of h and by having only to compute the equal-interval differences $\delta^k f$, which require no divisions, rather than the divided-differences. (The form of these equal-interval differences we have written are called *central differences*, so called because the subscript of the difference is the center of all the tabular arguments involved. These equal-interval differences can also be called *forward differences*, defined by

$$\Delta f(x_i) \triangleq f(x_{i+1}) - f(x_i) \tag{41}$$

or *backward differences*, defined by

$$\nabla f(x_{i+1}) \triangleq f(x_{i+1}) - f(x_i) \tag{42}$$

Let us calculate the number of operations required by the equal-interval difference formula given by equation 39, assuming the values $\delta^k f(x_i)/k!$ are already computed. The algorithm is as follows:

1. $s := (x - x_0)/h$
2. $\hat{f}(x) := f(x_0)$
3. factor $:= s$
4. for $k := 1$ by 1 until stopping criterion is met
   a. $\hat{f}(x) := \hat{f}(x) +$ factor $\times (\delta^k f(x_i)/k!)$ [for some i]
   b. factor $:=$ factor $\times (s \pm$ integer) [need not be computed in the last step]

Thus $2n$ M/D's and $2n$ A/S's are required to compute the approximation through $n + 1$ points, exactly as many as with divided-differences.

Similarly as with divided differences, n M/D's can be saved if n is known beforehand (see Problem 3.5).

To compute the n values $\delta^k f(x_i)/k!$ required by the equal-interval difference algorithm requires the calculation of $n(n + 1)/2$ differences and the factorials up to $n!$ and the division of at least the n differences to be used in the formula by the appropriate factorials. This all can be done with $2(n - 1)$ M/D's (or $(n - 1)[(n/2) + 1]$ if all differences must be divided by a factorial because different paths will be used for different evaluations) and $n(n + 1)/2$ A/S's. It is here that the savings in operations are made over the divided-difference method.

The efficiency analysis that we have done for the various methods of polynomial approximation with simple exact-matching constraints leads us to prefer, in most cases, the undivided-difference method when the tabular arguments are equally spaced and the divided-difference method when they are not. The error analysis, and thus the process by which one makes the decision on when to stop the path, will be put off until Chapters 5 and 6. But here we must indicate some of the problems with the accuracy of polynomial approximation with simple exact-matching constraints.

One problem is that if only a few points are used the accuracy may be low because the few constraints on the polynomial (exact matching at only a few points) are not enough to force it to follow the underlying function closely. But as the number of points is increased, if the tabular arguments are approximately equally spaced, as is most often the case, the approximating polynomial will often become very unsmooth (see Figure 3–7). That is, for an evaluation argument nearly equidistant between two adjacent tabular arguments, the unsmooth approximating function may be far from the smooth underlying function. This may be the case even if the tabular arguments are close together.

The inaccuracy of simple exact matching with a low-degree polynomial (through only a few points) can also be thought of as one of unsmoothness. Consider an approximation through four points of a table of n points with equally spaced arguments (see Figure 3–8). For an evaluation argument between x_1 and x_2 the approximating polynomial should pass through f at x_0, x_1, x_2, and x_3, but for an evaluation argument between x_2 and x_3 the polynomial should pass through f at x_1, x_2, x_3, and x_4, and so on. Thus the approximating function is not one polynomial but one polynomial for evaluation arguments between x_1 and x_2, another polynomial for those between x_2 and x_3, still another for those between x_3 and x_4, and so on. We say $\hat{f}(x)$ is a *piecewise* polynomial exact-matching approximation. At the tabular arguments this approximation has discontinuities of its first derivative, that is, it comes to a point and is thus very unsmooth. We would do better in approximating a smooth underlying function if our exact-matching approximation were smooth.

FIGURE 3–7
Unsmooth Polynomial Exact-Matching Approximation Through Many Points

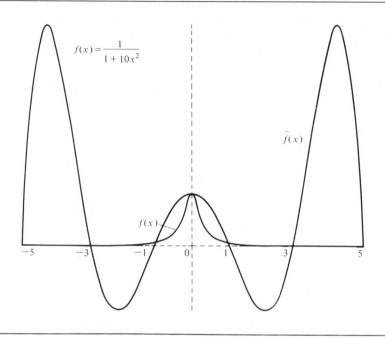

Smoothness Constraints: Splines.   To obtain a smooth piecewise exact-matching approximation we can add constraints to each piece that they match in slope, and also commonly in curvature, with the adjacent pieces at the tabular points at which they meet. Such approximations are called *splines*. We will develop the most common of these, which we will see is, under the appropriate definition of smoothness, the smoothest exact-matching approximation.

Assume we have a table that is to be fit by a polynomial spline in such a way that we have a different polynomial for each tabular interval (not necessarily of equal width). Assume further that we have fit the first m tabular intervals and wish to determine the polynomial for the next interval. We place four constraints upon this polynomial:

1. It must agree with the data point at the left end of the interval.
2. It must agree with the data point at the right end of the interval.
3. It must agree with the previous polynomial in slope (first derivative) at the left end of the interval.
4. It must agree with the previous polynomial in curvature (and thus second derivative) at the left end of the interval.

FIGURE 3–8
Piecewise Polynomial Exact-Matching Approximation

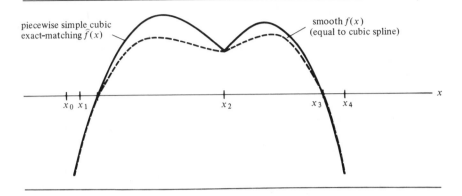

Since there are four constraints, a polynomial with four parameters, that is, a cubic polynomial, is required.

Let $p_{m+1}(x)$ be the cubic polynomial fit to the $(m + 1)$th interval $[x_m, x_{m+1}]$. If we let the second derivative at x_m be c_m and let c_m and c_{m+1} be parameters of $p_{m+1}(x)$, the specification of $p_{m+1}(x)$ becomes straightforward. Since $p_{m+1}(x)$ is a cubic polynomial, $p''_{m+1}(x)$ is a linear polynomial constrained to have the value c_m at x_m and c_{m+1} at x_{m+1}. Thus, $p''_{m+1}(x)$ is the exact-matching line,

$$p''_{m+1}(x) = c_m \frac{x_{m+1} - x}{x_{m+1} - x_m} + c_{m+1} \frac{x - x_m}{x_{m+1} - x_m} \qquad [43]$$

Integrating equation 43 twice produces

$$p_{m+1}(x) = \frac{c_m}{6} \frac{(x_{m+1} - x)^3}{x_{m+1} - x_m} + \frac{c_{m+1}}{6} \frac{(x - x_m)^3}{x_{m+1} - x_m}$$
$$+ A_{m+1}(x_{m+1} - x) + B_{m+1}(x - x_m) \qquad [44]$$

where the arbitrary linear term that is added is written in the form $A_{m+1} (x_{m+1} - x) + B_{m+1}(x - x_m)$ for computational simplicity in the solutions for A_{m+1} and B_{m+1}. These constants are determined by two constraints:

(i) $p_{m+1}(x_m) = y_m$ $\qquad\qquad\qquad\qquad\qquad\qquad$ [45]

which implies

$$\frac{c_m}{6}(x_{m+1} - x_m)^2 + A_{m+1}(x_{m+1} - x_m) = y_m \qquad [46]$$

so

$$A_{m+1} = \frac{y_m}{x_{m+1} - x_m} - \frac{c_m}{6}(x_{m+1} - x_m) \qquad [47]$$

and

(ii) $$p_{m+1}(x_{m+1}) = y_{m+1} \qquad [48]$$

which implies

$$\frac{c_{m+1}}{6}(x_{m+1} - x_m)^2 + B_{m+1}(x_{m+1} - x_m) = y_{m+1} \qquad [49]$$

so

$$B_{m+1} = \frac{y_{m+1}}{x_{m+1} - x_m} - \frac{c_{m+1}}{6}(x_{m+1} - x_m) \qquad [50]$$

The only constraint left to be satisfied, continuity of the first derivative at x_m,

$$p'_{m+1}(x_m) = p'_m(x_m) \qquad [51]$$

determines a relation among c_{m+1}, c_m, and c_{m-1}. Differentiating equation 44 produces

$$p'_{m+1}(x) = -\frac{c_m}{2}\frac{(x_{m+1} - x)^2}{x_{m+1} - x_m} + \frac{c_{m+1}}{2}\frac{(x - x_m)^2}{x_{m+1} - x_m} + \frac{y_{m+1} - y_m}{x_{m+1} - x_m}$$
$$- \frac{c_{m+1} - c_m}{6}(x_{m+1} - x_m) \qquad [52]$$

Equation 51 becomes

$$-\frac{c_m}{2}(x_{m+1} - x_m) + \frac{y_{m+1} - y_m}{x_{m+1} - x_m} - \frac{c_{m+1} - c_m}{6}(x_{m+1} - x_m)$$
$$= \frac{c_m}{2}(x_m - x_{m-1}) + \frac{y_m - y_{m-1}}{x_m - x_{m-1}} - \frac{c_m - c_{m-1}}{6}(x_m - x_{m-1}) \qquad [53]$$

Simplifying equation 53 produces the following relation between the second derivatives:

$$c_{m+1}(x_{m+1} - x_m) + 2c_m(x_{m+1} - x_{m-1}) + c_{m-1}(x_m - x_{m-1})$$
$$= 6\left[\frac{y_{m+1} - y_m}{x_{m+1} - x_m} - \frac{y_m - y_{m-1}}{x_m - x_{m-1}}\right] \qquad [54]$$

Equation 54 applied for $m = 1, 2, \ldots, N - 1$ can determine $N - 1$ of the c_i values, but equations 44, 47, and 50, which specify the N polynomials which make up the spline approximation, require

$N + 1$ c_i values, that is, at $x = x_i$ for $i = 0, 1, ..., N$. Two more constraints must be given to specify these two additional c_i values. There are an infinite number of possibilities for this pair of constraints. For example, any two of the c_i could be given. Alternatively, we could argue from our discussion above that for the polynomial, $p_1(x)$, used in the first tabular interval, constraints 3 and 4 specifying agreement with the slope and curvature of the previous polynomial at the left end of the interval were not applicable and one should specify the slope and curvature at x_0. These could be used to calculate c_1, and with c_0 and c_1, equation 54 could be used as a recurrence relation to compute the other $N - 1$ c_i values.

Still another possibility for the pair of additional constraints is to specify c_N and s_N and use the recurrence relation to work backwards. Among all the possibilities, the one normally used is to specify that $c_0 = c_N = 0$, that is, the curvature (and thus the second derivative) is zero at both endpoints, x_0 and x_N. The reason for this choice is that the spline thus produced is the smoothest of all exact-matching fits, where maximum smoothness is defined as minimum average second derivative and average is defined in the root mean square sense. We prove this result below.

THEOREM. The cubic polynomial spline fit with $c_0 = c_N = 0$ is the function among all twice-differentiable functions $f(x)$ which exact match at x_i, $0 \leq i \leq N$, for which the average

$$\left[\frac{1}{x_N - x_0} \int_{x_0}^{x_N} (f''(x))^2 \, dx \right]^{1/2}$$

is minimum.

Proof. The average above is minimum when $\int_{x_0}^{x_N} [f''(x)]^2 \, dx$ is minimum. Let $g(x)$ be the cubic spline fit in question, and let $f(x)$ be any other twice-differentiable function over $[x_0, x_N]$. Then

$$\int_{x_0}^{x_N} [f''(x)]^2 \, dx = \int_{x_0}^{x_N} [g''(x) + (f''(x) - g''(x))]^2 \, dx$$

$$= \int_{x_0}^{x_N} [g''(x)]^2 \, dx + 2 \int_{x_0}^{x_N} g''(x)[f''(x) - g''(x)] \, dx \quad [55]$$

$$+ \int_{x_0}^{x_N} [f''(x) - g''(x)]^2 \, dx$$

The middle integral on the right side of equation 55 can be integrated by parts to produce

$$I \triangleq \int_{x_0}^{x_N} g''(x)[f''(x) - g''(x)] \, dx$$

$$= g''(x)[f'(x) - g'(x)] \Big|_{x_0}^{x_N} - \int_{x_0}^{x_N} [f'(x) - g'(x)]g'''(x) \, dx \qquad [56]$$

$$= c_N[f'(x_N) - g'(x_N)] - c_0[f'(x_0) - g'(x_0)]$$

$$- \sum_{i=0}^{N-1} \int_{x_i}^{x_{i+1}} [f'(x) - g'(x)]K_i \, dx$$

since $g'''(x)$ is a constant (K_i) in $[x_i, x_{i+1}]$ because g is a cubic there. Since $c_N = c_0 = 0$,

$$I = - \sum_{i=0}^{N-1} K_i[f(x) - g(x)] \Big|_{x_i}^{x_{i+1}} = 0 \qquad [57]$$

because $f(x_i) = g(x_i)$ for all i by the exact-matching requirement on f and g. Thus, equation 55 becomes

$$\int_{x_0}^{x_N} [f''(x)]^2 \, dx = \int_{x_0}^{x_N} [g''(x)]^2 \, dx + \int_{x_0}^{x_N} [f''(x) - g''(x)]^2 \, dx \qquad [58]$$

Since the integrand of the second integral on the right side of equation 58 is always positive, we have

$$\int_{x_0}^{x_N} [f''(x)]^2 \, dx \geq \int_{x_0}^{x_N} [g''(x)]^2 \, dx \qquad [59]$$

Q.E.D.

Note that in equation 59 equality holds only if $f''(x) = g''(x)$ for all $x \in [x_0, x_N]$ except for some discrete values of x.

Thus, according to the above measure of smoothness, the cubic spline fit with $c_0 = c_N = 0$ is the *smoothest exact-matching fit* through the data points. To obtain this fit, we must use these initial conditions and the recurrence relation given by equation 54 to produce the values of the c_i needed in equation 44 to produce the desired cubic splines. The unknowns are $c_1, c_2, \ldots, c_{N-1}$. Equation 54 applied for $m = 1, 2, \ldots, N-1$ produces $N - 1$ linear equations in these $N - 1$ unknowns. Since the mth equation involves only c_m, c_{m-1}, and c_{m+1}, the matrix for these linear equations has only a diagonal, a superdiagonal, and a subdiagonal with nonzero elements. Such a matrix is called a *tridiagonal matrix*. The solution of linear equations involving a tridiagonal matrix is especially easy (see Problem 3.30).

The c_i values can be found once and for all for a given set of tabular data. With these values the approximation at any evaluation argument x can be computed (see Program 3–3, p. 95). We determine the interval $[x_m, x_{m+1}]$ in which x falls and thus determine m. Then we evaluate a form of equation 44 which has been algebraically simplified using the definitions $h \triangleq x_{m+1} - x_m$, $s \triangleq (x - x_m)/h$, $a \triangleq c_m h^2/6$, and $b \triangleq c_{m+1} h^2/6$.

As an example, consider the problem with $N = 3$ and tabular points $(1, 2)$, $(2, 1)$, $(3, 2)$, and $(4, 6)$. The linear equations produced from equation 54 with $m = 1$ and 2 and $c_0 = c_3 = 0$ are $4c_1 + c_2 = 12$ and $c_1 + 4c_2 = 18$, the solution of which is $c_1 = 2$, $c_2 = 4$. With these c_i, the approximation at $x = 1.5$ is produced as follows. 1.5 is in $[1, 2]$, that is, in the interval $[x_0, x_1] = [1, 2]$. The approximation thus comes from applying equation 44 with $m = 0$:

$$\hat{f}(1.5) = p_1(1.5)$$

$$= \frac{c_0}{6}(x_1 - 1.5)^3 + \frac{c_1}{6}(1.5 - x_0)^3 + A_1(x_1 - 1.5) + B_1(1.5 - x_0)$$

$$= 0 + \frac{0.125}{3} + \left(y_0 - \frac{c_0}{6}\right)0.5 + \left(y_1 - \frac{c_1}{6}\right)0.5$$

$$= \frac{0.125}{3} + 1 + \frac{1}{3} = 1.375$$

where A_1 and B_1 were computed from equations 47 and 50, respectively. Similarly $\hat{f}(2.4)$ would involve evaluating equations 44, 47, and 50 with $m = 1$.

Cubic spline approximation often produces far superior approximations than piecewise polynomial exact-matching approximations without constraints on derivative continuity (see Figures 3–8 and 3–9). Also, once the c_i are computed, it requires only 11 M/D's and 8 A/S's per evaluation. But it still does not approximate with adequate accuracy in cases where exact matching of any kind is contraindicated. If the data values have significant error (more than just representation error), the requirement that the approximating function matches the tabular points exactly makes it follow the error, producing an inaccurate approximation and one which is unsmooth since the error is unsmooth. To avoid this problem, when the data values have significant error we must not insist that the approximating function pass through the N data points (x_i, y_i) but only near enough to them to minimize some average of the differences $\hat{f}(x_i) - y_i$, where $\hat{f}$ has too few parameters for it to be made to pass through all of the data points.

Linear Least-Squares Approximation

There are many definitions of the average mentioned above, including the whole class†

---

† In least-squares and other "average" minimizing approximations, it is convenient to number N tabular points from 1 to N rather than $N + 1$ tabular points from 0 to N, as has been convenient for exact-matching approximation.

$$\left(\frac{\frac{1}{N} \sum\limits_{i=1}^{N} w_i |\hat{f}(x_i) - y_i|^p}{\sum\limits_{i=1}^{N} w_i} \right)^{1/p}$$

for some p and set of positive weights w_i, and the one to choose depends on the probability distribution of the error, a matter beyond the scope of this book. However, the most common definition is the one of the above class for $p = 2$, partially because analysis using it is most mathematically tractable and partially because it is the correct one for the most common error distribution, the normal distribution (see Pizer, 1975, pp. 358–361). For $p = 2$ we wish to minimize

$$\left(\frac{\frac{1}{N} \sum\limits_{i=1}^{N} \frac{1}{\sigma_i^2} (\hat{f}(x_i) - y_i)^2}{\sum\limits_{i=1}^{N} \frac{1}{\sigma_i^2}} \right)^{1/2}$$

where σ_i is the standard deviation of the error in y_i, a measure of the expected size of the error in y_i. We will assume here that all the data values have the same expected error size. Then to minimize the expression given above for $p = 2$ is the same as to minimize

$$\sum_{i=1}^{N} (\hat{f}(x_i) - y_i)^2$$

Doing this is called *least-squares approximation*.

We will restrict our attention to least-squares approximation when the approximating function $\hat{f}(x)$ is a linear combination of a set of n basis functions, $f^j(x)$:

$$\hat{f}(x) = \sum_{j=1}^{n} \hat{a}_j f^j(x) \qquad \textbf{[60]}$$

Examples of this case are polynomial least-squares approximation, for which†

$$f^j(x) = x^{j-1} \qquad \textbf{[61]}$$

and Fourier approximation, for which $f^j(x) = \sin(2k\pi x/N)$ or $\cos(2k\pi x/N)$ for some k. We will use the notation f_i^j for $f^j(x_i)$.

Least-squares approximation involves choosing the parameters a_j in $\sum_{j=1}^{n} a_j f^j(x)$ as the $\hat{a}_j$ that minimize the sum of squares,

---

† Note that in "x^{j-1}" the superscript is used to indicate exponentiation, whereas in "$f^j(x)$" the superscript is a counter among a set of functions.

$$S = \sum_{i=1}^{N} \left(\sum_{j=1}^{n} a_j f^j(x_i) - y_i \right)^2 \qquad [62]$$

This minimization is accomplished by setting the partial derivative of S with respect to each a_j equal to 0. Since from equation 62

$$\frac{\partial S}{\partial a_k} = 2 \sum_{i=1}^{N} \left(\sum_{j=1}^{n} a_j f_i^j - y_i \right) f_i^k \qquad [63]$$

we choose the $\hat{a}_j$, the best estimates of the a_j, so that

$$\sum_{i=1}^{N} \sum_{j=1}^{n} \hat{a}_j f_i^j f_i^k = \sum_{i=1}^{N} y_i f_i^k \qquad (k = 1, 2, ..., n) \qquad [64]$$

Reversing the order of summation on the left side of equation 64, we obtain

$$\sum_{j=1}^{n} \hat{a}_j \left(\sum_{i=1}^{N} f_i^j f_i^k \right) = \sum_{i=1}^{N} y_i f_i^k \qquad (k = 1, 2, ..., n) \qquad [65]$$

Equation 65 is a set of n linear equations in the n unknowns $\hat{a}_j$. It can be written using a matrix and vectors as

$$F\hat{a} = d \qquad [66]$$

where

$$F_{jk} = \sum_{i=1}^{N} f_i^j f_i^k \qquad (1 \leqslant k \leqslant n; \quad 1 \leqslant j \leqslant n) \qquad [67]$$

and

$$d_k = \sum_{i=1}^{N} y_i f_i^k \qquad (1 \leqslant k \leqslant n) \qquad [68]$$

Equations 65 are called the *normal equations*.

Given a set of tabular data $\{(x_i, y_i) \mid i = 1, 2, ..., N\}$, least-squares approximation thus involves a setup step in which the F matrix elements and d vector elements are evaluated and the linear equations given by equation 66 are solved for the $\hat{a}_j$, with $j = 1, 2, ..., n$. After this the approximation is calculated for any evaluation argument by using equation 60, which involves evaluating each basis function at the evaluation argument and calculating the weighted sum indicated.

For example, with $n = 2$, $f^1(x) = 1$, $f^2(x) = x$, and $N = 4$, with tabular data $(0, -4)$, $(1, 0)$, $(2, 4)$, and $(3, -2)$, the F elements are $F_{11} = 1^2 + 1^2 + 1^2 + 1^2 = 4$, $F_{12} = F_{21} = 1(0) + 1(1) + 1(2) + 1(3) = 6$, and $F_{22} = 0^2 + 1^2 + 2^2 + 3^2 = 14$, and the d elements are $d_1 = -4(1) + 0(1) + 4(1) + (-2)(1) = -2$, and $d_2 = -4(0) + 0(1) + 4(2) + (-2)(3) = 2$. Solving the equation

$$\begin{bmatrix} 4 & 6 \\ 6 & 14 \end{bmatrix} \begin{bmatrix} a_1 \\ a_2 \end{bmatrix} = \begin{bmatrix} -2 \\ 2 \end{bmatrix}$$

produces $a_1 = -2$, $a_2 = 1$, that is, $f(x) = x - 2$. Evaluating $\hat{f}(1.5)$ involves evaluating $f^1(1.5) = 1$ and $f^2(1.5) = 1.5$, and then computing $\hat{f}(1.5) = (-2)(1) + 1(1.5) = -0.5$.

Solving the normal equations is solving a set of linear equations, a matter covered in sections 3.4 and 6.4. We will see in Section 3.4 that this solution can be time consuming and in Section 6.4 that it can be very prone to error. This turns out to be especially the case of the normal equations for polynomial approximation (where the basis functions are given by equation 61) if the tabular arguments are approximately equally spaced. We would thus like to simplify this solution and improve its accuracy.

Solution of the normal equations is inefficient because in them the unknowns are coupled. That is, each equation involves all the unknowns. A further source of inefficiency is the fact that it is very time consuming to modify the number of basis functions, n, in the common case that n is not known beforehand but is to be determined by the quality of the approximation, increasing n until an adequate approximation is achieved (in the polynomial case n is the degree of the polynomial $+1$). Increasing n by 1 requires the normal equations to be completely re-solved after new coefficients have been computed.

Both of the above inefficiencies, as well as problems of inaccuracy of the solution of the normal equations, are avoided if the normal equations are uncoupled, that is, if each equation involves only one unknown. To do this we require that

$$F_{kj} = 0 \qquad \text{for} \quad k \neq j \tag{69}$$

so that each equation will be of the form

$$F_{kk} \, \hat{a}_k = d_k \tag{70}$$

From equation 67 zeroing F_{kj} for $k \neq j$ implies

$$\sum_{i=1}^{N} f^k(x_i) \, f^j(x_i) = 0 \tag{71}$$

In terms of the vectors $f^k = [f^k(x_1), \ f^k(x_2), \ ..., \ f^k(x_N)]^T$ and $f^j = [f^j(x_1), f^j(x_2), ..., f^j(x_N)]^T$, equation 71 is said to require that f^k and f^j be *orthogonal* according to the inner product $(u, v) = \Sigma_i u_i v_i$. If the f^j are mutually orthogonal, solving the normal equations involves computing

$$\hat{a}_k = \frac{d_k}{F_{kk}} \qquad (k = 1, 2, ..., n) \tag{72}$$

an efficient, accurate process. Furthermore, increasing n by 1 does not change d_k or F_{kk} for $k \leq n$, so it does not change $\hat{a}_k$ for $k \leq n$. Thus all that must be done, given the solution for the previous value of n, is to evaluate d_n and F_{nn} for the new value of n, divide these to produce $\hat{a}_n$, and add $\hat{a}_n \hat{f}^n(x)$ to the approximating function.

Thus we desire to choose the f^j as functions which are orthogonal (equation 71). For any nonorthogonal set of n linearly independent† functions, it is possible to find n linear combinations of these which are orthogonal. For example, a set of n orthogonal $(n-1)$th degree polynomials can be found for any set of tabular arguments $\{x_i\}$ (in fact, with the jth a polynomial of degree $j-1$ for each j), and this set can be used to compute efficiently and accurately the least-squares polynomial approximation of degree $n-1$. But this computation of the orthogonal functions is itself often time consuming and error prone, and it is better to have a set of basis functions which are known to be orthogonal over the tabular arguments.

Even with orthogonal polynomials high-degree polynomial least-squares approximation is flawed for approximately equally spaced tabular arguments because it often requires a large value of n to fit closely enough to the data points, producing inefficiency and frequently an unsmooth fit as with simple exact matching. (Note that a least-squares spline made of low-degree polynomials is a possible solution.)

In contrast, it can be shown that for equally spaced arguments a set of sine and cosine functions (sinusoids) produce a smooth, close fit for small n, and furthermore these functions are orthogonal for these arguments. If the N arguments have interval h, the basis functions are for even N

$$f^1(x) = 1$$

$$f^{2j}(x) = \cos\left[\frac{2j\pi(x - x_1)}{Nh}\right] \qquad \left(j = 1, 2, ..., \frac{N}{2}\right) \qquad [73]$$

$$f^{2j+1}(x) = \sin\left[\frac{2j\pi(x - x_1)}{Nh}\right] \qquad \left(j = 1, 2, ..., \frac{N}{2} - 1\right)$$

and for odd N

$$f^{2j+1}(x) = \cos\left[\frac{2j\pi(x - x_1)}{Nh}\right] \qquad \left(j = 0, 1, 2, ..., \frac{N-1}{2}\right) \qquad [74]$$

$$f^{2j}(x) = \sin\left[\frac{2j\pi(x - x_1)}{Nh}\right] \qquad \left(j = 1, 2, ..., \frac{N-1}{2}\right)$$

† The vectors which must be linearly independent are those obtained by evaluating each function at the set of tabular arguments.

In both cases it can be shown that

$$\sum_{i=1}^{N}(f^k(x_i))^2 = \begin{cases} N & (k=1) \\ N/2 & (1 < k \leq N-1) \end{cases}$$

For even N,

$$\sum_{i=1}^{N}(f^N(x_i))^2 = N$$

and for odd N,

$$\sum_{i=1}^{N}(f^N(x_i))^2 = \frac{N}{2} \qquad [75]$$

The approximation thus produced is said to constitute a *Fourier series* and is called a *Fourier approximation*.

An even smoother and more efficient fit can be obtained by producing an odd function from the y_i by shifting the x_i so that $x_1 = 0$, subtracting the straight line through (x_1, y_1) and (x_N, y_N), letting $y_{-i} = -y_i$ for $1 \leq i \leq N - 1$, and fitting a Fourier (sinusoidal) series to the $2(N - 1)$ points in the result (Pizer, 1975, pp. 398–409). The modified data set, since it is odd, will be fit by a linear combination of sine functions alone (no cosines), and it is shown in Pizer (1975) that such an approximation, fully specified in Programs 3–4 (p. 96), normally requires fewer basis functions than the general Fourier approximation to produce an adequate fit. This approximation, produced by subtraction of the linear tendency and odd reflection, is of the form

$$\hat{f}(x) = y_1 + \frac{y_N - y_1}{N-1}\left(\frac{x-x_1}{h}\right) + \sum_{j=1}^{n}\hat{a}_j \sin\left[\frac{j\pi(x-x_1)}{(N-1)h}\right] \qquad [76]$$

where the $\hat{a}_j$ are given by

$$\hat{a}_j = \frac{2}{N-1}\sum_{i=2}^{N-1}\left\{y_i - \left[y_1 + \frac{y_N-y_1}{N-1}(i-1)\right]\right\}\sin\left[\frac{j\pi(x_i-x_1)}{(N-1)h}\right] \qquad [77]$$

To summarize, least-squares approximation is useful when the tabular data values have significant error. Although such approximation can be done using any set of basis functions f^j that produce linearly independent vectors $[f^j(x_1), f^j(x_2), \ldots, f^j(x_N)]^T$, probably the most effective least-squares approximation is the Fourier approximation, especially with linear tendency subtraction and odd reflection. The coefficients of this approximation can often be computed most efficiently by an algorithm called the *Fast Fourier transform* (Pizer, 1975, pp. 409–414), which is discussed in the last pages of this chapter. However, a full specification of this algorithm is beyond the scope of this book.

It should be noted that least-squares approximation for which the number of basis functions used, n, is equal to the number of tabular points, N, is an exact-matching approximation, as the n coefficients $\hat{a}_j$ are determined by the n constraints given by the tabular points to make the sum

$$S = \sum_{i=1}^{N} (\hat{f}(x_i) - y_i)^2$$

achieve its minimum value of zero. Thus, for example, we can produce a Fourier exact-matching approximation by the above least-squares approach.

We have said that least-squares approximation should be used when the data points have significant error. When they do not, exact matching is usually preferable. Among exact-matching approximations, the most efficient is piecewise low-degree simple polynomial exact matching, usually obtained using a difference algorithm, but this often produces unsatisfactory accuracy because of unsmoothness. Somewhat more time consuming but usually producing a much better fit is the cubic spline approach. Fourier exact matching also usually produces a better fit than piecewise polynomial approximation using differences but often not as good as that produced by cubic splines. Furthermore, the Fourier algorithm is less efficient than the cubic splines algorithm in both computing its coefficients and evaluating $\hat{f}(x)$. Examples of the quality of fit given by each of these approximation techniques are given in Figure 3–9. Note that all the techniques do badly for extrapolation, that is, when the evaluation argument is outside the range of tabular arguments.

3.2 Operators on Functions

We are often interested in the result of an operator applied to a function, for example, the derivative of a function (produced by the differentiation operator), the integral of a function (produced by the integration operator), or the zero of a function (produced by the equation solution operator). In some of these cases the result is a number (for example, in definite integration) and in others it is a function (for example, in differentiation). In the latter case we usually want to evaluate the resulting function at a given argument, in which case we will let the operator in question include the evaluation step. Thus, let us assume that we wish to compute the number

$$y = T(f) \tag{78}$$

where T is an operator and f is a function, and that we can not compute $T(f)$ analytically or can not do so efficiently.

The general approach is to approximate f by a function $\hat{f}$ to which we can easily apply T analytically, and then to apply T to $\hat{f}$:

$$\hat{y} = T(\hat{f}) \qquad [79]$$

We say we *approximate and operate*. We have already seen this technique for the evaluation operator: to evaluate f at x_{eval} we approximated f by an easily evaluated $\hat{f}$, and then we evaluated $\hat{f}$ at x_{eval}. The same approach works for almost every operator.

Thus numerical differentiation of f at x_{deriv} [calculating $f'(x_{deriv})$] is accomplished by approximating f by a function $\hat{f}$ which is easily differ-

FIGURE 3–9

Approximation by Various Algorithms

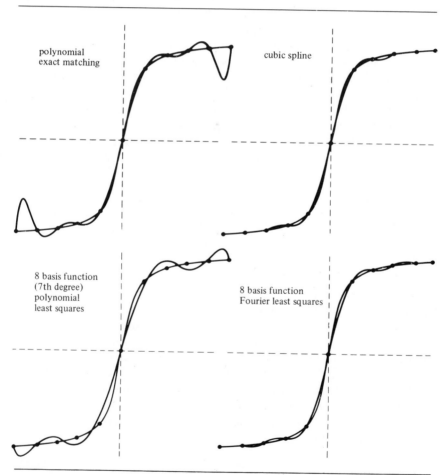

polynomial
exact matching

cubic spline

8 basis function
(7th degree)
polynomial
least squares

8 basis function
Fourier least squares

entiated and whose derivative is easily evaluated, and then $f'(x_{\text{deriv}})$ is approximated by $\hat{f}'(x_{\text{deriv}})$.

The basis functions and approximation criterion used must be chosen not only so that T may be applied easily but also so that $T(\hat{f})$ is a good approximation to $T(f)$. For example, in numerical differentiation where f is given as a table of measured values (with measurement error!), piecewise simple polynomial exact matching is a poor approximation technique because, although polynomials are easily differentiated and the result evaluated, a small error in the tabular data can cause a large error in the derivative (see Figure 3–10). Significantly improved results will be obtained using an approximation technique which gives a smoother approximation, such as cubic splines or Fourier least-squares approximation. The latter method should do better with inaccurate tabular values. However, all approximation techniques result in a derivative value approximation that is rather sensitive to error, so numerical differentiation should be avoided altogether if possible.

As another example of the use of the approximate and operate approach, let us consider the definite integral operation:

$$T(f) = \int_a^b f(x)\,dx \qquad\qquad [80]$$

Here we approximate $\int_a^b f(x)\,dx$ by $\int_a^b \hat{f}(x)\,dx$, where $\hat{f}$ is chosen to be easily integrated. This approach is called *numerical integration* or *quadrature*.

FIGURE 3–10
Sensitivity of Numerical Differentiation Based on Piecewise Polynomial
Exact Matching

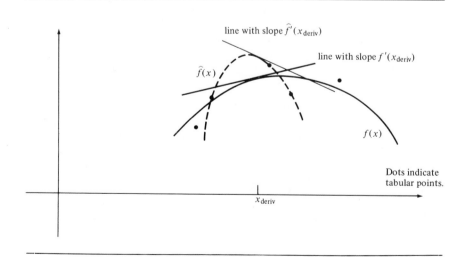

Numerical integration is more accurate than numerical differentiation because it tends to cancel errors in the approximation to the function (see Figure 3–11). When one considers that integration is the inverse of differentiation, this behavior seems reasonable. Since differentiation maps small approximation errors into large errors, integration maps large approximation errors into small errors.

Numerical integration is necessary both when the function is not available analytically and when the function is available analytically but the integral can not be obtained analytically. In fact, the second case is the more common one. Therefore, we shall discuss methods that assume that the argument values to be used can be specified by the methods. In the case of a fixed table, these methods may not be applicable.

Newton-Cotes Methods for Numerical Integration

Because of the good error behavior of integration, we can expect that a method based on simple polynomial exact-matching approximation will be satisfactory. The Newton-Cotes methods involve the integration of simple equal-interval exact-matching polynomial approximation formulas. That is, to evaluate

$$I \triangleq \int_a^b f(x)\, dx \qquad\qquad [81]$$

FIGURE 3–11
The Accuracy of Numerical Integration: Cancellation of Approximation Errors

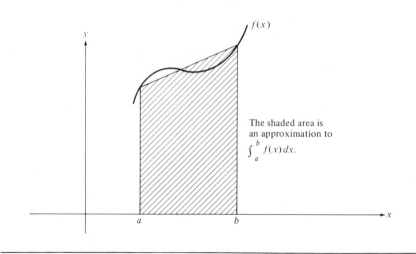

The shaded area is an approximation to $\int_a^b f(x)\,dx.$

we divide the interval $[a, b]$ into n equal intervals, defining

$$x_0 = a; \quad x_n = b; \quad h = \frac{b - a}{n}; \quad x_i = x_0 + ih \quad \text{(for } n \geqslant 1\text{)} \quad [82]$$

and we fit an nth-degree polynomial at these arguments and integrate that polynomial to produce the approximation $\hat{I}$.

FIGURE 3–12
Numerical Integration Using the Rectangular Rule. (a) Based on left endpoint of interval; (b) based on midpoint of interval.

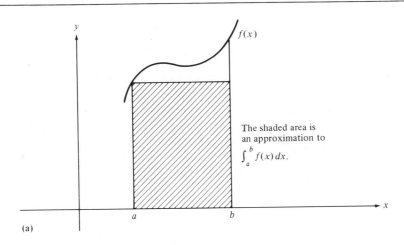

(a)

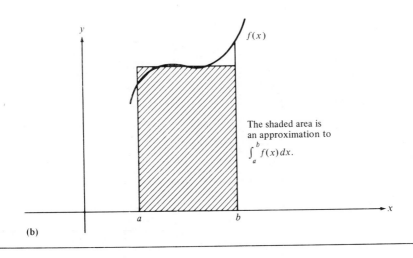

(b)

For $n = 0$ the approximating polynomial is the constant $f(x_0)$. The approximation to the integral is the area of the rectangle of height $f(x_0)$ and width $b - a = x_1 - x_0 \triangleq h$ (see Figure 3–12a), so

$$\hat{I} = (b - a)f(x_0) \qquad \qquad \textbf{[83]}$$

This integration method is called the *rectangular rule*. It can be improved by using a value of the constant (the height of the rectangle) which is more likely to be closer to the average value of $f(x)$ in $[a, b]$ than is the value of f at one end of the interval. The value at the center of the interval, $f((a + b)/2)$, is such a value. This version of the rectangular rule (see Figure 3–12b), given by

$$\hat{I} = (b - a)f\left(\frac{a + b}{2}\right) = hf\left(\frac{x_0 + x_1}{2}\right) \qquad \qquad \textbf{[84]}$$

is therefore an improvement over the previous version.

We could expect further improvement by using a higher-degree polynomial approximation. For $n = 1$ the polynomial fitted is a straight line and the approximation to the integral is the area of the trapezoid under this straight line (see Figure 3–13). This integration method is called the *trapezoidal rule*. Mathematically, it involves integrating the first two terms of the Newton divided-difference formula:

FIGURE 3–13
Numerical Integration Using the Trapezoidal Rule

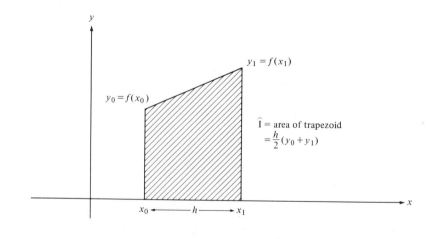

$$\hat{I} = \int_{x_0}^{x_1} (f(x_0) + (x - x_0)f[x_0, x_1]) \, dx$$

$$= (x_1 - x_0)f(x_0) + \frac{(x_1 - x_0)^2}{2} f[x_0, x_1]$$

$$= (x_1 - x_0)\left(f(x_0) + \frac{x_1 - x_0}{2} \frac{f(x_1) - f(x_0)}{x_1 - x_0}\right) \qquad [85]$$

$$= \frac{x_1 - x_0}{2}(f(x_0) + f(x_1))$$

$$= \frac{h}{2}(f(x_0) + f(x_1))$$

where $x_0 = a$, $x_1 = b$, and $h = x_1 - x_0$. Notice that the right side of equation 85 is precisely the area of a trapezoid with height h and parallel sides of lengths $f(x_0)$ and $f(x_1)$.

If $n = 2$, that is, if we divide the integration range into two intervals, fit a parabola, and integrate it, we have

$$\hat{I} = \int_{x_0}^{x_2} (f(x_0) + (x - x_0)f[x_0, x_1]$$

$$+ (x - x_0)(x - x_1)f[x_0, x_1, x_2]) \, dx \qquad [86]$$

Rewriting $x - x_1$ as $(x - x_0) - h$, where

$$h = \frac{x_2 - x_0}{2} = x_1 - x_0 \qquad [87]$$

we have

$$\hat{I} = (x_2 - x_0)f(x_0) + \frac{(x_2 - x_0)^2}{2} f[x_0, x_1]$$

$$+ \left(\frac{(x_2 - x_0)^3}{3} - h\frac{(x_2 - x_0)^2}{2}\right) f[x_0, x_1, x_2]$$

$$= (x_2 - x_0)\left(f(x_0) + h\frac{f(x_1) - f(x_0)}{h} + \frac{1}{3}h^2 \frac{f(x_2) - 2f(x_1) + f(x_0)}{2h^2}\right) \qquad [88]$$

$$= (x_2 - x_0)\left(\frac{1}{6}f(x_0) + \frac{2}{3}f(x_1) + \frac{1}{6}f(x_2)\right)$$

$$= \frac{h}{3}(f(x_0) + 4f(x_1) + f(x_2))$$

The integration method given by equation 88 is called *Simpson's rule*. Higher-order rules are developed in the same way, but we will see later that they are not normally used because of considerations of accuracy. Accuracy *is* increased by using a finer subdivision of $[a, b]$; but instead of using a higher-degree polynomial approximation for this increased number of tabular arguments, a piecewise low-degree approximation is used for the integrand. Thus with an nth degree rule we let $N = kn$ for some integer k, $x_0 = a$, and $x_N = b$, and we rewrite

$$\int_{x_0}^{x_N} f(x)\,dx = \int_{x_0}^{x_n} f(x)\,dx + \int_{x_n}^{x_{2n}} f(x)\,dx + \cdots + \int_{x_{(k-1)n}}^{x_N} f(x)\,dx \quad [89]$$

Then we approximate $f(x)$ in each "panel" $[x_{jn}, x_{(j+1)n}]$ by an nth degree exact-matching polynomial through f at x_{jn}, x_{jn+1}, x_{jn+2}, ..., $x_{(j+1)n}$ and integrate these approximations.

For example, for $n = 1$ this method consists of summing the trapezoidal approximations over $[x_0, x_1]$, $[\dot{x}_1, x_2]$, ..., and $[x_{N-1}, x_N]$; that is from equation 85

$$\hat{I} = \frac{h}{2}(f(x_0) + f(x_1)) + \frac{h}{2}(f(x_1) + f(x_2))$$
$$+ \cdots + \frac{h}{2}(f(x_{N-1}) + f(x_N)) \quad [90]$$
$$= \frac{h}{2}(f(x_0) + 2f(x_1) + 2f(x_2) + \cdots + 2f(x_{N-1}) + f(x_N))$$

This is the so-called *composite trapezoidal rule* (see Program 3–5, p. 98).

Similarly the composite Simpson's rule from equation 89 is

$$\hat{I} = \frac{h}{3}(f(x_0) + 4f(x_1) + f(x_2)) + \frac{h}{3}(f(x_2) + 4f(x_3) + f(x_4))$$
$$+ \cdots + \frac{h}{3}(f(x_{N-2}) + 4f(x_{N-1}) + f(x_N))$$
$$= \frac{h}{3}(f(x_0) + 4f(x_1) + 2f(x_2) + 4f(x_3)$$
$$+ \cdots + 2f(x_{N-2}) + 4f(x_{N-1}) + f(x_N)) \quad [91]$$

where N is a multiple of 2 (see Program 3–6, p. 99).

For illustration, we apply the composite Simpson's rule with $N = 4$ to $I = \int_0^\pi \sin(x)\,dx$ (integrating, we see $I = 2$). Then $h = (\pi - 0)/4 = \pi/4$, so $\hat{I} = (\pi/12)\,(\sin(0) + 4\sin(\pi/4) + 2\sin(\pi/2) + 4\sin(3\pi/4) + \sin(\pi)) = 2.005$, which is not a bad approximation.

The composite trapezoidal rule has the advantage that, if the result for a given value of N is not accurate enough, N can be doubled with a computation involving only the present result and the additional tabular function values. It is left to Problem 3.19 to show that if $\hat{I}_N$ is the result of applying the trapezoidal rule with N panels,

$$\hat{I}_{2N} = \frac{1}{2}\left\{\hat{I}_N + h \sum_{i=1}^{(b-a)/h} f\left(a + \left(i - \frac{1}{2}h\right)\right)\right\} \qquad [92]$$

where $h = (b - a)/N$, the panel width in the N interval rule. It is also shown there that Simpson's rule with $2N$ intervals can be computed from the trapezoidal rule results as

$$\hat{I}_{2N}^{\text{Simp}} = \hat{I}_{2N}^{\text{trap}} + \frac{\hat{I}_{2N}^{\text{trap}} - \hat{I}_N^{\text{trap}}}{3} \qquad [93]$$

The composite Newton-Cotes rules are used more frequently than are the simple rules because of the better error properties of the former. For the same reason the composite Simpson's rule is better than the composite rectangular and trapezoidal rules, but later we shall see better methods than any of these.

The basic point of this section has not been to develop methods of numerical differentiation and integration but to illustrate the application of the approach, approximate and operate. It is hard to overemphasize this approach — it is the most common in all of numerical computing. So when we try to solve equations $f(x) = 0$, we approximate $f(x)$ by $\hat{f}(x)$ and apply the equation solution operator to $\hat{f}$, which has been chosen so that this solution is easily found; when we try to find optima of functions we approximate them and find the optima of the approximation; etc., etc. Whatever the operator on a function, we approximate the result by approximating the function and applying the operator to the approximation.

3.3 Solution of Differential Equations

A common problem in numercial computing is to evaluate at a point or points a function specified by a differential equation: an equation in which some derivative of a function is given in terms of lower-order derivatives of the function, the function itself, and the argument value, and the object is to find the function. Multidimensional differential equations, called *partial differential equations,* are solved by extensions of the techniques used to solve single variable equations, called *ordinary differential equations.* Equations involving derivatives higher than the first also are solved by extension or application of techniques for solving equations involving derivatives of order no higher than the first, called *first-order differential equations.* Thus, here we will cover only first-order ordinary differential equations with initial values:

$$y'(x) = f(x, y) \qquad [94]$$

$$y(x_0) = y_0 \qquad\qquad [95]$$

Equation 94 gives a direction for each point in the (x, y) plane. Starting at any point in the plane, we can move an infinitesimal distance in the specified direction, recompute the function at the new x value to get a new direction, move an infinitesimal distance in that direction, and so on, defining a function (see Figure 3–14, in which a finite step width, h, is used; in the correct solution, $h \to 0$). If we start at the same x but a different y, a different function is produced. In fact, the second function can not coalesce with or cross the first function. Let us assume either was possible and did occur. Let (x_c, y_c) be the point in common. Then using equation 94, we could extend y back to $x = x_0$. Its value there could not be both the y_0 given by the first condition and that given by the second condition.

What we have said is that equation 94 gives a so-called direction field on the space, defining a family of functions $y(x)$, and equation 95 picks out a particular member of the family (see Figure 3–15).

The method of numerical solution of ordinary differential equations is very much like that specified above for infinitesimal distances, except that in real numerical solution we must use noninfinitesimal distances h instead of infinitesimal distances δ. Thus, the numerical solution of first-order ordinary differential equations involves using x_i, y_i, $y_i' = f(x_i, y_i)$, and possibly x, y, and y' at previous values of x to compute $y(x_{i+1}) \equiv y_{i+1}$ (see Figure 3–14). This is an extrapolation process, and extrapolation is very prone to error. The error problem is aggravated by

FIGURE 3–14
Solution of a Differential Equation as Extrapolation

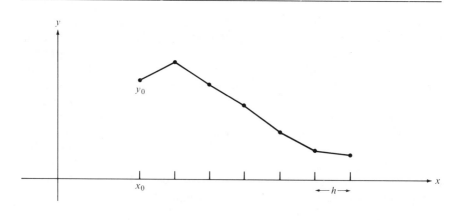

FIGURE 3–15
A Family of Functions Defined by a Differential Equation

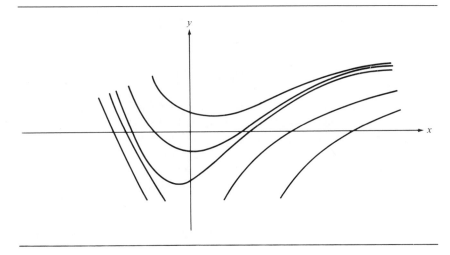

the fact that y_{i+1} is used to compute y_{i+2}, y_{i+2} is used to compute y_{i+3}, and so on, thereby propagating error. We must be very careful to choose h small enough so that the extrapolation error is small.

We will see that to compute $y(x_{\text{goal}})$, we normally take a number of steps between x_0 and x_{goal}. Any error that we make at an intermediate step will change the member of the family of solutions on which we are moving. After this point, unless we use information about the function at points before the point (value of x) at which we stand, we can do no better than stay on the new member of the family.

3.3.1 REJECTION OF LONG TAYLOR SERIES

Given equation 94 with the function f given analytically, we could compute formulas for all derivatives of y by analytically differentiating equation 94 and use these derivatives in a Taylor formula to compute $y(x)$ at any desired value of x for which the Taylor series converges.

Differentiating equation 94, we obtain

$$y''(x) = \frac{d}{dx}y'(x) = \frac{\partial f}{\partial x}(x, y) + \frac{\partial f}{\partial y}(x, y)\frac{dy}{dx}$$

$$= \frac{\partial f}{\partial x}(x, y) + f(x, y)\frac{\partial f}{\partial y}(x, y) \qquad [96]$$

Differentiating again, we obtain

$$y'''(x) = \frac{d}{dx} y''(x) = \frac{\partial^2 f}{\partial x^2}(x, y) + \frac{\partial f}{\partial x}(x, y) \frac{\partial f}{\partial y}(x, y)$$

$$+ 2f(x, y) \frac{\partial^2 f}{\partial x \partial y}(x, y) + [f(x, y)]^2 \frac{\partial^2 f}{\partial y^2}(x, y) \quad [97]$$

$$+ f(x, y) \left[\frac{\partial f}{\partial y}(x, y) \right]^2$$

and so on. In general, $y^{(n)}(x)$ depends on all partial derivatives of f of order $n - 1$ or less. Thus, evaluating a Taylor series using terms of degree up to the nth requires the evaluation of n derivatives of order $n - 1$, $n - 1$ derivatives of order $n - 2$, and so on; that is, it requires evaluating $n(n + 1)/2$ functions. If f is complicated, as it often is when we can not solve the equation analytically, evaluation of so many derivatives is very time consuming. (Furthermore, the user must do the analytic differentiation required unless a formal differentiation program is used.) Thus, instead of using a high-degree Taylor series to approximate over a relatively large distance, it is more efficient to divide the interval $[x_0, x_{goal}]$ into a number of pieces and successively use lower-degree Taylor series over each interval. That is, one can achieve the same overall error with less work by using low-degree Taylor series and small intervals and applying the Taylor series to move from x_0 to x_1, from x_1 to x_2, and so on, through from x_{N-1} to $x_N = x_{goal}$, rather than using one high-degree Taylor series to move from x_0 to x_{goal} in one step. This statement is true even when one takes into account the fact that in the former method one has to be concerned about the propagation of error from step to step. Therefore, we will not consider methods that use derivatives of y higher than the first.

3.3.2 RECURRENT METHODS

The methods that we will use, then, all consist of dividing the problem into a sequence of steps, and in each step applying the same operator to the result of the previous step, that is, recurrently. A step of a method for solving differential equations solves for y_{i+1}, given y_i and $y_i' = f(x_i, y_i)$. It is useful to write the ideal step of the method:

$$y(x_{i+1}) = y(x_i) + \int_{x_i}^{x_{i+1}} y'(x) \, dx = y_i + \int_{x_i}^{x_{i+1}} f(x, y(x)) \, dx \quad [98]$$

By writing the problem in this form, we can solve the problem by numerical integration and thus take advantage of the favorable error properties of numerical integration. The problem could also be cast in terms of numerial differentiation, but this would be a bad choice.

We have reduced our problem to finding the best numerical integration method for numerically integrating $\int_{x_i}^{x_{i+1}} f(x, y(x))\, dx$, where we know f and x throughout the integration interval but we do not know y throughout the interval. The solution of differential equations can be thought of as a combination of extrapolation, with its poor error properties, and numerical integration, with its relatively good error properties.

Since at the outset of the ith step the only value of y we know in the interval is y_i, we can use the left endpoint form of the rectangular rule for integration (Newton-Cotes rule for $n = 0$; see Figure 3–12a). In doing so, we approximate f by $f(x_i, y_i)$ throughout the whole interval. Assume that the interval width h is constant from step to step. Then the rectangular rule applied to equation 98 produces

$$\hat{y}_{i+1} = y_i + hf(x_i, y_i) \tag{99}$$

The method which moves from step to step using equation 99 to give the computed value of y_{i+1} given y_i is called *Euler's method*.

Euler's method can also be written

$$\hat{y}_{i+1} = y_i + hy'_i \tag{100}$$

(see Program 3–7, p. 100). In this form it is simply a Taylor series truncated after the first-degree term. It can be graphically interpreted as moving a step-length h from y_i in the direction of the tangent to $y(x)$ at x_i (see Figure 3–16). Notice that h may be negative if we wish to move in a negative direction from x_0.

We have argued in Section 3.2 that using the midinterval value in

FIGURE 3–16
A Step of Euler's Method

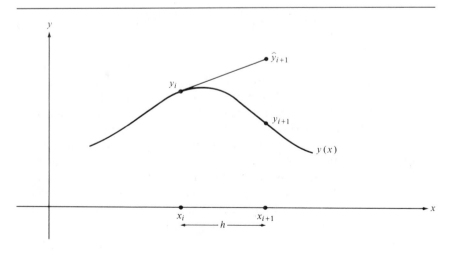

the rectangular rule for integration gives better results than using the value at the left end of the interval. If we try to apply this idea to the integration in equation 98, we find we need $y(x_{i+1/2})$, a value we do not have when we try to integrate. But if we consider the interval of width $2h$, $[x_{i-1}, x_{i+1}]$, we can recast equation 98 as

$$y_{i+1} = y_{i-1} + \int_{x_{i-1}}^{x_{i+1}} y'(x)\, dx = y_{i-1} + \int_{x_{i-1}}^{x_{i+1}} f(x, y(x))\, dx \qquad [101]$$

so using the midinterval rectangular rule we can approximate y_{i+1} by

$$\hat{y}_{i+1} = y_{i-1} + 2hf(x_i, y_i) \qquad\qquad [102]$$

Since for all steps but the first we have both y_i and y_{i-1}, we can use this *modified Euler's method* for these steps, with improved results over the simple Euler's method.

We might do even better if we used the trapezoidal rule, based on a first-degree polynomial approximation, instead of the rectangular rule, based on a zeroth-degree approximation. Applying the trapezoidal rule to equation 98, we obtain

$$\hat{y}_{i+1} = y_i + h\left(\frac{f(x_i, y_i) + f(x_{i+1}, y_{i+1})}{2}\right) \qquad [103]$$

Equation 103 is called *Heun's method*. The difficulty is that we do not know y_{i+1}, so we can not compute $f(x_{i+1}, y_{i+1})$. But we can approximate y_{i+1} by using the modified Euler's method. If we use this approximate value of y_{i+1} in equation 103, we might produce a better value for y_{i+1}:

$$y_{i+1}^{(c)} = y_i + h\left(\frac{f(x_i, y_i) + f(x_{i+1}, y_{i+1}^{(p)})}{2}\right) \qquad [104]$$

where $y_{i+1}^{(p)}$ is the value of y_{i+1} predicted by the modified Euler's method. We say we use the modified Euler's method as a *predictor* for y_{i+1} (except at the first step, where another method must be used) and Heun's method as a *corrector* for y_{i+1} (see Program 3–8, p. 101). Note that each step consists of the calculation of a predictor and a corrector.

Other methods for the solution of differential equations are based on other integration rules. In general, we will see that we must be concerned not only with the approximation error in these rules but also with the propagation of this error from step to step. In many cases the error in successive steps can interact to cause a small error at one step to produce violent error oscillations or growth in the successive results, a behavior called *instability*.

3.4 Solution of Equations

It is often necessary to find the value or values of x for which

$$f(x) = 0 \qquad [105]$$

(the roots of the equation) for a given function f. In the case of more than one variable it may be similarly necessary to find the values of the n variables $x_1, x_2, ..., x_n$ which make the n functions $f_1(x_1, x_2, ..., x_n)$, $f_2(x_1, x_2, ..., x_n)$, ..., $f_n(x_1, x_2, ..., x_n)$ all zero. This system of equations can be written in vector form as

$$\mathbf{f(x)} = \mathbf{0}, \qquad [106]$$

where $\mathbf{f} \triangleq [f_1, f_2, ..., f_n]^T$ and $\mathbf{x} \triangleq [x_1, x_2, ..., x_n]^T$. Since the algorithms for the solution of multidimensional equations are often generalizations of algorithms for one equation in one unknown, we will treat the latter first.

3.4.1 SOLUTION OF ONE EQUATION IN ONE UNKNOWN

We need to find the root $x = \xi$ for which $f(\xi) = 0$. If $f(x)$ is a polynomial of degree 4 or less (or the result of dividing such a polynomial by a function which does not have a root at the roots of f in which we are interested), then there exist methods of direct solution, as long as we assume that methods of computing the $1/2$, $1/3$, and $1/4$ powers of numbers are available. In particular, the solution of the linear equation

$$a_1 x + a_0 = 0 \qquad [107]$$

is simply computed as

$$\xi = \frac{-a_0}{a_1} \qquad [108]$$

If f is not a polynomial of degree 4 or less, there is no general closed form method for finding its root(s) and we are forced to use an iterative method in which first the root is approximated, then an algorithm is applied to improve the approximation, then the algorithm is reapplied to the result to obtain still a better approximation, and so on. We call the sequence of approximations the *iterates* $x_0, x_1, x_2, \ldots$.

The accuracy of the roots computed by these iterative methods is a matter for Chapter 9. But there will clearly be inadequate accuracy if the sequence of iterates does not converge to the root. Here we will need to discuss whether such convergence occurs.

The efficiency of root-finding methods depends on the number of iterations required and the amount of computation per iterative step. The number of iterations required depends on the rate of convergence. Though this matter will be treated in more detail in Chapter 9, we will treat it at the intuitive level here. The existence of convergence and its rate usually depend not only on the method but also on the shape of the function f. Thus, a skill in sketching the graphs of functions is useful. We will not attempt to develop this skill here, but will refer the student to Pizer (1975, Section 3.2).

As to the efficiency of each iterative step, normally each step requires one or more evaluations of the function f or a related function. These evaluations are generally the dominant part of the computation. Thus, we will often specify the efficiency of the iterative step in terms of the number of function evaluations it requires.

In the following we will assume that we are searching for roots of multiplicity 1, that is, if ξ is the root,

$$0 < \lim_{x \to \xi} \left| \frac{f(x)}{x - \xi} \right| < \infty$$

Algorithms for multiple roots, that is, for which

$$0 < \lim_{x \to \xi} \left| \frac{f(x)}{(x - \xi)^n} \right| < \infty \qquad \text{for} \quad n \neq 1$$

will be treated only in passing.

Root-Trapping Methods

Bisection Method. The first approach to finding the root of $f(x) = 0$ is to find an interval in which the root is known to be, then subdivide the interval and determine which subinterval contains the root, and iterate this process. If applicable, such a procedure always produces convergence to the root. The simplest form of this "divide and conquer" root-trapping approach is to find two points x^+ and x^- such that $f(x^+) > 0$ and $f(x^-) < 0$. Since f is assumed to be continuous, there must be a root of f in the interval $[x^+, x^-]$.† We choose x as the midpoint of the interval and evaluate $f(x)$. If $f(x) > 0$, the root must be in the interval $[x, x^-]$. We can replace x^+ with x, resulting in an interval that is half the size of the previous interval. If $f(x) < 0$, we can replace x^- with x, again resulting in an interval that is half the size of the previous interval. Thus, at each iteration the root is trapped in an interval that is half the size of that produced by the previous iteration. We continue until the interval is smaller than the tolerance within which we wish to know the root, or until $f(x)$ is so small that the arithmetic error in $f(x)$ may be as large as $f(x)$, so we can not trust the sign of the computed value of $f(x)$.

† The notation $[a, b]$ will be used to denote the closed interval with endpoints a and b, with no requirement that $a \leq b$.

This technique for finding the root of a nonlinear equation is called the *bisection method*. It is carefully specified in Program 3–9 (p. 102). We start with two argument values straddling the root. These two values, which we call x_0^+ and x_0^-, must satisfy the relations $f(x_0^+) > 0$ and $f(x_0^-) < 0$. The existence of such points is necessary for applicability of the bisection method. The points exist if $f(x)$ crosses the x axis at $x = \xi$, which is certainly the case if ξ is a single root.

At the $(i+1)$th iteration, the algorithm begins with two interval endpoints, which we will call x_i^+ and x_i^-, and computes a new iterate, which we will call x_{i+1}, as the interval midpoint. For example, let $f(x) = x^2 - 2$, $x_0^+ = 2$, and $x_0^- = 1$. Then $x_1 = 1.5$, and since $f(1.5) > 0$, $x_1^+ = 1.5$ and $x_1^- = x_0^- = 1$. Then $x_2 = 1.25$, and since $f(1.25) < 0$, $x_2^- = 1.25$ and $x_2^+ = x_1^+ = 1.5$. Further iterations continue the convergence to $\xi = 1.414$.

An advantage of the bisection method is that if appropriate starting points are found the sequence of iterates always converges to the desired root. Let us analyze its efficiency. As far as computation per iterative step, the method is quite efficient. At each step the major cost is the improvement, which requires only one addition and one division beyond a single evaluation of the function f.

From the point of view of the number of iterative steps required to achieve a desired accuracy, we see that the maximum error magnitude is halved at each iteration. Such convergence is relatively slow.

Method of False Position. How can we improve the rate of convergence of the bisection method? Can we take advantage of our knowledge of the function f to make a better estimate of the root in the interval in question than the midpoint of the interval? We will do so by applying our standard technique — approximate and operate — on which many of the methods for solving nonlinear equations are based: Using the present estimate(s) of the root, we will approximate the nonlinear equation by a linear equation [approximate $f(x)$ by a straight line] and use the root of our approximation to produce a better estimate of the root of the nonlinear equation. That is, we can fit a straight line between the points $(x_i^+, f(x_i^+))$ and $(x_i^-, f(x_i^-))$ to obtain a linear approximation to the function f in the interval (see Figure 3–17), and we can use the root of the line as an approximation of the root of f. Then we can proceed as with the bisection method, replacing the appropriate one of x_i^+ and x_i^- with the new estimate of the root. If f is reasonably well approximated by the straight line, we can expect this method to produce faster convergence than the bisection method because we are choosing the successive new interval endpoints in a more educated manner.

This technique is called the *method of false position*, which is specified in Program 3–10 (p. 103). As with the bisection method, we start with two values x_0^+ and x_0^- straddling the root such that $f(x_0^+) > 0$ and

FIGURE 3–17
A Step of the Method of False Position

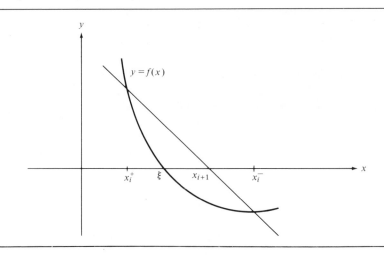

$f(x_0^-) < 0$. At the $(i+1)$th iteration, the algorithm takes the interval endpoints, x_i^+ and x_i^-, and computes the new iterate x_{i+1} as the zero of the line going through f at x_i^+ and x_i^- :

$$x_{i+1} = \frac{x_i^+ f(x_i^-) - x_i^- f(x_i^+)}{f(x_i^-) - f(x_i^+)} \qquad [109]$$

Before replacing one of the interval endpoints with x_{i+1}, we must check whether we have converged to within our desired tolerance. Because the interval width need not approach zero with this method, the test that we used with the bisection method is not applicable. Here are the two most common convergence tests, used with all methods which follow in this section:

1. Test whether the successive iterates are relatively close enough together, that is, whether

$$\frac{|x_{i+1} - x_i|}{|x_{i+1}|} < \varepsilon \qquad [110]$$

2. Test whether

$$|f(x_{i+1})| < \delta \qquad [111]$$

Each of these tests can produce undesirable results with appropriately pathological functions. Test 1 is most commonly used because it depends less on our accuracy in calculating f. Of course, we could require that the convergence satisfy both tests before we stopped.

As an example of the application of the method, again let $f(x) = x^2 - 2$, $x_0^+ = 2$, and $x_0^- = 1$. Then $x_1 = (2(-1) - 1(2))/(-1 - 2) = 4/3 \approx 1.33$. Since $f(4/3) < 0$, $x_1 = 4/3$ and $x_1^+ = x_0^+ = 2$. Then $x_2 = 1.4$, and since $f(1.4) < 0$, $x_2 = 1.4$ and $x_2^+ = x_1^+ = 2$. And so on. As we hoped, x_2 from the method of false position (1.4) is closer to the root than is x_2 from the bisection method (1.25).

It can be proved that the sequence of iterates produced by the method of false position always converges to a root of $f(x) = 0$ (Pizer, 1975, pp. 192–194). But we must analyze the speed of convergence. The amount of computation per step is only slightly more time consuming than that of the bisection method. Like bisection, it requires the evaluation of f at only one new point per iterative step (the evaluation of f at the other interval endpoint has been done in a previous step). However, the evaluation must be somewhat more accurate for the method of false position because the value itself is used rather than simply its sign.

As for the number of iterations required, even though the method of false position is designed to produce faster convergence than the bisection method, for some functions it does not. Since the method of false position depends on approximating f by a straight line, we might correctly predict that when the approximation is a good one, convergence will be fast (see Figure 3–18a). We can see that the convergence is particularly slow when the interval endpoint on one side of the root becomes frozen far from the root, producing a straight-line approximation that does not improve quickly (see Figure 3–18b). And it can be shown

FIGURE 3–18

Speed of Convergence of Method of False Position. (*a*) f fit well by line; (*b*) f fit poorly by line.

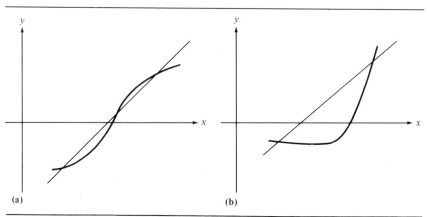

that freezing one endpoint always happens after some iteration if $f''(\xi) \neq 0$ (see Problem 9.3).

Newton's Method

To improve the method of false position, we need to find a linear approximation that is assured of becoming close to f as the iterates approach the root. A reasonable approximating line is the tangent to f at the most recent iterate, x_i (see Figure 3–19). Using the root of this line as the next iterate, x_{i+1}, produces

$$f'(x_i) = \frac{f(x_i) - 0}{x_i - x_{i+1}}$$ [112]

which leads to the iterative step

$$x_{i+1} = x_i - \frac{f(x_i)}{f'(x_i)}$$ [113]

This iterative method is called the *Newton–Raphson method*, or just *Newton's method* (see Program 3–11, p. 105).

Note that with Newton's method the root is no longer trapped in an interval at each iterative step. Unlike the methods of bisection and false position, it is not a root-trapping method. But like the method of false position and other methods we will see, it is based on the idea of

FIGURE 3–19
Newton's Method

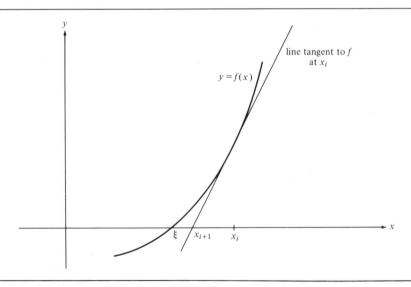

at each step approximating f by a straight line and letting its root be the next iterate.

We will see that unlike the root-trapping methods, Newton's method is not assured of convergence. But because the linear approximation on which it is based improves as the iterates approach the root, when it does converge it usually does so very quickly. And we will see in Chapter 9 that there always exists some interval about a single root for which Newton's method does converge.

When does Newton's method not converge? Examining equation 113, we see there is difficulty if $f'(x_i) = 0$. Thinking of the iterative step graphically, we can see there is trouble when $f'(x_i) = 0$. In that case the approximating straight line is parallel to the x axis, so it has no root. Trouble also occurs when $f'(x_i)$ is approximately 0. In that case the value x_{i+1} may be in the wrong direction from x_i (see Figure 3–20a), an os-

FIGURE 3–20
Newton's Method When $f'(x_i) \approx 0$

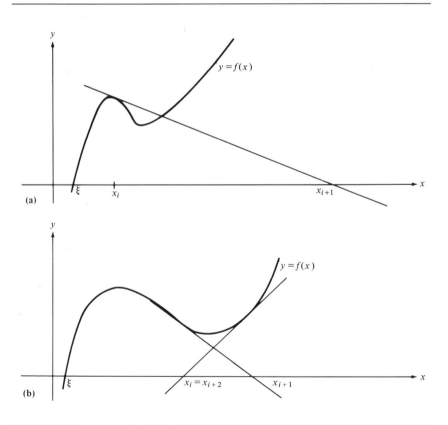

cillation may take place about the zero derivative (see Figure 3–20b), or the "improvement" in the root estimate may be a great overcorrection. We must be very careful when considering application of Newton's method to a function which has zero or near-zero derivatives between the root and any point that may be reached by our iteration. In fact, we must establish beforehand that this behavior does not occur. Certainly, a function that has many oscillations between our estimate of the root and the root itself is a clear case where Newton's method should not be used.

On the other hand, if the function $f(x)$ has no near-zero derivative near the root, Newton's method is a most effective method. For example, consider the operation of taking the square root of a. This involves solving the equation

$$f(x) = x^2 - a = 0,$$ [114]

which is interpreted graphically in Figure 3–21. We see that for any initial guess greater than 0, there are no near-zero derivatives in the region of interest. Thus Newton's method, given by

$$x_{i+1} = \frac{1}{2}\left(x_i + \frac{a}{x_i}\right)$$ [115]

converges quickly. Therefore, this method is used for finding square roots in most square root subroutines commonly provided. For example, if we apply this method to the equation used to illustrate previous methods, $x^2 - 2 = 0$, with $x_0 = 2$, we see $x_1 = (1/2)(2 + 2/2) = 1.5$, $x_2 = (1/2)(1.5 + 2/1.5) = 1.42$, and so forth, a rapid convergence.

Geometric interpretation of Newton's method also provides another sufficient condition for convergence when the method is used: $f(x)\,f''(x) >$

FIGURE 3–21
Graphic Interpretation of $x^2 - a = 0$

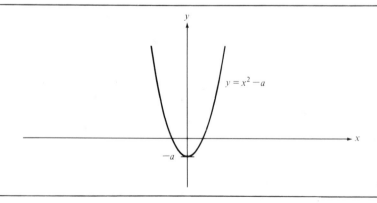

$y = x^2 - a$

0 in $[x_0, \xi]$. Sketches on which this condition is based and the formal proof of its sufficiency are left to Problem 3.26.

To use Newton's method, we must evaluate $f'(x)$ analytically and provide the formula for f' to the computer. If we have no way of formally differentiating f — for example, if we have a subroutine for f but no analytic expression for f — Newton's method can not be applied. Even if we can differentiate f, at each iteration we require two function evaluations, one of f and one of f', as compared to one function evaluation for the method of false position. If the time required for the evaluation of $f'(x_i)$ is about the same as that for $f(x_i)$, one step of Newton's method requires approximately the same amount of computation as two steps of false position. Although Newton's method will normally require fewer steps for convergence than the method of false position, the total time needed to obtain a solution of the required accuracy may be less with false position than with Newton's method (although this will not normally be the case).

We should note that many common functions have derivatives that involve the same function or functions; the derivatives can be simply computed from the values of the original function. Two examples follow:

$$\frac{d}{dx} e^x = e^x \qquad [116]$$

$$\frac{d}{dx} \tan(x) = \sec^2(x) = 1 + \tan^2(x) \qquad [117]$$

In other cases, the derivative can be written in terms of the original function but in a way that requires an amount of computation which although not great is not negligible. For example,

$$\frac{d}{dx} \sin(x) = (1 - \sin^2(x))^{1/2} \qquad [118]$$

Especially in the first cases, the extra computation involved in evaluating the derivative at each step does not significantly increase computation time once we have evaluated the function. In any case, the computation time per iterative step must be taken into account when deciding whether Newton's method should be used.

Before moving on to a method that converges almost as quickly as Newton's method without requiring two function evaluations per step, let us note the general form of Newton's method:

$$x_{i+1} = g(x_i) \qquad [119]$$

Such an iteration is called a *Picard iteration*. After one interval endpoint becomes frozen, the method of false position is also a Picard iteration. Also, there are an infinite number of ways in which a given equation

$f(x) = 0$ can be rewritten in the form $x = g(x)$, and any of these can be made into a Picard iteration: $x_{i+1} = g(x_i)$. We will analyze Picard iterations in detail in Chapter 9. Here, it is worth noting that if a Picard iteration converges (and it may not), it must converge to a point ξ for which $\xi = g(\xi)$. This point ξ is called a *fixed point* of the iteration.

Secant Method

To return to our survey of methods, we wish to find a method which like Newton's method produces a better and better approximating line as $x_i \to \xi$ but which requires only one function evaluation per step. The extra function evaluation of Newton's method comes from having to evaluate $f'(x_i)$, which is the slope of the tangent line. If instead of using this tangent we use the line (the secant) through the two most recent iterates, x_i and x_{i-1}, this line will also get better and better as x_i and thus x_{i-1} approach the root. The equation defining the iterative step is produced in the same way as that producing Newton's method (equation 112), replacing the slope of the tangent $f'(x_i)$ by the slope of the secant $(f(x_i) - f(x_{i-1}))/(x_i - x_{i-1})$:

$$\frac{f(x_i) - f(x_{i-1})}{x_i - x_{i-1}} = \frac{f(x_i) - 0}{x_i - x_{i+1}} \qquad [120]$$

From equation 120 follows the step of the *secant method* (see Figure 3–22), which is not a Picard iteration since x_{i+1} depends on both x_i and x_{i-1}:

$$x_{i+1} = \frac{x_{i-1}f(x_i) - x_i f(x_{i-1})}{f(x_i) - f(x_{i-1})} \qquad [121]$$

The same formula is used in the method of false position. The only difference is that, after computing x_{i+1}, in the secant method we always discard x_{i-1} (see Program 3–12, p. 106), whereas in the method of false position we discard either x_i or x_{i-1}, depending on the sign of f at these points. The secant method always uses the most recent information and thus can not freeze one point. Because of this difference, the secant method does not always converge, but when it does converge it converges faster than the method of false position. It often requires more iterative steps than Newton's method to achieve a desired accuracy, but usually it requires less computation time because of its more efficient steps (with one fewer function evaluation).

When does the secant method get in trouble? Generally when Newton's method does: around zero derivatives.

To illustrate the secant method, we again use $x^2 - 2 = 0$ with $x_0 = 0$ and $x_1 = 2$. Then

FIGURE 3-22
Secant Method

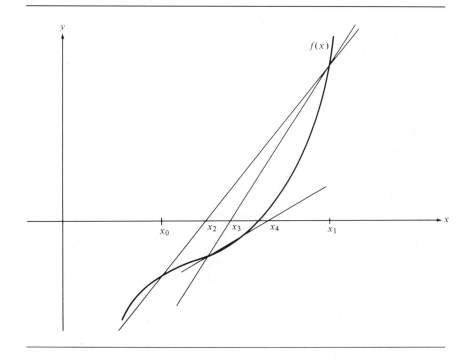

$$x_2 = \frac{0(2) - 2(-2)}{2 - (-2)} = 1$$

$$x_3 = \frac{2(-1) - 1(2)}{-1 - 2} = \frac{4}{3}$$

$$x_4 = \frac{1(-2/9) - (4/3)(-1)}{-2/9 - (-1)} = 1.43$$

and so forth, a fairly fast convergence, but not as fast as with Newton's method.

Hybrid Methods

We have seen that the root-trapping methods always converge but usually slowly. We have seen other methods that usually converge much more quickly but sometimes do not converge. We can combine these methods to produce a hybrid that always converges, and usually quickly, by using the result of the fast method as long as that result is in the interval in which the root is trapped, and using the result of the always-convergent method when it is not. The interval is made smaller at each step by taking the new iterate in place of the endpoint for which the function f evaluated at the iterate has the same sign. For example, we can produce a secant-false position hybrid (see Program 3–13, p. 107) that is quickly convergent and is also always-convergent if false position is applicable.

3.4.2 SOLUTION OF *N* EQUATIONS IN *N* UNKNOWNS

As with one equation in one unknown, the problem of solving many equations in many unknowns divides into two cases: where direct solution is possible and where it is not. The most common situation where direct solution is possible is when the equations are linear: no variable appears in a term in a more complicated way than multiplied by a constant. With most nonlinear equations only an iterative approach is possible.

Root-trapping methods do not generalize to solving many nonlinear equations in many unknowns. Thus methods involve either (1) Picard iteration obtained by solving each of the equations for a different unknown, or (2) iteration obtained by approximating and operating: approximating the nonlinear equations by linear equations and solving these to obtain an improved estimate of the solution. The latter technique is generally more successful.

We will therefore be interested in the solution of linear equations not only in its own right but as a subroutine for solving nonlinear equations. We have also seen methods of function approximation that require the solution of linear equations as a subroutine, namely, least-squares approximation and cubic splines.

Systems of n linear equations in n unknowns,

$$
\begin{aligned}
A_{11}x_1 + A_{12}x_2 + \cdots + A_{1n}x_n &= b_1 \\
A_{21}x_1 + A_{22}x_2 + \cdots + A_{2n}x_n &= b_2 \\
\vdots \qquad \vdots \qquad\qquad \vdots \qquad \vdots & \\
A_{n1}x_1 + A_{n2}x_2 + \cdots + A_{nn}x_n &= b_n
\end{aligned}
\tag{122}
$$

are usually written as

$$
Ax = b
\tag{123}
$$

where A is an $n \times n$ matrix and x and b are n-vectors (n-element arrays of one dimension). It can be shown that there exists a unique solution to this equation if $\det(A)$, the determinant of A, is nonzero. In high school we learned Cramer's rule for solving nonsingular systems of linear equations:

$$Ax = b \Rightarrow x_i = \frac{\det(A^i)}{\det(A)}$$

where A^i is the matrix obtained by replacing the ith column of A with b. This method involves evaluating $(n + 1)$ $n \times n$ determinants. To evaluate a determinant by expansion on a row or column involves more than $n!$ multiplications, a number which grows very rapidly with n. Let us examine a method which requires $n^3/3 + O(n^2)$ multiplications. This method, called *Gaussian elimination*, results from trying to transform the original system of equations into another system with the same solution but computationally easier to solve than the original one.

Assume we are given the following system of simultaneous equations, which is written in both element and matrix-and-vector notations:

$$x_1 - x_2 + x_3 = 2 \qquad \begin{bmatrix} 1 & -1 & 1 \\ 2 & 1 & 1 \\ 4 & 2 & 1 \end{bmatrix} x = \begin{bmatrix} 2 \\ 7 \\ 11 \end{bmatrix} \qquad \begin{matrix} [124] \\ \\ [125] \\ \\ [126] \end{matrix}$$

$$2x_1 + x_2 + x_3 = 7$$

$$4x_1 + 2x_2 + x_3 = 11$$

This set of equations has the solution $x_1 = 1$, $x_2 = 2$, $x_3 = 3$. We can accomplish the desired transformation by a multiple-stage process. In stage 1, we eliminate the first variable from equations 2 through n. In stage 2, we eliminate the second variable from equations 3 through n. In stage 3, we eliminate the third variable from equations 4 through n, and so on, until we arrive at a set of equations where the nth equation involves only x_n, the $(n-1)$th equation involves x_{n-1} and x_n, and so on. Once the equations are in this form, we can solve the nth equation for x_n. Substituting its value in the $(n-1)$th equation, we can solve for x_{n-1}, and so on, until we solve for x_1 in the first equation after substituting the already computed values for x_2 through x_n.

Consider how we can apply this idea to equations 124, 125, and 126. If we subtract twice equation 124 from equation 125, we have not changed the solution, and we obtain

$$3x_2 - x_3 = 3 \qquad\qquad [127]$$

an equation in which x_1 has been eliminated. Similarly, if we subtract four times equation 124 from equation 126, we obtain

$$6x_2 - 3x_3 = 3 \qquad\qquad [128]$$

Thus we have completed stage 1, producing a new set of equations, made up of equations 124, 127, and 128, that has the same solution as the original set, equations 124, 125, and 126, but in which x_1 has been eliminated from all but the first equation. This transformed set of equations can be written

$$\begin{bmatrix} 1 & -1 & 1 \\ 0 & 3 & -1 \\ 0 & 6 & -3 \end{bmatrix} x = \begin{bmatrix} 2 \\ 3 \\ 3 \end{bmatrix}$$

Stage 2 involves subtracting twice equation 127 (the new second equation) from equation 128 (the new third equation) to obtain

$$-1x_3 = -3 \qquad\qquad [129]$$

and replacing equation 128 (the present third equation) by this new third equation, equation 129, to produce the following transformed set, which has the same solution as the original set but in which x_2 has been eliminated from all equations below the second equation:

$$\begin{bmatrix} 1 & -1 & 1 \\ 0 & 3 & -1 \\ 0 & 0 & -1 \end{bmatrix} x = \begin{bmatrix} 2 \\ 3 \\ -3 \end{bmatrix}$$

It is simple to solve for x_3, x_2, and x_1 in our transformed equations, 124, 127, and 129, as follows. The last equation, equation 129, implies $x_3 = 3$. Substituting $x_3 = 3$ into the second equation, equation 127, we compute $3x_2 - 3 = 3$, so $x_2 = 2$. Substituting $x_3 = 3$ and $x_2 = 2$ into the first equation, equation 124, we obtain $x_1 - 2 + 3 = 2$, so $x_1 = 1$.

Let us describe the elimination process in general. The elimination proceeds at the ith stage with the last $n - i + 1$ equations in $n - i + 1$ unknowns (the first $i - 1$ unknowns have been eliminated from these equations; in matrix terms the first $i - 1$ columns of A are zero below the diagonal). The ith unknown is eliminated (the ith column is zeroed below the diagonal) by subtracting an appropriate constant times the first equation of the group from the second, where the constant is chosen to make the resulting coefficient of the ith unknown equal to zero, then subtracting a different constant times the first equation from the third equation, choosing the constant to eliminate the ith unknown, and so on. This process results in an upper triangular matrix (only 0's below the diagonal) and is called the *triangularization* step of the solution. It is followed by the *substitution* step, in which for each equation we substitute the values of the variables already evaluated and solve for the remaining variable in the equation. (See Program 3–14, p. 109, for full specification of the algorithm.)

Let us first analyze this Gaussian elimination algorithm with respect to efficiency — the number of multiplications and divisions required.

The ith stage of triangularization requires, for each of $n - i$ rows, one division to determine the multiplying constant and $n - i + 1$ multiplications to determine the resulting elements in the row that are not necessarily equal to zero. Thus the total number of multiplications and divisions required in the triangularization is

$$\sum_{i=1}^{n-1} (n - i)(n - i + 2)$$

Using the formulas

$$\sum_{i=1}^{m} i = \frac{m(m + 1)}{2} \qquad \text{[130]}$$

and

$$\sum_{i=1}^{m} i^2 = \frac{m(m + 1)(2m + 1)}{6} \qquad \text{[131]}$$

we find that $n^3/3 + O(n^2)$ multiplications and divisions are required; $n^3/3 + O(n^2)$ additions and subtractions are also required.

The back substitution involves n stages, where the ith stage requires $i - 1$ multiplications and 1 division. The total number of multiplications and divisions required by this back substitution is

$$\sum_{i=1}^{n} i = \frac{n(n + 1)}{2} = \frac{n^2}{2} + O(n)$$

Thus the complete Gaussian elimination process requires $n^3/3 + O(n^2)$ multiplications and divisions. This is much less than the approximately $1.7(n + 1)!$ M/D's required by Cramer's method.

In order to achieve greater efficiency, we can try to solve the system of linear equations by an iterative method, even though it can be solved directly. We try the Picard approach: we solve equation 1 for x_1, equation 2 for x_2, etc., producing

$$x_j = \frac{1}{A_{jj}} \left(b_j - \sum_{\substack{k=1 \\ k \neq j}}^{n} A_{jk} x_k \right) \qquad \text{for } j = 1, 2, ..., n \qquad \text{[132]}$$

and then we use this as a basis for iteration:

$$x_j^{i+1} = \frac{1}{A_{jj}} \left(b_j - \sum_{\substack{k=1 \\ k \neq j}}^{n} A_{jk} x_k^i \right) \qquad \textbf{[133]}$$

where x_k^i is the estimate of x_k at the ith iterative step. This method is called the *Jacobi method* (see Program 3–15, p. 111).

Each application of equation 133 requires n M/D's and $n - 1$ A/S's. One iterative step requires n such applications and thus n^2 M/D's and $n(n - 1)$ A/S's. Therefore, if appropriate accuracy can be obtained by the Jacobi method in fewer than $n/3$ iterative steps, efficiency will have been gained over Gaussian elimination. Furthermore, we will see later that accuracy will also likely have been gained.

For sparse matrices A, ones for which most of the elements are zero, the Jacobi method is particularly efficient as compared to Gaussian elimination. Every zero element, A_{jk} for $j \neq k$, saves a multiplication every iterative step. Thus a matrix with elements which are 90% zero will require less than 10% of the work per iteration than if all the elements were nonzero. In contrast, in the triangularization step of Gaussian elimination each subtraction of a multiple of one equation from another changes many zero elements to nonzero, so there is little saving due to sparsity. Thus, for large, sparse matrices, the Jacobi method has a major advantage over Gaussian elimination.

The difficulty with the Jacobi method is that it often fails to converge. We can show that a sufficient condition for convergence when using the Jacobi method is that the diagonal coefficient in each row of A is in magnitude greater than the sum of the magnitudes of the remaining elements in the row. Such a matrix is said to be *row diagonally dominant*. Another sufficient condition for convergence is that A is a matrix in which the diagonal element in each column is larger in magnitude than the sum of the magnitudes of the remaining elements in that column (is *column diagonally dominant*). Note that the diagonal dominance conditions are sufficient but not necessary. However, matrices that are strongly nondiagonally dominant are not likely to produce convergence.

It is often effective to modify the original set of equations to produce a matrix which is, if not diagonally dominant, as close to diagonally dominant as possible. For example, consider the following set of equations, which has the solution $x_1 = x_3 = 1$, $x_2 = 2$:

$$\begin{aligned} x_1 + 6x_2 + 2x_3 &= 15 \\ x_1 + x_2 - 6x_3 &= -3 \\ 6x_1 + x_2 + x_3 &= 9 \end{aligned} \qquad [134]$$

Setting up the Jacobi method by solving the ith equation for x_i, we obtain

$$\begin{aligned} x_1 &= 15 - 6x_2 - 2x_3 \\ x_2 &= -3 - x_1 + 6x_3 \\ x_3 &= 9 - 6x_1 - x_2 \end{aligned} \qquad [135]$$

If we guess $x^0 = 0$, we obtain

$$x^1 = \begin{bmatrix} 15 \\ -3 \\ 9 \end{bmatrix} \qquad x^2 = \begin{bmatrix} 15 \\ 36 \\ -78 \end{bmatrix}$$

The method is clearly diverging. On the other hand, if we switch equations 1 and 3 of the original set and then switch equations 2 and 3 of the result, we obtain

$$
\begin{aligned}
6x_1 + x_2 + x_3 &= 9 \\
x_1 + 6x_2 + 2x_3 &= 15 \\
x_1 + x_2 - 6x_3 &= -3
\end{aligned} \qquad [136]
$$

This result is both row and column diagonally dominant. Solving equation i for x_i, we obtain

$$
\begin{aligned}
x_1 &= \frac{3}{2} - \left(\frac{1}{6}\right)x_2 - \left(\frac{1}{6}\right)x_3 \\[6pt]
x_2 &= \frac{5}{2} - \left(\frac{1}{6}\right)x_1 - \left(\frac{1}{3}\right)x_3 \qquad [137] \\[6pt]
x_3 &= \frac{1}{2} + \left(\frac{1}{6}\right)x_1 - \left(\frac{1}{6}\right)x_2
\end{aligned}
$$

Guessing $x^0 = 0$, we obtain

$$
x^1 = \begin{bmatrix} \dfrac{3}{2} \\[4pt] \dfrac{5}{2} \\[4pt] \dfrac{1}{2} \end{bmatrix} \qquad
x^2 = \begin{bmatrix} 1 \\[4pt] \dfrac{25}{12} \\[4pt] \dfrac{7}{6} \end{bmatrix} \qquad
x^3 = \begin{bmatrix} \dfrac{23}{24} \\[4pt] \dfrac{35}{18} \\[4pt] \dfrac{73}{72} \end{bmatrix}
$$

We are clearly converging.

Thus we see that by switching the rows or columns of a matrix we can sometimes produce either row or column diagonal dominance or close to it. Furthermore, we can scale the rows or columns by appropriate factors to improve the chance that switching the rows or columns will produce diagonal dominance.

The Jacobi method is an example of a *method of simultaneous displacements*. That is, at the end of each iteration, we replace all of the elements of x^i simultaneously with the elements of x^{i+1}. But as soon as we have computed x_1^i, we presumably have a better estimate of x_1; we would do well to use it immediately and not wait until the end of the iterative step. If we use the computed values as soon as we have them, we have a *method of successive displacements*, known as the *Gauss-Seidel method* (see Program 3–16, p. 113). For example, applying the Gauss-Seidel method to equation 136 involves solving the ith equation for x_i, for $1 \le i \le n$, just as in the Jacobi method, to produce equation 137, and then iterating as follows. With the starting guess $x^0 = 0$, we compute a new estimate for x_1 from the first equation (of 137), producing $x_1^1 = 3/2$. Now using as our estimate $\hat{x} = (3/2, 0, 0)^T$, we compute a new estimate for x_2 from the second equation, producing $x_2^1 = 5/2 - (1/6) 3/2 - (1/3) 0 = 9/4$. Using as our new estimate $\hat{x} = (3/2, 9/4, 0)^T$, we compute a new estimate for x_3 from equation 3, thereby producing $x_3^1 = 1/2 + (1/6) 3/2 + (1/6) 9/4 = 9/8$. Our estimate of x after the first iteration, $x^1 = (3/2, 9/4, 9/8)^T$, is considerably closer to the solution $(1, 2, 1)^T$ than was the result of the first iteration of the Jacobi method. Using this estimate, we return to the first equation (of 137) to compute x^2, and so forth.

The formal equation for a stage of the Gauss-Seidel method is

$$x_j^{i+1} = \frac{b_j}{A_{jj}} - \sum_{k=1}^{j-1} \frac{A_{jk}}{A_{jj}} x_k^{i+1} - \sum_{k=j+1}^{n} \frac{A_{jk}}{A_{jj}} x_k^i \qquad [138]$$

Compare this equation to equation 133 (and Program 3–16 to Program 3–15). It requires the same number of arithmetic operations per iterative step as the Jacobi method and has the same efficiency advantages with sparse matrices. Because we are using the new values immediately, we can expect that if the Jacobi method converges the Gauss-Seidel method will also converge and faster. We can also expect that if the Jacobi method diverges our new values will be worse sooner, so we can expect the Gauss-Seidel method to diverge more quickly. This is in fact most often the case. Generally, we should use the Gauss-Seidel method instead of the Jacobi method. In the case of convergence, fewer iterative steps will be required; in the case of divergence, we will find out sooner that the method is diverging. For completeness, we should note that there exist matrices for which the Jacobi method converges but the Gauss-Seidel method does not, and vice versa. Furthermore, even if A is diagonally dominant, there exist cases for which the Jacobi method converges faster (in the limit) than the Gauss-Seidel method does. It can be shown, however, that if after each row of A has been divided by its diagonal element all of the off-diagonal elements have the same sign, the predicted behavior must hold (Varga, 1962, Chapter 3).

For most matrices A neither the Gauss-Seidel nor the Jacobi method converges, and thus they are seldom used. But sparse matrices are often diagonally dominant, because the jkth element represents the effect of the jth component of a large system on the kth and a component more strongly affects itself than other elements. Thus, exactly where it has the most advantage in efficiency, the Gauss-Seidel method is most likely to converge, so it is often used with large sparse matrices.

There exist iterative methods for the solution of linear equations which always converge, but they usually converge slowly and thus are only used where the increased accuracy that they can provide over Gaussian elimination is necessary.

Systems of n nonlinear equations in n unknowns often fail to have solutions or have nonunique solutions, and if they have a solution it is very hard to find iterative methods that converge. A Picard iteration, like the Jacobi method, can be obtained by solving the ith equation for the ith unknown and using this to produce an iteration (see Pizer, 1975, Section 3.5.1), and improvement can be gained over the simultaneous displacements form of this iteration by using the successive displacements form. More time consuming and more likely to converge to the desired root, but still unlikely, are the n-dimensional analogs of Newton's method and the secant method.

3.5 Summary and Complements

A major object of this chapter has been to present some of the common problems that must be solved numerically and the most straightforward numerical methods for their solution. There has been no attempt to cover all problems or all classes of solutions to any problem, let alone all numerical methods in any class. For example, no attempt has been made to discuss the problem of function optimization, the class of splines where the pieces join at places (called knots) other than the tabular arguments, nor predictor–corrector methods of choice for the solution of differential equations, despite the usefulness of all of these. The student who wishes exposure to these or other problems that we have not covered or greater exposure to the problems that we have covered before moving on to generalization from them and analysis of them is referred to any of the large number of numerical methods books (see especially Hornbeck, 1975).

In fact, few of the methods presented in this chapter are methods of choice as they stand, and fewer of the programs presented should be used as they stand, their presentation having been only for the purpose of clarifying the algorithm. The methods presented were chosen because they illustrate concerns which we must treat to improve the methods and they exemplify classes of methods to which these improvements must apply. The application of these improvements to the methods that have been presented will produce methods of choice.

A most important result of the presentation and development of these methods has been the identification of four of the five basic computing strategies that we will cover: approximate and operate, divide and conquer, recurrence, and fixed-point iteration. Understanding these strategies will allow the reader to develop or understand solutions to numerical problems not treated in this book. In all cases the problem to be solved involves applying an operator, but we find that it either can not be applied directly or can not be applied efficiently and accurately.

The strategy of approximate and operate is used when the operator in question is to be applied to a function. In this circumstance we approximate the function and apply the operator to the approximation. The approximating function is chosen so that the operator can be efficiently applied to it and so that adequate accuracy in the result is achieved.

The strategy of recurrence involves dividing the action of the original operator into a number of steps, each involving the application of the same operator. This step operator produces, from the result(s) of the previous step(s), an intermediate result to the original problem, such that a correct answer for each intermediate result can be analytically defined. The step operator is chosen so that even its repetitive application is more efficient than the original operator and so that it produces adequate accuracy.

The strategy of fixed-point iteration involves finding an estimate improvement operator that has as a fixed point the result of the original operator. The improvement operator is chosen to be efficient and to produce a more accurate estimate of the desired result, given some number, n, of previous estimates. Methods based on this strategy operate by first accepting n initial estimates and then repetitively applying the improvement operator to the last n estimates produced, until adequate accuracy is achieved.

The divide and conquer strategy is applied to an operator T on an object with n components, and it involves dividing these into two or more parts, often each with $n/2$ components, applying the operator to each of the halves separately, and then somehow combining the results of the two halves. The application of the operator to the halves is accomplished by the same approach, except for cases with an appropriately small n, for which a special, normally simple, method is provided. We have seen only a sort of degenerate form of divide and conquer, in the root-trapping methods for solving one nonlinear equation in one unknown. In this case the components are n appropriately small subdivisions of the interval in which the root is trapped, and the operator T is root finding, that is, determining in which subdivision, if any, the root falls. The divide and conquer step consists of applying the root-finding operator T to each half of the interval and combining the results. The degeneracy comes from the fact that we know that the root does not fall in one of the halves, so both one of the applications of the root-finding operator and the combination step are trivial.

Nonetheless, with most applications of divide and conquer, such degeneracy does not occur. In these methods the operator must be applied to each of the halves of the components, and each of these applications is again accomplished by dividing and conquering. For example, the Fast Fourier transform computes the n coefficients of the exact-matching Fourier series through n points by finding the $n/2$ coefficients of the exact-matching Fourier series through the $n/2$ points with odd index, finding the $n/2$ coefficients of the exact-matching Fourier series through the $n/2$ points with even index, and finding each of the n desired coefficients as a weighted sum of one of the computed coefficients from the first group and one from the second group. Furthermore, the coefficients of the Fourier series for the points with even index are computed using the same approach, and similarly for the Fourier series for the points with odd index. This recursion continues until the case of a Fourier series through one point is arrived at, a trivial operation.

The time required to apply the divide and conquer algorithm when the division is into two parts consists of the time required to apply the algorithm to each of the halves plus the time to combine the two sets of results. Thus, if $t(n)$ is the time required to compute with n components,

$$t(n) = t\left(\frac{n}{2}\right) + t\left(\frac{n}{2}\right) + c(n) \qquad [139]$$

where $c(n)$ is the cost of combining the results of T applied to each half of the n components to produce the result of applying T to the n components. The recurrence relation 139 can be solved to determine the efficiency of the method. For example, if, as with the Fast Fourier transform

$$c(n) = kn \qquad [140]$$

for some constant k, then

$$t(n) = O(n \log n) \qquad [141]$$

REFERENCES

Hornbeck, R. W., *Numerical Methods*. Quantum Publ., New York, 1975.

Pizer, S. M., *Numerical Computing and Mathematical Analysis*. Science Research Associates, Chicago, 1975.

Varga, R. S., *Matrix Iterative Analysis*. Prentice-Hall, Englewood Cliffs, N.J., 1962.

PROGRAM 3-1
Barycentric Form of Lagrange Interpolation

```
{ ********************************************************************}
{* Barycentric form of Lagrange Method for polynomial interpolation  *}
{* Set in an enclosing block is                                      *}
{*     n, the degree of the approximating polynomial.                *}
{* Given are                                                         *}
{*     (x(j),y(j)),  the data points (0<=j<=n); and                  *}
{*     xeval,  the evaluation argument.                              *}
{* The approximation at xeval is calculated.                         *}
{ ********************************************************************}

FUNCTION b_l_approx (
    x,y: real_array;        {data points}
    xeval: real             {evaluation argument}
                   ): real;

        {real_array is defined as ARRAY[0:n] of real}

VAR
    a: real_array;          {constants in summed terms}
    num,denom: real;        {numerator, denominator of approximation}
    term: real;             {term of sum}
    i,j: integer;           {subscripts}

BEGIN

    {Compute constants in summed terms}
      FOR j:=0 TO n DO
        BEGIN
          a[j] := 1;
          FOR i := 0 TO j-1 DO
            a[j] := a[j]/(x[j]-x[i]);
          FOR i := j+1 TO n DO
            a[j] := a[j]/(x[j]-x[i])
        END;

    {Compute approximation at x=xeval}
      {Compute numerator and denominator sums
        num :=0;
        denom :=0;
        FOR j:=0 TO n DO
          BEGIN
            term := a[j]/(xeval-x[j]);
            denom := denom + term;
            num := num + y[j]*term
          END;

    {Compute final result}
      b_l_approx := num/denom

    END; {b_l_approx}
```

Newton Divided-Difference Interpolation

```
{**************************************************************************}
{* Newton divided difference method for polynomial interpolation        *}
{* Assumed set in an enclosing block is                                 *}
{*      n, the number of first differences.                             *}
{* Given are                                                            *}
{*      x(j), 0<=j<=n, such that x(j)<x(j+1),  the n+1 data             *}
{*          arguments;                                                  *}
{*      p, the last column in the difference table where signs          *}
{*          do not fluctuate near the path to be used;                  *}
{*      xeval,  the evaluation argument; and                            *}
{*      tol,  the error tolerance.                                      *}
{* Also assumed available is a subroutine, f(i,diff), which gives the   *}
{* divided difference of order "diff" where "i" is the index of the     *}
{* least data argument to be used to find that difference.              *}
{**************************************************************************}

FUNCTION Newton(
    x: real_array;       {n+1 data arguments}
    p: integer;          {last useful difference column}
    xeval: real;         {evaluation argument}
    tol: real            {error tolerance}
            ):real;

    {"real_array" has been defined as ARRAY[0..n] OF real}
    {"infinity" has been defined as a very large number}

VAR

    index: integer;      {index of lowest data argument used}
    diff: real;          {difference between argument and xeval}
    diff_order: integer; {order of highest divided difference used}
    newterm: real;       {new term in approximating sum}
    prevsize,prevx: real; {previous summand size and tabular argument}
    factor: real;        {product of (x-x(j)) used}
    lower_x,higher_x: real;{next lower and higher arguments that have}
                          {  not yet been used (or +/- infinity)    }
    approx: real;        {intermediate estimation}

BEGIN

    {Find the closest tabular argument to xeval}
       diff := abs(xeval-x[0]);
       index := 0;
       WHILE (abs(xeval-x[index+1]) < diff) and (index<n) DO
          BEGIN
             index := index + 1;
             diff := abs(xeval-x[index])
          END;
```

```
    IF (index=n-1) and (abs(xeval-x[n])<diff)
        THEN index :=n;

{Initialize for approximation computation}
    diff_order := 0;
    newterm := f(index,diff_order);
    prevx := x[index];
    IF index >0
        THEN lower_x := x[index-1]
        ELSE lower_x := -infinity;
    IF index < n
        THEN higher_x := x[index+1]
        ELSE higher_x := infinity;
    factor := 1;
    approx := 0;

{Add each new term and find the next}
    REPEAT
      {Add new term}
        approx := approx + newterm;
      {Set up for next term}
        prevsize := abs(newterm);
        factor := factor *(xeval-prevx);
        diff_order := diff_order +1;
      {Find next closest argument to xeval}
        IF (xeval - lower_x < higher_x - xeval)
          THEN BEGIN
            index := index-1;
            prevx := x[index];
            IF index > 0
              THEN lower_x := x[index-1]
              ELSE lower_x := -infinity
            END
          ELSE BEGIN
            prevx := x[index + diff_order];
            IF index+diff_order <n
              THEN higher_x := x[index + diff_order]
              ELSE higher_x := infinity
            END;
      {Compute new term}
        newterm := factor*f(index,diff_order)
    UNTIL ((abs(newterm) >= prevsize) OR (diff_order >= p)
              OR (abs(newterm) < tol ));

    IF abs(newterm) < prevsize
        THEN Newton := approx + newterm
        ELSE Newton := appro
    END; {Newton}
```

PROGRAM 3-3
Cubic Spline Exact-Matching Approximation

```
{*******************************************************************}
{* Cubic Spline Exact-Matching Approximation                      *}
{* Set in an enclosing block is                                   *}
{*      n, the number of intervals.                               *}
{* Given are                                                      *}
{*      (x(i),y(i)), 0<=i<=n, n+1 data points such that           *}
{*          x(i-1)<x(i) for 1<=i<=n;                              *}
{*      c(i), 0<=i<=n, n+1 second derivative values at the        *}
{*          x(i); and                                             *}
{*      xeval, an evaluation argument.                            *}
{* Most likely, the x(i), the y(i), and two of the c(i) values were *}
{* given to a previous program and it computed the remaining c(i). *}
{*******************************************************************}

FUNCTION c_s_approx(
   x,y: real_array;     {data arguments and values}
   c:   real_array;     {second derivative values}
   xeval: real          {evaluation argument}
               ): real;

   {real_array is defined as ARRAY[0:n] of real}

VAR
   h: real;             {width of particular interval}
   s,scomp: real;       {fraction of interval and its complement}
   a,b: real;           {coefficients in evaluation formula}
   factor: real;        {factor for coefficients}
   i: integer;          {subscript indicating interval containing xeval}

BEGIN

   {Find interval in which xeval falls}
   i:=0;
   REPEAT i:=i+1 UNTIL ((xeval<x[i]) OR (i=n)) ;

   {Evaluate polynomial for interval (x(i-1),x(i)) at xeval}
   h := x[i]-x[i-1];
   s := (xeval-x[i-1])/h;
   scomp := 1-s;
   factor := h*h/6;
   a := c[i-1]*factor;
   b := c[i]*factor;

   {Final result}
   c_s_approx := (a*scomp*scomp + y[i-1] -a)*scomp + (b*s*s + y[i] -b)*s

END; {c_s_approx}
```

PROGRAM 3-4a

Fourier Least-Squares Approximation with Subtraction of Linear Tendency and
Odd Reflection: Coefficient Calculation

```
{******************************************************************************}
{* Coefficient calculation for Fourier least-squares approximation          *}
{* with linear tendency subtraction                                         *}
{* Defined in an enclosing block are                                        *}
{*      n,  the number of data points;                                      *}
{*      m,  the number of sine terms in the approximation; and              *}
{*      sintab(i,j), 1<=i<=n, 1<=j<=m, table of sinusoid values             *}
{*            where value (i,j) is sin(j*pi*i/(n-1)).                        *}
{* Given are                                                                *}
{*      y(i), 1<=i<=n, data values at equally spaced arguments.             *}
{* Output values are                                                        *}
{*      a(j), 1<=j<=m, the sinusoid coefficients in the                     *}
{*            approximation; and                                            *}
{*      line0 and line1, line coefficients.                                 *}
{******************************************************************************}
PROCEDURE coeff_calc(
    y: real_array;              {data values}
    VAR a: sin_array;           {approximation coefficients}
    VAR line0,line1: real       {line coefficients in approximation}
                    );

    {real_array is defined as ARRAY[1..n] of real}
    {sin_array is defined as ARRAY[1..m] of real}

VAR
    afactor: real;              {factor in coefficient calculation}
    i,j: integer;               {subscripts}

BEGIN

  {Compute line coefficients}
  line0 := y[1];
  line1 := (y[n]-y[1])/(n-1);

  {Subtract linear tendency from data values}
  FOR i := 2 TO n-1 DO
     y[i] := y[i] - (line0 + line1*(i-1));

  {Compute sinusoid coefficients}
  afactor := 2/(n-1);
  FOR j:=1 TO m DO
    BEGIN
      a[j] := 0;
      FOR i:=1 TO n-2 DO
          a[j] := a[j] + y[i+1]*sintab[i,j];
      a[j] := afactor*a[j]
    END

END; {coeff_calc}
```

Fourier Least-Squares Approximation with Subtraction of Linear Tendency and
Odd Reflection: Evaluation of Approximation at Argument

```
{*****************************************************************************}
{* Fourier least squares approximation with linear tendency           *}
{*       subtraction - evaluation of approximation at argument         *}
{* Set in an enclosing block are                                       *}
{*       n, the number of data points; and                            *}
{*       m, the number of sine terms in the approximation.            *}
{* Given are                                                           *}
{*       a(j), 1<=j<=m, coefficients of sine terms of the             *}
{*            approximation;                                           *}
{*       line0 and line1, line coefficients;                          *}
{*       x1, the first tabular argument;                              *}
{*       x, the evaluation argument; and                              *}
{*       h, the tabular argument interval width.                      *}
{* The approximation at x is computed.                                 *}
{*****************************************************************************}

FUNCTION f_ls_approx(
     line0,line1: real;            {line coefficients}
     a: sin_array;                 {sine term coefficients}
     x1,x: real;                   {first and evaluation arguments}
     h: real                       {tabular argument interval}
                    ): real;

     {sin_array is defined as ARRAY[1:m] of real}

CONST
     pi = 3.141592;

VAR
     s: real;                      {argument in units of h}
     sintable: sin_array;          {table of sin(j*pi*(x-x1)/((n-1)*h))}
     costable: sin_array;          {table of cos(j*pi*(x-x1)/((n-1)*h))}
     approx: real;                 {temporary approximation result}
     j: integer;                   {subscript}

BEGIN
   {Compute line terms}
   s := (x-x1)/h;
   approx := line1*s + line0;

   {Compute table of sines at s}
   s := pi*s/(n-1);
   sintable[1] := sin(s);
   costable[1] := cos(s);
   FOR j:=2 TO m DO
     BEGIN
       sintable[j] := sintable[j-1]*costable[1]+costable[j-1]*sintable[1]
       costable[j] := costable[j-1]*costable[1]-sintable[j-1]*sintable[1]
     END;

   {Compute sine terms}
   FOR j:=1 TO m DO
     approx := approx + a[j] * sintable[j];

   {Final result}
   f_ls_approx := approx

END; {f_ls_approx}
```

PROGRAM 3-5
Trapezoidal Rule for Numerical Integration

```
{***********************************************************************}
{* Trapezoidal rule for numerical integration of f(x) over an         *}
{*    interval, xmin to xmax                                          *}
{* Given are                                                          *}
{*    f, the function to be integrated;                               *}
{*    n, the number of tabular intervals; and                         *}
{*    xmin and xmax, the limits of integration.                       *}
{* An approximation to the integral is computed.                      *}
{***********************************************************************}

FUNCTION trap_int(
    FUNCTION f(z:real) :real;    {function to be integrated}
    n: integer;                  {number of tabular intervals}
    xmin,xmax: real              {integration limits}
               ): real;

VAR
    h: real;                 {tabular interval width}
    x: real;                 {tabular argument}
    func_sum: real;          {weighted sum of f(x) values}
    i: integer;              {loop counter}

BEGIN

    {Initialize}
    h := (xmax-xmin)/n;
    func_sum := 0.5*(f(xmin) + f(xmax));
    x := xmin;

    {Compute sum of tabular values}
    FOR i:=1 TO n-1 DO
      BEGIN
         x := x+h;
         func_sum := func_sum + f(x)
      END;

    {Normalize integral value}
    trap_int := h*func_sum

END; {trap_int}
```

PROGRAM 3-6
Simpson's Rule for Numerical Integration

```
{******************************************************************************}
{* Simpson's rule for numerical integration of f(x) over an interval,      *}
{*    xmin to xmax                                                          *}
{* Given are                                                               *}
{*    f, the function to be integrated,                                    *}
{*    n, the number of tabular intervals (n is assumed even); and          *}
{*    xmin and xmax, the limits of integration.                            *}
{* An approximation to the integral is computed.                           *}
{******************************************************************************}

FUNCTION Simp_int(
    FUNCTION f(z:real) :real;    {function to be integrated}
    n: integer;                  {number of tabular intervals}
    xmin,xmax: real              {integration limits}
                ): real;

VAR
    m: integer;            {number of panels}
    h: real;               {tabular interval width}
    x: real;               {tabular argument}
    oddsum,evensum: real;  {sums of odd and even terms}
    i: integer;            {loop counter}
BEGIN

    {Initialize}
    h := (xmax-xmin)/n;
    m := n DIV 2;
    oddsum := 0;
    evensum := 0;
    x := xmin;

    {Compute sums of even and odd terms}
    FOR i:=1 TO m-1 DO
        BEGIN
            x := x+2*h;
            oddsum := oddsum + f(x-h);
            evensum := evensum + f(x)
        END;
    oddsum := oddsum + f(x+h);

    {Weight sums and normalize integral value}
    Simp_int := (h/3)*(f(xmin) + 4*oddsum + 2*evensum + f(xmax))

END; {Simp_int}
```

PROGRAM 3-7
Euler's Method for Solving First–Order
Ordinary Differential Equation

```
{******************************************************************}
{* Euler's method for solving a first-order differential          *}
{*     equation: y'(x)=yprime(x,y) and y(x0)=y0                    *}
{* Defined in an enclosing block is                               *}
{*     n, the number of steps.                                    *}
{* Given are                                                      *}
{*     yprime(x,y), the derivative of y(x);                       *}
{*     x0 and y0, the initial x and y values; and                 *}
{*     h, the step size.                                          *}
{* Output parameters are                                          *}
{*     y(i), 0<=i<=n, the solution values.                        *}
{******************************************************************}

PROCEDURE Euler(
  FUNCTION yprime(x:real; y:real) :real;
                    {the derivative of y}
  x0,y0: real;       {initial x and y values}
  h: real;           {step size}
  VAR y: real_array  {solution values}
             );

  {real_ array is defined as ARRAY[0..n] of real}


VAR
  xi: real;          {tabular argument}
  i: integer;        {subscript used with solution array}

BEGIN

  {Initialize}
    y[0] := y0;
    xi := x0;

  {Recurrently compute y[i] values}
    FOR i:=0 TO n-1 DO
        BEGIN
          y[i+1] := y[i] + h*yprime(xi,y[i]);
          xi := xi + h
        END

END; {Euler}
```

100

PROGRAM 3-8
Modified Euler-Heun Predictor-Corrector Method for Solving First-Order
Ordinary Differential Equation

```
{*******************************************************************}
{* Modified Euler-Heun method for solving a first-order ordinary    *}
{*    differential equation:                                        *}
{*    y'(x)=yprime(x,y) and y(x0)=y0                                 *}
{* Defined in an enclosing block is                                 *}
{*    n, the number of steps.                                       *}
{* Given are                                                        *}
{*    yprime(x,y), the derivative of y(x);                          *}
{*    x0 and y0, the initial x and y values; and                    *}
{*    h, the step size.                                             *}
{* Output parameters are                                            *}
{*    y(i), 0<=i<=n, the solution values.                           *}
{*******************************************************************}

PROCEDURE Euler_Heun(
    FUNCTION yprime(x:real; y:real) :real;
                        {the derivative of y}
    x0,y0: real;        {initial x and y values}
    h: real;            {step size}
    VAR y: real_array   {solution values}
            );

    {real_ array is defined as ARRAY[0..n] of real}


VAR
    xi: real;           {tabular argument}
    i: integer;         {subscript used with solution array}
    lfderiv,rtderiv: real;
                        {value of derivative at left and right ends}
                        {  of the interval}
    pred: real;         {approximate value at right end}

BEGIN

    {Initialize}
      y[0] := y0;

    {Compute first step using simple Euler predictor}
      lfderiv := yprime(x0,y0);
      pred := y0 + h*lfderiv;
      rtderiv := yprime(x0+h,pred);
      y[1] := y0 + h*(lfderiv + rtderiv)/2;
      xi := x0 +h;

    {Recurrently compute y[i] values}
      FOR i:=1 TO n-1 DO
        BEGIN
          lfderiv := yprime(xi,y[i]);
          pred := y[i-1] + 2*h*lfderiv;
          rtderiv := yprime(xi+h,pred);
          y[i+1] := y[i] + h*(lfderiv + rtderiv)/2;
          xi := xi + h
        END

END; {Euler_Heun}
```

PROGRAM 3–9
Bisection Method for Solving Nonlinear Equations

```
{*****************************************************************************}
{* Bisection method for finding roots of equations                         *}
{* Given are                                                               *}
{*    f, the function for which the root is desired;                       *}
{*    lftguess and rtguess, estimates of the root such that                *}
{*          lftguess < root and rtguess > root; and                        *}
{*    tol, a threshold on the error bound on the root estimate.            *}
{* Assumed available is a sign function, "sign(x)", which returns          *}
{* the integer -1 for negative x and 1 otherwise, where x is real.         *}
{*****************************************************************************}
FUNCTION bisect_root(
    FUNCTION f(x:real):real;      {function for which root is desired}
    lftguess,rtguess: real;       {left and right root estimates}
    tol: real                     {error tolerance}
            ):real;
VAR
  lftsign: integer;    {sign of f(lftguess)}
  xnew: real;          {new root estimate}
BEGIN

  {Initialize}
    lftsign := sign(f(lftguess));

  {Iterate improvement until desired accuracy is achieved}
    REPEAT
      {Compute improved estimate of root}
        xnew := (lftguess + rtguess)/2;
      {Replace the guess at which f agrees in sign with the sign}
      {  of xnew with the value of xnew}
        IF sign(f(xnew)) = lftsign
          THEN lftguess := xnew
          ELSE rtguess := xnew;
    UNTIL (abs((rtguess-lftguess)/2) <= tol);

  {Return final result}
    bisect_root := xnew

END; {bisect_root}
```

PROGRAM 3–10
Method of False Position for Solving Nonlinear Equations

```
{**********************************************************************}
{* Method of false position for finding roots of equations            *}
{* Given are                                                           *}
{*    f, the function for which the root is desired;                   *}
{*    lftguess and rtguess, estimates of the root such that            *}
{*           lftguess < root and rtguess > root; and                   *}
{*    tol, a threshold on relative difference between succesive        *}
{*        root estimates.                                              *}
{* Assumed available is a sign function, "sign(x)", which returns      *}
{* an integer, -1 for negative x and 1 otherwise, where x is real.     *}
{**********************************************************************}

FUNCTION f_p_root(
   FUNCTION f(x:real):real;      {function for which root is desired}
   lftguess,rtguess: real;       {left and right root estimates}
   tol: real                     {error tolerance}
            ):real;

VAR
   new_root: real;       {most recent root estimate}
   old_root: real;       {next most recent root estimate}
   lftval,rtval,newval: real;
                         {f(lftguess),f(rtguess),f(new_root)}
   slope: real;          {slope of line through f at lftguess and rtguess
   lftsign: integer;     {sign of f(lftguess)}

BEGIN

   {Initialize}
      old_root := lftguess;
      new_root := rtguess;
      lftval := f(lftguess);
      rtval := f(rtguess);
      lftsign := sign(lftval);

   {Iterate improvement until desired accuracy is achieved}
      WHILE abs((old_root - new_root)/old_root) > tol DO
         BEGIN
            {Compute improved estimate of root}
              old_root := new_root;
              slope := (lftval-rtval)/(lftguess-rtguess);
              new_root := lftguess - lftval/slope;
            {Replace the guess at which f agrees in sign with}
            {  new_root with the value of new_root}
              newval := f(new_root);
              IF sign(newval) = lftsign
                THEN
                  BEGIN
                    lftguess := new_root;
                    lftval := newval
                  END
                ELSE
                  BEGIN
                    rtguess := new_root;
                    rtval := newval
                  END
         END;
```

```
    {Return final result}
      f_p_root := new_root;
END; {f_p_root}
```

PROGRAM 3–11

Newton's Method for Solving Nonlinear Equations

```
{*****************************************************************}
{* Newton's method for finding roots of equations for one equation  *}
{*      in one unknown                                              *}
{* Given are                                                        *}
{*    f, the function for which the root is desired;                *}
{*    fprime, the function giving the derivative of f;              *}
{*    xguess, a guess at the root; and                              *}
{*    tol, a threshold on the relative difference between           *}
{*         successive root estimates.                               *}
{*****************************************************************}

FUNCTION Newton_root(
   FUNCTION f(z: real): real;      {function to solve for root}
   FUNCTION fprime(z: real): real;{derivative of f}
   xguess: real;                  {guess at root}
   tol: real                      {threshold for stopping iteration}
          ): real;

VAR
   xnew: real;                    {improved estimate of root}

BEGIN

   {Make initial root improvement}
     xnew := xguess - f(xguess)/fprime(xguess);

   {Iterate improvement until required accuracy is achieved}
     WHILE (abs((xnew-xguess)/xnew) > tol) DO
       BEGIN
         {Compute improved estimate of root}
           xguess := xnew;
           xnew := xguess - f(xguess)/fprime(xguess);
       END;

   {Return final result}
     Newton_root := xnew;

END; {Newton_root}
```

Secant Method for Solving Nonlinear Equations

```
{*****************************************************************************}
{* Secant method of finding roots of equations for one equation in         *}
{*     one unknown                                                          *}
{* Given are                                                                *}
{*     f, the function for which the root is desired;                       *}
{*     newguess and oldguess, two guesses at the root; and                  *}
{*     tol, a threshold on the relative difference between                  *}
{*         successive root estimates.                                       *}
{*****************************************************************************}


FUNCTION secant_root(
    FUNCTION f(z: real): real;    {function to solve for a root}
    newguess,oldguess: real;      {two root guesses}
    tol: real                     {threshold for stopping iteration}
            ): real;

VAR
    newval,oldval: real;          {value of f at newguess and oldguess}
    xnew: real;                   {improved root estimate}
    slope: real;                  {slope of secant}

BEGIN

    {Initialize}
      oldval := f(oldguess);

    {Iterate improvement until required accuracy is achieved}
      REPEAT
        newval := f(newguess);
        {Compute slope of secant through newguess and oldguess}
          slope := (newval-oldval)/(newguess-oldguess);
        {Compute improved root estimate}
          xnew := newguess - newval/slope;
        {Make newguess and oldguess the two most recent root estimates}
            oldguess := newguess;
            newguess := xnew;
            oldval := newval
      UNTIL (abs((newguess-oldguess)/oldguess) <= tol);

    {Return final result}
      secant_root := xnew

END; {secant_root}
```

PROGRAM 3-13
Hybrid Secant-False Position Method

```
{************************************************************************}
{* Hybrid secant-false position method for finding roots of one        *}
{*    equation in one unknown: f(x)=0                                   *}
{* Given are                                                            *}
{*    f, the function for which the root is desired;                    *}
{*    lftguess and rtguess, root estimates where lftguess < root        *}
{*        and rtguess > root; and                                       *}
{*    tol, a threshold on the relative difference between root          *}
{*        estimates.                                                    *}
{* Assumed available is a sign function, "sign(x)", which returns       *}
{* an integer, -1 for negative x and 1 otherwise, where x is real.      *}
{************************************************************************}


FUNCTION hybrid_root(
   FUNCTION f(z:real): real;        {function to solve for root}
   lftguess,rtguess: real;          {closest estimates to left and right}
                                    {  of root                          }
   tol: real                        {threshold for stopping iteration}
            ): real;

VAR
   xnew: real;                      {new root estimate}
   xold,xprev: real;                {next two most recent root estimates}
   lftval,rtval: real;              {f(lftguess), f(rtguess)}
   newval,oldval: real;             {f(xnew), f(xold)}
   preval: real;                    {f(xprev)}
   slope: real;                     {slope of line through f}
   lftsign: integer;                {sign of f(lftguess)}

BEGIN

   {Initialize}
      xold := lftguess;
      xprev := rtguess;
      lftval := f(lftguess);
      rtval := f(rtguess);
      lftsign := sign(lftval);
      oldval := lftval;
      preval := rtval;

   {Iterate improvement until desired accuracy is achieved}
   REPEAT
      {Compute improved estimate of root by secant method}
         slope := (oldval-preval)/(xold-xprev);
         xnew := xold - oldval/slope;
      IF ((xnew > rtguess) OR (xnew < lftguess))
         THEN BEGIN
            {Secant result no good; use false position}
               slope := (lftval-rtval)/(lftguess-rtguess);
               xnew := lftguess - lftval/slope
         END;
      {Update recent root estimates and corresponding function values}
         xprev := xold;
         xold := xnew;
         preval := oldval;
         newval := f(xnew);
         oldval := newval;
```

```
                    {Take xnew in place of the one of lftguess and rtguess}
                    {    for which f agrees in sign with f at xnew}
                      IF sign(newval) = lftsign

                      THEN BEGIN
                        lftguess := xnew;
                        lftval := newval
                        END
                      ELSE BEGIN
                        rtguess := xnew;
                        rtval := newval
                        END
          UNTIL (abs((xold-xprev)/xold) <= tol);

{Return final result}
  hybrid_root := xnew
```

PROGRAM 3-14

Elementary Algorithm for Gaussian Elimination

```
{***************************************************************************}
{* Solve Ax = b; set of n linear equations in n unknowns                 *}
{* Set in an enclosing block is                                          *}
{*     n, number of given equations and unknowns.                        *}
{* Given are                                                             *}
{*     A, the n x n matrix of coefficients; and                         *}
{*     b, an n-vector (right-hand-side of the equation to be            *}
{*        solved).                                                       *}
{* Output parameter is                                                   *}
{*     x, the solution vector.                                           *}
{***************************************************************************}

     PROCEDURE Gauss_elim(
       A: real_matrix;              {coefficients}
       b: real_n_vector;            {right-hand-side of equation}
       VAR x: real_n_vector         {solution vector}
                 );

          {real_matrix is defined as ARRAY[1..n,1..n] OF real}
          {real_n_vector is defined as ARRAY[1..n] OF real}

VAR
  amult: real;          {row multiplier}
  i,j,k: integer;       {row and column indicators}

BEGIN

  {Triangularize}
  {Eliminate i-th column below i-th row}


  FOR i := 1 TO n-1 DO
    FOR j := i+1 TO n DO
      BEGIN
        {Compute multiplier for j-th row}
          amult := A[j,i]/A[i,i];
        {Compute nonzero elements of j-th row}
          FOR k:= i+1 TO n DO
            A[j,k] := A[j,k] - amult*A[i,k];
        {Compute new b[j]}
          b[j] := b[j] - amult*b[i]
      END;

{Back substitute}
  FOR i:=n DOWNTO 1 DO
    BEGIN
      x[i] := b[i];
      {Subtract terms in already computed x[j]}
        FOR j:=i+1 TO n DO
          x[i] := x[i] - A[i,j]*x[j];
      {Solve for x[i]}
      x[i] := x[i]/A[i,i]
    END

END; {Gauss_elim}
```

PROGRAM 3-15

Jacobi Method for Solution of Linear Equations

```
{***************************************************************************}
{* Solve Ax = b, a set of n linear equations in n unknowns  by the      *}
{*      Jacobi method                                                    *}
{* Set in an enclosing block is                                          *}
{*      n, number of given equations and unknowns.                       *}
{* Given are                                                             *}
{*      A, the n x n matrix of coefficients;                             *}
{*      b, an n-vector (right-hand-side of the equation to be solved);   *}
{*      xold, a guess at the solution vector; and                        *}
{*      tol, a tolerance for stopping iteration. The test for            *}
{*         convergence is one of many possibilities:  Is the             *}
{*         size of the vector of differences between the old             *}
{*         and new vectors small enough relative to the size of          *}
{*         the new vector?  The size of a vector is taken to be          *}
{*         the maximum element magnitude.                                *}
{* Output parameter is                                                   *}
{*      xnew, the solution vector.                                       *}
{***************************************************************************}
    PROCEDURE Jacobi(
      A: real_matrix;                {coefficients}
      b: real_n_vector;             {right-hand-side of equation}
      xold: real_n_vector;          {old solution vector}
      tol: real;                    {threshold for stopping iteration}
      VAR xnew: real_n_vector       {new solution vector}
                  );

        {real_matrix is defined as ARRAY[1..n,1..n] OF real}
        {real_n_vector is defined as ARRAY[1..n] OF real}

VAR
  C: real_matrix;       {normalized "A" matrix}
  d: real_n_vector;     {normalized "b" vector}
  maxnew,new,maxdif,diff: real;
                        {used for stopping criteria}
  j,k: integer;         {row and column indicators}

BEGIN

  {Normalize matrix}
    FOR j:=1 TO n DO
      BEGIN
        FOR k:=1 TO j-1 DO
            C[j,k] := A[j,k]/A[j,j];
        FOR k:=j+1 TO n DO
            C[j,k] := A[j,k]/A[j,j];
        d[j] := b[j]/A[j,j]
      END;

  {Iterate improvement until required accuracy is achieved}
    REPEAT
      {Compute new estimate}
        maxnew := 0;    maxdif := 0;
        FOR j:=1 TO n DO
          BEGIN
            xnew[j] := d[j];
            FOR k:=1 TO j-1 DO
              xnew[j] := xnew[j] - C[j,k]*xold[k];
            FOR k:=j+1 TO n DO
```

```
            xnew[j] := xnew[j] - C[j,k]*xold[k];
       {Find maximum absolute difference between old}
       {  and new elements}
          diff := abs(xnew[j]-xold[j]);
          IF diff > maxdif THEN maxdif := diff;
       {Find maximum new element value}
          new := abs(xnew[j]);
          IF new > maxnew THEN maxnew := new
      END;
   {Let present estimate be improved estimate}
     FOR j:=1 TO n DO
        xold[j] := xnew[j]
   UNTIL (maxdif/maxnew <= tol)

END; {Jacobi}
```

PROGRAM 3–16

Gauss-Seidel Method for Solution of Linear Equations

```
{*****************************************************************************}
{* Solve Ax = b, a set of n linear equations in n unknowns  by      *}
{*    the Gauss-Seidel method                                        *}
{* Set in an enclosing block is                                      *}
{*    n, number of given equations and unknowns.                     *}
{* Given are                                                         *}
{*    A, the n x n matrix of coefficients;                           *}
{*    b, an n-vector (right-hand-side of the equation to be          *}
{*        solved);                                                   *}
{*    x, a guess at the solution vector which changes as the         *}
{*        guess improves and holds the solution at the time          *}
{*        of return; and                                             *}
{*    tol, a tolerance for stopping iteration. The test for          *}
{*        convergence is one of many possibilities:  Is the          *}
{*        size of the vector of differences between the old          *}
{*        and new vectors small enough relative to the size of       *}
{*        the new vector?  The size of a vector is taken to be       *}
{*        the maximum element magnitude.                             *}
{*****************************************************************************}

PROCEDURE Gauss_Seidel(
  A: real_matrix;              {coefficients}
  b: real_n_vector;            {right-hand-side of equations}
  VAR x: real_n_vector;        {solution vector}
  tol: real                    {used to stop iteration}
                );

      {real_matrix is defined as ARRAY[1..n,1..n] OF real}
      {real_n_vector is defined as ARRAY[1..n] OF real}

VAR
  C: real_matrix;        {normalized "A" matrix}
  d: real_n_vector;      {normalized "b" vector}
  xcomp: real;           {new element estimate}
  maxnew,new,maxdif,diff: real;
                         {used for stopping criterion}
  j,k: integer;          {row and column indicators}

BEGIN

  {Normalize matrix}
    FOR j:=1 TO n DO
      BEGIN
        FOR k:=1 TO j-1 DO
            C[j,k] := A[j,k]/A[j,j];
        FOR k:=j+1 TO n DO
            C[j,k] := A[j,k]/A[j,j];
        d[j] := b[j]/A[j,j]
      END;

  {Iterate improvement until required accuracy is achieved}
    REPEAT
      {Compute new estimate}
        maxnew := 0;     maxdif := 0;
        FOR j:=1 TO n DO
          BEGIN
            xcomp := d[j];
            FOR k:=1 TO j-1 DO
              xcomp := xcomp - C[j,k]*x[k];
```

```
          FOR k:=j+1 TO n DO
            xcomp := xcomp - C[j,k]*x[k];
            {Find maximum absolute difference between}
            {  old and new elements}
              diff := abs(xcomp-x[j]);
              IF diff > maxdif THEN maxdif := diff;
            {Use new value immediately}
              x[j] := xcomp;
            {Find maximum new element value}
              new := abs(xcomp);
              IF new > maxnew THEN maxnew := new
          END
    UNTIL (maxdif/maxnew <= tol)

END; {Gauss_Seidel}
```

PROBLEMS

3.1 Let $f(x) = \sum_{j=0}^{n} a_j x^j / (x^m + \sum_{k=0}^{m-1} b_k x^k)$. Assume that we are given N pairs (x_i, y_i) such that $f(x_i) = y_i$, with $i = 1, 2, \ldots, N$, where the x_i are distinct. Finally, assume that $N > n + m$.

(a) Give conditions on the x_i and y_i such that there exists a unique set of values of the a_j, for $j = 0, 1, \ldots, n$, and the b_k, for $k = 0, 1, \ldots, m - 1$, so that $f(x)$ passes through these data points.

(b) For $n = 0$, $m = 1$, and $N = 2$ give an example of data points such that no $f(x)$ of the form given passes through them.

*(c) Show that if $m \geqslant 1$, if a function $f(x)$ of the form given passes through the data points, it is unique.

3.2

(a) Assume we are given the value of $f(x)$ at n values of x ($x = x_1, x_2, \ldots, x_n$) and $f'(x_k)$ for some k between 1 and n. Develop an interpolation polynomial of minimum degree which agrees with these $n + 1$ values.

(b) Assume we are given $\{(x_i, y_i, y_i') | i = 0, 1, \ldots, N\}$, a set of argument values, and for each argument the value of a function f and a value for its derivative at that argument. Find a $(2N+1)$th-degree polynomial $\hat{p}(x)$ such that for each i, $\hat{p}(x_i) = y_i$ and $\hat{p}'(x_i) = y_i'$. Interpolation using this polynomial is called Hermite interpolation.

3.3 Let $p_i(x) = y_i$ and $p_{i j_1 j_2 \cdots j_n k}(x) = [(x_k - x)p_{i j_1 j_2 \cdots j_n}(x) - (x_i - x) \, p_{j_1 j_2 \cdots j_n k}(x)] / (x_k - x_i)$.

(a) Prove by induction that $p_{01 \cdots N}(x) = \hat{p}(x)$, where $\hat{p}(x)$ is the polynomial determined by the Lagrange method passing through (x_0, y_0), $(x_1, y_1), \ldots, (x_N, y_N)$. This method for determining $\hat{p}(x)$ is called iterated linear interpolation.

(b) Show that iterated linear interpolation is less efficient than the Newton divided-difference method.

3.4 Carry out the following problem, first using Lagrange interpolation (in barycentric form) and then using the divided-difference method. Compute $\hat{f}(2.16)$ using the accompanying table for f, first using a quadratic polynomial and then using a cubic polynomial. Compare your answers to the correct value of this function e^x: $f(2.16) = 8.6711$.

---

* This problem is challenging.

x	$f(x)$
2.00	7.3891
2.10	8.1662
2.25	9.4877
2.30	9.9742

3.5   Given tabular data produced by the algorithm developed in Problem 2.1, evaluate at 4(0.2)5 a four-point exact-matching piecewise cubic approximation to these data, using what you consider to be the most efficient procedure for the data. State why you chose the method you did, and determine (analytically) the number of M/D's and A/S's required overall.

3.6   Consider the problem of developing a spline approximation for which $\hat{f}(x) = p_{m+1}(x)$ if $x \in [x_m, x_{m+1}]$ and $p_{m+1}(x)$ is a quadratic that exact matches at x_m and x_{m+1} and agrees with p_m and p_{m+2} in slope at the endpoints x_m and x_{m+1}, respectively, for $m = 0, 1, ..., N-1$. Letting $s_m \overset{\Delta}{=} \hat{f}'(x_m)$, find

(a)   an expression for $p_{m+1}(x)$ in terms of s_m, s_{m+1}, x_m, x_{m+1}, y_m, and y_{m+1};
(b)   an equation relating s_m and s_{m+1}; and
(c)   how many additional constraints must be specified to fully specify $\hat{f}(x)$ on $[x_0, x_N]$.

3.7   The accompanying table gives at 4(0.2)5 the values of the underlying function given in Problem 2.1 and of three different polynomial exact-matching approximations. Discuss the quality of each of the three approximations as objectively and as quantitatively as you can.

x	Underlying function	9th degree polynomial through −4(1)5	3rd degree piecewise fit through 4 points	Cubic spline through −4(1)5 with $c_0 = c_N = 0$
4.0	−0.19253	−0.19253	−0.19253	−0.19253
4.2	−0.15434	−0.04220	−0.15407	−0.15377
4.4	−0.11597	0.13602	−0.11547	−0.11519
4.6	−0.07743	0.28750	−0.07685	−0.07672
4.8	−0.03877	0.30405	−0.03832	−0.03835
5.0	0	0	0	0

3.8   Given the data $\{(1, 1.3), (2, 1.8), (3, 2.9), (5, 5.1)\}$, where all of the y_i have the same expected error size and the underlying function is assumed to be a straight line, compute the least-squares line.

3.9 Let $f^1(x) = 1$ and $f^2(x) = 1/x$. Find a least-squares fit $\hat{f}(x) = \hat{a}_1 f^1(x) + \hat{a}_2 f^2(x)$ with respect to the data points $\{(1, 1), (2, 2), (4, 5)\}$, where it is assumed the expected size of error in each y_i is the same.

3.10 In fitting a least-squares line to the points $\{(x_i, y_i)|1 \leq i \leq N\}$ when there are errors of the same magnitude in the x_i and the y_i, it is reasonable to minimize the sum of the squares of the perpendicular distances from the data points to the fitted line. Give equations for the slope and y intercept of the fitted line as a function of the x_i and the y_i.

3.11 Assume we wish to construct a piece of a least-squares cubic spline over $\{x_j, x_{j+1}, \ldots, x_{j+k-1}\}$ that agrees at x_j in value, slope, and curvature with a piece over a previous interval. Assume the previous piece has been calculated, so that $\hat{f}(x_j)$, $\hat{f}'(x_j)$, and $\hat{f}''(x_j)$ can be assumed to be given. Find expressions for the coefficients of the piece of the cubic spline, $\hat{f}(x)$, to be fit over the above argument points.

3.12 Let $\{f^j(x)|i = 1, 2, \ldots, N\}$ be functions that are orthogonal with respect to the tabular arguments $x_1, x_2, \ldots, x_N$, and let $\hat{a}_j$, for $j = 1, 2, \ldots, N$, be the least-squares coefficients of the respective $f^j(x)$ when fitting to the data $\{(x_i, y_i)|i = 1, 2, \ldots, N\}$, where each y_i has equal expected error size. Prove Parseval's theorem, which states that

$$\sum_{i=1}^{N} y_i^2 = \sum_{j=1}^{N} \left(\hat{a}_j^2 \sum_{i=1}^{N} (f^j(x_i))^2 \right)$$

3.13 Let $z_i = x_i h + c$, $1 \leq i \leq N$, where $x_{i+1} - x_i = 1$ and N is even. Consider the following least-squares problems:

(1) Data: $\{(x_i, y_i)|1 \leq i \leq N\}$.
   Basis functions: $\{f^k(x)|1 \leq k \leq N\}$, a set of orthogonal functions over $\{x_i|1 \leq i \leq N\}$.
(2) Data: $\{(z_i, y_i)|1 \leq i \leq N\}$.
   Basis functions: $\{g^k(z)|1 \leq k \leq N\}$, where $g^k(z) = f^k((z - c)/h)$.
(a) Show that the g^k, $1 \leq k \leq N$, are orthogonal over $\{z_i|1 \leq i \leq N\}$.
(b) Show that the coefficient of f^k in the least-squares fit in problem 1 is equal to the coefficient of g^k in problem 2, for $1 \leq k \leq N$.
(c) If the f^k are the functions defined by equation 73 and $x_i = i - M$, give an expression for g_i^k.

3.14 At $4(0.2)5$ evaluate the two-term Fourier least-squares approximation, with subtraction of linear tendency and odd reflection, of the data generated by the algorithm produced in Problem 2.1 for $x = 0(1)5$. (Note that subtraction of linear tendency has no effect here and odd reflection produces the function tabulated in Problem 2.1.) Compare quantitatively the approximation obtained to that from four-point piecewise cubic approximation, as listed in Problem 3.7.

3.15 Consider the problem of fitting a Fourier approximation to $f(x)$ = $1/(1 + x^2)$ from the function values at the tabular arguments x_i = 0, 1, 2, ..., 11.

(a) Do an ordinary Fourier approximation (least-squares approximation with the orthogonal basis functions given in equation 73 and no reflection or linear tendency subtraction) to these data for n = 10. This will produce coefficients $\hat{a}_j$ for j = 1, 2, ..., 10 and an approximation $\hat{f}(x)$ = $\sum_{j=1}^{10} \hat{a}_j f^j(x)$. Note that since $c_j \triangleq (\sum_{i=0}^{11}[f^j(i)]^2)^{1/2}$ is a measure of the "size" of f^j, $\hat{a}_j c_j$ is a measure of the size of the coefficient of a normalized (unit size) f^j. Thus the rate at which $|\hat{a}_j c_j|/|\hat{a}_1 c_1|$ gets small with j determines the ability of the method to approximate $f(x)$ well with only a few terms. Therefore, compute $|\hat{a}_{10} c_{10}|/|\hat{a}_1 c_1|$. As another measure of quality of fit, also compute $|\hat{f}(10.5) - f(10.5)|$.

(b) Consider the set of data produced by reflecting $f(x)$ evenly about the origin and using the data from the resulting function at x_i = -10, -9, ..., -1, 0, 1, ..., 11. [For the particular function $f(x)$ used here, which is itself even, this is equivalent to evaluating $f(x)$ at x_i = -10, -9, ..., 11.] Fitting a Fourier approximation to these even data produces zero ceofficients for all of the sine terms, thus producing a pure cosine approximation:

$$\hat{f}(x) = \sum_{j=1}^{n} \hat{a}_j \cos\left[\frac{(j - 1)\pi x}{N}\right]$$

with

$$\hat{a}_j = \frac{1}{c_j}\left(f(0) + (-1)^{j-1}f(N) + 2\sum_{i=1}^{10} f(i) \cos\left[\frac{(j - 1)\pi i}{N}\right]\right),$$

where here N = 11. For this approximation c_j = $\sqrt{11}$ if $2 \le j \le 10$ and c_j = $\sqrt{22}$ if j = 1 or j = 11. Compute the a_j for j = 1, 2, ..., 10. Then compute $|\hat{a}_{10} c_{10}|/|\hat{a}_1 c_1|$ and $|\hat{f}(10.5) - f(10.5)|$ for this approximation.

(c) Compute the coefficients $\hat{a}_j$ (given by equation 77) of the Fourier approximation with linear tendency subtraction and odd reflection. Here c_k is constant with k, so compute $|\hat{a}_{10}|/|\hat{a}_1|$ and $|\hat{f}(10.5) - f(10.5)|$ for this approximation [equation 76 gives $\hat{f}(x)$].

(d) Compare each of the two measures of quality of approximation across the three approximation methods applied in parts a–c. Which method is to be preferred for this function?

3.16 Consider the set of functions $\{f^j\}$ such that the jth member of the set is alternately $+1$ or -1, j times over $[0, 2^N]$:

$$f^j(x) = (-1)^i \quad \text{for} \quad a_{ji} \le x < a_{j,i+1} \quad (i = 0, 1, ..., j - 1)$$

where a_{j0} = 0 and a_{jj} = 2^N (j = 1, 2, ...). So $f^1(x)$ = 1.

(a) For $j = 2, 3, 4, 5$ choose the a_{ji} values so that the set $\{f^j\}$ are orthogonal over the continuous interval $[0, 2^N]$ according to the inner product $(g, h) = \int_0^{2^N} g(x)h(x)\, dx$. These f^j are called *Walsh functions.* Normalize the f^j.

(b) For $j = 2, 3, 4, 5$ choose the a_{ji} values so that the set $\{f^j\}$ are orthogonal over a given set of not necessarily equally spaced points x_i, $i = 1, 2, \ldots, 2^N$. Normalize the f^j.

(c) Give a method to find a least-squares approximation to $\{(x_i, y_i)|$ $1 \leq i \leq 2^N\}$ for $x_i = i$ in terms of the first $m \ll 2^N$ of the f^j found in part b.

(d) Discuss the suitability of the f^j for smoothing.

3.17 An astronomer is interested in the function expressing the distance of a given point on the earth to that of a given point on the sun as a function of day of the year. He has measurements for this distance at intervals of 14.6 days through the year (365 days).

(a) He wants a fairly accurate value of $f(x)$ for any $x = $ the day of the year, and it is suggested to him that he has the choice between the following approximating functions:

$$\hat{f}_1(x) = a_0 + a_1 x + a_2 x^2 + a_3 x^3 + a_4 x^4$$

$$\hat{f}_2(x) = b_0 + b_1 \sin\left(\frac{2\pi x}{365}\right) + b_2 \cos\left(\frac{2\pi x}{365}\right) + b_3 \sin\left(\frac{4\pi x}{365}\right) + b_4 \cos\left(\frac{4\pi x}{365}\right)$$

Which of these should he choose? Why?

(b) If he chooses $\hat{f}_2(x)$, he can rewrite this function as

$$\hat{f}_2(x) = c_0 + c_1 \sin\left(\frac{2\pi x}{365}\right) + c_2 \sin^2\left(\frac{2\pi x}{365}\right) + c_3 \cos\left(\frac{2\pi x}{365}\right)$$
$$+ c_4 \cos\left(\frac{2\pi x}{365}\right) \sin\left(\frac{2\pi x}{365}\right)$$

Given values for c_0, c_1, c_2, c_3, c_4, and $2\pi x/365$, a sine subroutine which requires 100 arithmetic operations, a cosine subroutine which requires 100 operations, and a square root subroutine which requires 25 operations, how many operations would be required to evaluate this expression most efficiently?

(c) What approximation method would you suggest the astronomer use (whichever function he uses): **(i)** a single exact-matching fit; **(ii)** a piecewise exact-matching fit, using only exact-matching constraints; **(iii)** a spline exact-matching fit; or **(iv)** a single least-squares fit? Why?

(d) Our astronomer really wants to know the first day of the year on which the distance in question is equal to its average over the year $\frac{1}{365} \int_0^{365} f(x)\, dx$. Briefly how would you tell him to find this day?

3.18   Determine numerically $\int_0^1 e^{-x}\, dx$

(a)   using the trapezoidal rule with 7 points (6 panels);
(b)   using the trapezoidal rule with 66 points (65 panels);
(c)   using Simpson's rule with 7 points (3 panels).

Design your algorithms to minimize the total number of arithmetic operations required. Compare (by quantitative discussion) the answers to the correct answer of $1 - e^{-1} = 0.632121$.

3.19   Let $\hat{I}_N$ be the result of the trapezoidal rule applied with N panels:

$$\hat{I}_N = \frac{h}{2}\left[f(x_0) + 2 \sum_{i=1}^{N-1} f(x_i) + f(x_N) \right]$$

where h is one Nth of the integration interval. Show that

$$\hat{I}_{2N} = \frac{1}{2}\left[\hat{I}_N + h \sum_{i=1}^{(b-a)/h} f\left(a + \left(i - \frac{1}{2}\right)h \right) \right]$$

3.20   Let $f(x) = 0$ be an equation with only one root. To find the root of $f(x) = 0$, where f is given analytically, two alternative approaches are suggested:

(a)   Find three points, x_0, x_1, and x_2, near the root, and for $i = 2, 3,$ ... let x_{i+1} be a root of the quadratic fitted through f at x_i, x_{i-1}, and x_{i-2}, namely, that root which is closest to the mean of these three points.
(b)   Noting that the root $= f^{-1}(0)$ and beginning with two points, x_0 and x_1, near the root (where f can be assumed to be monotonic, so f^{-1} exists), for $i = 1, 2, ...$ approximate $x = f^{-1}(y)$ by $q_i(y)$, the ith degree polynomial through the tabular points (y_0, x_0), (y_1, x_1), ..., (y_i, x_i), where $y_i = f(x_i)$, and let $x_{i+1} = q_i(0)$.

Each method uses the idea of "approximate and operate." In part a the approximation is by a quadratic, and the operation is root finding. In part b the approximation is by an ith degree polynomial for $i = 1, 2,$ ..., and the operation is evaluation at $y = 0$. Discuss the strengths and weaknesses of each method, comparing them when such comparison is relevant.

3.21   A step of a proposed method for the solution of first-order ordinary initial value differential equations

$$y'(x) = f(x, y)$$
$$y(x_0) = y_0$$

operates as follows:

$$y^{(p)}_{i+1/2} = y_i + \frac{h}{2}f(x_i, y_i)$$

$$y^{(p)}_{i+1} = y_{i+1/2} + \frac{h}{2}f(x_{i+1/2}, y^{(p)}_{i+1/2})$$

$$\hat{y}_{i+1} = y_i + \frac{h}{6}[f(x_i, y_i) + 4f(x_{i+1/2}, y^{(p)}_{i+1/2}) + f(x_{i+1}, y^{(p)}_{i+1})]$$

(a)   Assume we have a general subroutine *G* written to apply this method to an arbitrary function *f* given by a subroutine *F*. Assume the calling and return processes from *F* together take a time equivalent to 20 arithmetic operations. Let *F* compute the function $x^3y + 9x^2 - 1$. How many arithmetic operations should that computation take? Note that calling *F* will require that number plus 20.

(b)   Assume that the differential equation solver *G(F)* defined in part a is applied to call the function *F* defined in part a. Further assume that to achieve a given accuracy this method requires an interval width *h* which is 4/3 times that required by the modified Euler-Heun method:

$$y^{(p)}_{i+1} = y_{i-1} + 2Hf(x_i, y_i)$$

$$\hat{y}_{i+1} = y_i + \frac{H}{2}[f(x_i, y_i) + f(x_{i+1}, y^{(p)}_{i+1})]$$

where *H* is the interval required by the modified Euler-Heun method. Assuming $H \ll |x_{goal} - x_0|$, which method of the two suggested would you use? Explain.

(c)   Interpret the method proposed in part a. That is, on what approximations of the function $y'(x)$ is it based?

3.22   Assume we have a table of the function $f(x)$ for a set of equally spaced values of *x* on the interval $[0,100]$. Assume we know that these $f(x)$ values have an average error of 5%. We need to find the smallest positive value of *x* for which $f''(x) + f'(x) = 0$. Assuming that we can localize the region of the required value of *x*, suggest a method by which this problem might be solved, and discuss the reasons for your choice. You need not give the algorithm in the detail of a program, nor do you need to give mathematical formulas, but you should indicate of what subproblems your method consists and how you propose to solve them.

3.23   A function *f* is given by

$$f(x) = e^{-x}g(x) + 0.01\,e^{-2x} + 0.02$$

where $g(x)$ is a function not known analytically but for which a sketch is known well enough to tell that Newton's method and the secant method applied to *f* will converge to a root of interest for a reasonable starting guess. Assume we have two subroutines called G and GPRIME, which supply values of $g(x)$ and $g'(x)$, respectively, for any given x. We also have a subroutine EXP to evaluate e^x for any given x. These subroutines are known to require 500, 200, and 50 arithmetic operations respectively.

(a)   Write a description of a program segment which accomplishes one iteration of Newton's method for this function as efficiently as possible. (Don't seek to be general purpose, just efficient.)
(b)   Write a description of a program segment which accomplishes one iteration of the secant method for this function as efficiently as possible. (Don't seek to be general purpose, just efficient.)
(c)   Assume the secant method requires three iterations to make the improvement in the root estimate made by two iterations of Newton's method. Which method would you use for this problem? Explain quantitatively.

3.24   We wish to solve the differential equation $y'(x) = -2xy^2$, $y(0) = 1$ at $x = 2$. Both of the following methods require 10 function evaluations:

(a)   Euler's method with $h = 0.2$;
(b)   the Euler-Heun method with $h = 0.4$.

Execute both methods, and compare the relative errors in the two results at $x = 2$. Note that the solution to the differential equation is

$$y(x) = \frac{1}{1 + x^2}$$

3.25   Assume we wish to solve a first-order ordinary differential equation with initial value:

$$y'(x) = f(x, y) \qquad y(x_0) = y_0$$

(a)   If the Heun corrector is to be used, the following two predictors will be approximately equivalent in that the Heun corrector applied to either will produce approximately the same accuracy for each step.
(i) Modified Euler predictor:

$$y_{i+1}^{(p)} = y_{i-1} + 2hf(x_i, y_i)$$

(ii) Simple Euler predictor corrected by the Heun formula to produce a better *predictor:*

$$y_{i+1}^{(p1)} = y_i + hf(x_i, y_i)$$

$$y_{i+1}^{(p)} = y_i + \frac{h}{2}[f(x_i, y_i) + f(x_{i+1}, y_{i+1}^{(p1)})]$$

For $i > 0$ (steps beyond the first), why is predictor (i) preferable? For $i = 0$, why is predictor (ii) preferable?

(b) Consider the following two candidates for recurrent algorithms.

(i) Use the three-term Taylor series for $y(x)$ to produce

$$\hat{y}_{i+1} = y_i + hy_i' + \frac{h}{2}y_i''$$

$$= y_i + hf(x_i, y_i) + \frac{h^2}{2}\left[\frac{\partial f}{\partial x}(x_i, y_i) + f(x_i, y_i)\frac{\partial f}{\partial y}(x_i, y_i)\right]$$

(ii) Use the modified Euler-Heun algorithm:

$$\hat{y}_{i+1} = y_i + \frac{h}{2}[f(x_i, y_i) + f(x_{i+1}, y_{i+1}^{(p)})]$$

where for $i \geqslant 1$ the modified Euler predictor is used and for $i = 0$ the simple Euler predictor is used (not its corrected version described in part a).

Assume that evaluating f requires M arithmetic operations and evaluating $\partial f/\partial x$ and $\partial f/\partial y$ together requires αM operations over and above those required for evaluating f. Assume that M is much larger than the number of additional arithmetic operations required by a single step of either method, given the values of f, $\partial f/\partial x$, and $\partial f/\partial y$. Assume further that to obtain equivalent accuracy at x_{goal} the interval width h_1 used in method (i) must be $h_2/\sqrt{2}$, where h_2 is the interval width used in method (ii). Show that method (i) is preferable if $\alpha < 0.414 = \sqrt{2} - 1$ and method (ii) is preferable if $\alpha > 0.414$.

(c) Let $f(x, y) = x^2y^2[\log(x) + \log(y)] + 3xy - 2$, where log is base e. Assume that evaluating log requires 45 arithmetic operations. What is α, where α was defined in part b?

3.26 Assume ξ is a root of $f(x)$, and $f(x)$ is twice differentiable on $[x_0, \xi]$. Assume $f(x)f''(x) > 0$ if $x \in [x_0, \xi)$.

(a) There are four possible general shapes which f can have on $[x_0, \xi]$. Sketch them. Argue geometrically that in each case Newton's method with starting value x_0 will converge to ξ.

(b) Prove that Newton's method with starting value x_0 will converge to ξ.

3.27 Consider the equation $x - e^{1-x}[1 + \ln(x)] = 0$. It has roots at $x = \xi_1 = 1$ and $x = \xi_2 \approx 0.5$. For the starting value, $x = 0.8$, set up Newton's method and execute two steps to slide-rule accuracy. To which root is the iteration converging? How can you get Newton's method to converge to the other root?

3.28 The function $f(x) = 0.2 \tan^{-1}(x - 10) + e^{-x} \cos(4x) - 0.25$ on $0 \leq x \leq 20$ is sketched as shown here. Based on this sketch (and realizing

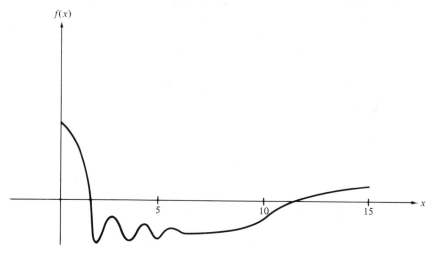

that it is a sketch, but one showing all the (2) roots on [0, 20]), give any fears you have about convergence and speed of convergence to the desired root in each of the following cases.

Desired root = the one between $x = 10$ and $x = 15$:

(a) Newton's method with $x_0 = 9$
(b) Newton's method with $x_0 = 15$
(c) Secant method with $x_0 = 9$, $x_1 = 15$
(d) Method of false position with $x_0 = 9$, $x_1 = 15$

Desired root = the one between $x = 0$ and $x = 5$:

(e) Newton's method with $x_0 = 5$
(f) Secant method with $x_0 = 0$, $x_1 = 5$
(g) Method of false position with $x_0 = 0$, $x_1 = 5$

3.29 This problem will probably be done most easily by hand calculation. Consider the following set of linear equations:

$$3x_1 + x_2 + x_3 = 5$$
$$x_1 + 3x_2 - x_3 = 3$$
$$3x_1 + x_2 - 5x_3 = -1$$

(a) Solve these equations by simple Gaussian elimination.
(b) Switch the second and third equations. Call the result the modified set of equations. What goes wrong when you try to solve the modified set by simple Gaussian elimination?

(c) Solve the original set of equations by doing three iterations of the Gauss-Seidel method with $x^{(0)} = 0$. Is the iteration converging or diverging?

(d) Solve the modified set of equations by doing three iterations of the Gauss-Seidel method with $x^{(0)} = 0$. Is the iteration converging or diverging?

3.30 Consider the linear equations $Ax = b$, where A is an $n \times n$ tridiagonal matrix, that is, $A_{ij} = 0$ if $|i - j| > 1$.

(a) Show that with $2(n - 1)$ multiplications and divisions we can triangularly factorize the tridiagonal matrix A into a lower triangular matrix L with only a diagonal and a subdiagonal and an upper triangular matrix U with ones on the diagonal and only a superdiagonal besides.

(b) Once we have done this factorization, how many multiplications and divisions are required to solve $LUx = b$ for a given b?

3.31

(a) If A is $n \times n$ and B is $n \times p$, show $AX = B$ can be solved by elimination in $pn^2 + n^3/3 + O(n^2)$ multiplications and divisions. Thus if B is $n \times n$, the solution can be found by elimination in $\frac{4}{3}n^3 + O(n^2)$ multiplications and divisions.

(b) If $B = I$, show that the solution can be found by elimination in "only" n^3 multiplications and divisions.

3.32 Assume that you are given $\{(x_i, y_i)|i = 1, 2, ..., N\}$, where each y_i has normally distributed error with standard deviation $= 20\%$ of the mean and where the x_i are more or less evenly distributed in $[0, 1000]$. Assume further that we know that these data come from an underlying function $f(x)$ defined on $[0, \infty]$ which decreases monotonically to zero in both value and slope magnitude as $x \to \infty$. Say that we wish the value $v = \int_0^\infty f(x)\,dx$. Specify the method you would use to determine v and the reasons for your choices. Be brief, but give enough detail so that there is no question about the steps of which your method consists.

3.33 Assume we are designing an airfoil and we know the shape of the bottom of the foil. Assume the top of the airfoil has the shape $f(x)$, where x is horizontal distance along the airfoil as in the accompanying diagram.

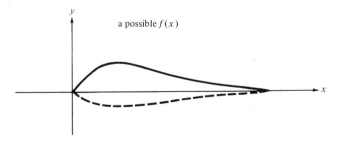

a possible $f(x)$

Assume we have a set of equations which we can solve analytically to give $f(x)$ for any x but for each x a different analytic solution is required which takes us 1 day of work to do without a blunder. We wish to know the point of the maximum air flow across the airfoil, and we know that this point occurs when $f'(x) = -[f''(x)]^2$. Give a method for finding this point. Justify your decisions.

CHAPTER FOUR

Algorithms and Operators

In Chapter 3 we covered only part of the numerical problems that arise, and we presented only a small fraction of the numerical methods available to solve the problems we covered. Not only are improvements available to the methods presented, but for each problem there are many different algorithms to solve it, each best in different circumstances. Instead of trying to understand the strategies of development, the accuracy, and the efficiency of each of these methods separately, we would do well to find a general point of view that will allow us to develop and analyze all or at least large classes of these methods together. This generalization is the subject of this chapter.

This approach will allow us not only to produce general techniques of development and analysis of numerical methods but also to see fundamental concepts that will inform the design of computer hardware and software. All of these steps will be taken in Chapters 5 through 9.

4.1 Vector Spaces and Operators

4.1.1 OPERATORS

The common characteristic of all numerical algorithms and the mathematical actions that they approximate is that they accept input data which are selected from a *domain D* of possible inputs and produce an output selected from a *range R* of possible outputs. The mapping between these domain elements and range elements is an *operator* defined by the algorithm or mathematical action.

The operator itself may be composed of component operators. For example, Newton's method (Section 3.4) can be viewed as an operator S mapping an initial iterate x_0 into a root ξ of a function f, or it can be viewed as repeated application of a root improvement operator T, first upon x_0, then upon the result, then upon that result, etc., so that

$$x_n = T(x_{n-1}) \qquad (n = 1, 2, ...) \tag{1}$$

or

$$x_n = T(T(\; ... \; (T(x_0)) \; ... \;)) \tag{2}$$

and

$$\xi = \lim_{n \to \infty} x_n \tag{3}$$

Furthermore, the operator T can be viewed as composed of an operator U which determines the values of the function f and its derivative f' at a point x, and an operator V which determines the new iterate, so that

$$U(x_{n-1}) \triangleq (x_{n-1}, f(x_{n-1}), f'(x_{n-1})) \tag{4}$$

$$V(x_{n-1}, f(x_{n-1}), f'(x_{n-1})) \triangleq x_{n-1} - \frac{f(x_{n-1})}{f'(x_{n-1})} = x_n \tag{5}$$

and

$$T(x_{n-1}) \triangleq V(U(x_{n-1})) \tag{6}$$

These various operators for Newton's method can be illustrated schematically as in Figure 4–1.

Thus, algorithms can generally be represented in increasing detail by compositions of operators operating upon successive operands. This detailing can proceed, if necessary, down to the primitive operations of a machine itself, or even its microcode. The framework that allows the study of properties of algorithms in general depends on our view of algorithms as operators, that is, mappings from one set of objects to another.

FIGURE 4–1
Illustrating Operators in Newton's Method

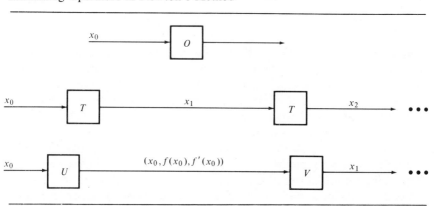

The *operands* of the operators (algorithms and the corresponding mathematical actions) described in Chapter 3 have been diverse. They have included real numbers, polynomials, n-tuples of real numbers, and functions of a real variable. It will be shown that these objects, and many others common to numerical algorithms, have certain mathematical properties in common. Mathematicians call objects with these properties a vector space.

4.1.2 VECTOR SPACES

A *vector space* is a triple consisting of a set and two operators, say $\langle V, +, \cdot \rangle$, and is defined with respect to a field F whose elements are called *scalars* (usually F is the set of real numbers). The elements of the set V are called *vectors;* the operator $+$ is called *vector addition* and operates on vectors (we write $x + y$); and the operator $\cdot$ is called *scalar multiplication,* combining scalars and vectors and producing vectors (we write $\alpha \cdot x$). They must satisfy the following properties:

1. The set is closed under vector addition and multiplication by scalars.
2. Vector addition is commutative $[x + y = y + x]$.
3. Vector addition is associative $[x + (y + z) = (x + y) + z]$.
4. There is a unique vector, denoted 0, for which $x + 0 = x$.
5. For every vector x there is a unique inverse vector under addition, denoted $-x$, for which $x + (-x) = 0$.
6. Multiplication by two scalars is associative $[\alpha \cdot (\beta \cdot x) = (\alpha \cdot \beta) \cdot x]$.
7. Multiplication by the unit scalar of F is an identity operation $[1 \cdot x = x]$.
8. Multiplication is distributive over addition of both vectors and scalars $[\alpha \cdot (x + y) = \alpha \cdot x + \alpha \cdot y; (\alpha + \beta) \cdot x = \alpha \cdot x + \beta \cdot x]$.

A test of these properties would show that each of the following useful sets are vector spaces, using the normal addition and scalar multiplication:

1. the set of all polynomials
2. the set of polynomials of degree less than or equal to n, for any n
3. the set of Fourier functions, $\sin(2\pi k\nu x)$ and $\cos(2\pi k\nu x)$ for all integers k, on any fundamental frequency ν
4. the set of all continuous functions of one variable
5. the set of solutions to a homogeneous equation $Ax = 0$
6. the set of solutions to a homogeneous differential equation $y' - f(x)y = 0$, for any continuous f
7. the set of n-tuples of real numbers, for any n; we will call this common set the *n-vectors,* and we will write it L_n
8. the set of real numbers!

On the other hand, many desirable sets are *not* vector spaces (why?):

1. the set of all polynomials of exact degree n, where n is an integer
2. the set of floating point numbers represented in some computer (IBM S/360, for example)
3. the set of solutions to a nonhomogeneous equation $Ax = b$, where $b \neq 0$

The notion of vector space is an abstraction of the vector notion found in mechanics, and the properties guarantee that many of the concepts associated with such vectors are carried over, as we shall see below. Furthermore, many other sets that are not themselves vector spaces are nevertheless subsets of vector spaces and may still be usefully treated in many situations using these concepts.

In particular, the normal algebraic notions of linear combination, linear independence, spansion, basis, and dimension can be defined for a vector space. Consider a set $\{x^1, x^2, ...\}$, a subset of elements of a vector space X. Then any sum formed as $\Sigma_i \alpha_i x^i$ is called a linear combination of the vectors $x^1, x^2, ...$. If one of the members of the set $\{x^1, x^2, ...\}$, say without loss of generality x^m, can be written as a linear combination of the other members ($x^m = \Sigma_{i \neq m} \beta_i x^i$), then we say that x^m is linearly dependent on the other vectors of the set and we say that the set $\{x^1, x^2, ...\}$ is *linearly dependent*. We say that a set is *linearly independent* if a linear combination $\Sigma_i \alpha_i x^i$ of its elements vanishes only if all of the α_i are zero. One can readily prove that a set is linearly dependent if and only if it is not linearly independent.

A set of vectors $\{x^1, x^2, ...\}$ is said to *span* a vector space X if each $x \in X$ can be represented as a linear combination of $\{x^1, x^2, ...\}$. A set of vectors $\{x^1, x^2, ...\}$ is called a *basis* for a vector space X if it is linearly independent and spans X. For example, in the vector space of polynomials of degree less than or equal to n, the set of polynomials $\{1, t, t^2, t^3, ..., t^n\}$ is a basis. A basis is not necessarily unique, since $\{1, 1 + t, 1 + t + t^2, ..., 1 + t + \cdots + t^n\}$ is also a basis for the aforementioned polynomial space. However, a theorem can be proved that if any basis for a vector space has a finite number of elements n, the number of elements in all bases for the space must be identical. Furthermore, it can be shown that any set of n linearly independent vectors in this vector space form a basis.

If a vector space has a finite basis, the number of elements in that basis is called the *dimension* of the space. If some basis has an infinite number of vectors, all bases will do so, and the dimension of the space is said to be infinite. For example, the vector space consisting of all polynomials has basis $\{1, t, t^2, t^3, ...\}$ and therefore has infinite dimension.

In this text we will usually restrict ourselves to consideration of only finite-dimensional vector spaces. In this case, any vector space can be placed in one-to-one correspondence to a vector space of n-tuples, where n is the dimension of the space. Thus, at any point in a treatment applicable to vector spaces one can, without losing generality, think of it in terms of a space of n-tuples or $1 \times n$ matrices.

4.2 Measures of Size of and Angle Between Vectors

4.2.1 NORMS, INNER PRODUCTS, AND ORTHOGONALITY

We are often concerned with the *size* of an object such as an output or an output error. We prefer that the size of the object is a single quantity even if the object is more complex. However, if all we know is that the output is a member of a vector space, it is not intuitively obvious how such a quantity might be represented. Traditionally, if the vector space is the set of real n-tuples $(x_1, x_2, ..., x_n)^T$, one measure of its size is its root sum of squares $(\Sigma_{i=1}^{n} x_i^2)^{1/2}$, otherwise known as its Euclidean length or Euclidean norm, since in two or three dimensions this represents length in Euclidean geometry. But what does one do if the vector space is a polynomial space or a set of functions? Furthermore, even for n-vectors we often prefer to measure the size of the vector by the absolute value of its largest element in magnitude, or perhaps by the sum of the magnitudes of the elements.

In order to preserve such power and flexibility and still be able to refer to a size generically, we define a *norm* to be any real-valued function having certain properties which will assure that it captures our idea of size. Thus a *norm* in a vector space X is a real-valued function on X, denoted $\|x\|$ for $x \in X$, for which the following statements apply:

1. The size (norm) of the **0** vector is 0: $\|\mathbf{0}\| = 0$.
2. The size (norm) of every nonzero vector is > 0: $x \neq \mathbf{0} \Rightarrow \|x\| > 0$.
3. The multiplication of a vector by a scalar multiplies the size (norm) of that vector by the magnitude of the scalar: $\|\lambda x\| = |\lambda| \|x\|$.
4. The triangle inequality holds: $\|x + y\| \leq \|x\| + \|y\|$; adding two vectors cannot produce a vector of a size larger than the sum of sizes of the two vectors.

It can be readily demonstrated that the "ℓ_p norms," defined on a vector space of n-tuples $x = (x_1, x_2, ..., x_n)^T$ as

$$\|x\|_p = \left(\sum_{i=1}^{n} |x_i|^p \right)^{1/p} \quad \text{for} \quad p \geq 1 \qquad [7]$$

satisfy these properties and are therefore norms. The norm ℓ_2 is merely the Euclidean norm defined above. It is so common that we will define the notation $|x|$ to mean $\|x\|_2$. The norm ℓ_1 is just the sum of magnitudes of the elements, while the norm ℓ_∞ (known as the Tchebycheff norm) is equal to the magnitude of the element of greatest magnitude (see Problem 4.2).

In a vector space of polynomials (or that of continuous functions) q on the interval $[a, b]$, the following might be acceptable norms:

$$\|g\|_p = \left(\int_a^b |g(x)|^p \, dx \right)^{1/p} \qquad \text{for} \quad p \geq 1 \qquad [8]$$

These are known as the $\mathscr{L}_p$ norms. They have many properties similar to the ℓ_p norms, including that the $\mathscr{L}_\infty$ norm of a function is its maximum magnitude over $[a, b]$. Many other means for constructing norms exist, and some of these will be introduced throughout this text.

One's choice of norm often depends upon one's objective in an analysis. For example, if worst case components of an error are your principal concern, you would choose an ℓ_∞ norm, whereas if you prefer equal weight to be applied to each error component you would choose an ℓ_1 norm for error measurement. Sometimes the choice of norm is made on the basis of mathematical convenience: some norms are more easily manipulated or are more compatible with a particular theoretical framework than others. Other times when the norm must be evaluated, the norm is chosen on the basis of computational convenience.

What allows one to have some freedom in the choice of norms is that it can be proved, at least for finite dimensional vector spaces, that over all vectors in the vector space the ratio of two norms is bounded both above and below by strictly positive numbers. Thus, it cannot be the case that one norm will indicate that the size of a vector is small while another norm will indicate that it is large.

Just as a generalization of the notion of size will be found useful, so will a mechanism that measures the angle between two vectors. In two- or three-dimensional Euclidean space we are familiar with the so-called *scalar product,* which consists of the product of the length of the first vector and the length of the projection of the second vector on the first. We write

$$x \cdot y = |x| \, |y| \cos (\theta_{xy}) \qquad [9]$$

where θ_{xy} is the angle formed between the two vectors and $x \cdot y$ is a notation for the scalar product, a scalar valued function of x and y. For three vectors $x = (x_1, x_2, x_3)^T$ and $y = (y_1, y_2, y_3)^T$ one can show that

$$x \cdot y = x_1 y_1 + x_2 y_2 + x_3 y_3 = x^T y \qquad [10]$$

Thus, the angle between the vectors is given by

$$\theta = \cos^{-1}\left(\frac{x \cdot y}{|x|\,|y|}\right) \tag{11}$$

where

$$|x| \triangleq \left(\sum_{i=1}^{n} x_i^2\right)^{1/2} = (x \cdot x)^{1/2} \tag{12}$$

that is, vector size is defined as the square root of the scalar product of the vector with itself.

Again, there are properties of the scalar product that we would like to have in analysis, even if we do not wish to or are unable to use the scalar product itself. Thus, we define an *inner product* on a vector space V to be a scalar valued (real or complex) function on $V \times V$, denoted (x, y) for $x, y \in V$, and for which

1. $(x, y) = \overline{(y, x)}$ $\tag{13}$

   where $\bar{\alpha}$ denotes the complex conjugate of α. (The function is symmetric for real vectors.)

2. $(\lambda x, y) = \lambda(x, y)$ $\tag{14}$

3. $(x + y, z) = (x, z) + (y, z)$ $\tag{15}$

4. $(x, x) > 0$ if $x \neq 0$

From this definition one can show (see Problem 4.4) that for any two vectors x, y and any inner product (x, y)

   a. $(x, \lambda y) = \bar{\lambda}(x, y)$ $\tag{16}$

   b. $(x, x) = 0$ if and only if $x = 0$ $\tag{17}$

One can also show that for any inner product, $(x, x)^{1/2}$ has the properties of a norm. It is called the "norm induced by the inner product."

A further derived property of an inner product (see Problem 4.6) is that the magnitude of an inner product of two vectors is always bounded by the product of the induced norms of the vectors, with equality occurring only when the vectors are collinear:

$$|(x, y)| \leq \|x\|\,\|y\| \tag{18}$$

where $\|x\| = (x, x)^{1/2}$. This relation, called the *Cauchy-Schwartz inequality,* lends substance to the interpretation of inner product as the product of the size of one vector and the size of a projection of the second on the first, even where we are not sure what might be meant by the notion of "projection."

The Cauchy-Schwartz inequality implies that $-1 \leq (x, y)/(\|x\| \|y\|) \leq 1$, with the value being ± 1 only when x is a scalar multiple of (in the same "direction" as) y. Therefore $(x, y)/(\|x\| \|y\|)$ can be interpreted as the cosine of a generalized angle between the vectors x and y.

By our definition, the following functions are inner products:

1. On the space of n-vectors, $(x_1, \ldots, x_n)^T$, $(y_1, \ldots, y_n)^T$, the scalar product

$$(x, y) = \sum_{i=1}^{n} x_i y_i \qquad [19]$$

This inner product induces the ℓ_2 norm, which can be shown to be the only ℓ_p norm that is induced by an inner product.

2. On the space of n-vectors,

$$(x, y) = \sum_{i=1}^{n} w_i x_i y_i \qquad [20]$$

for any set of strictly positive weights w_i.

3. On the space of polynomials defined over $[a, b]$

$$(p, q) = \int_a^b p(\xi) q(\xi) \, d\xi \qquad [21]$$

4. On an n-dimensional space of functions, none of which besides the zero function is zero at all of a set of distinct points $\{\xi_i | i = 1, 2, \ldots, n\}$ in the domain of these functions,

$$(f, g) = \sum_{i=1}^{n} f(\xi_i) g(\xi_i) \qquad [22]$$

There are many others which are occasionally useful.

If we imagine that the inner product of two vectors resembles a process of projecting one on the other and then multiplying sizes, we can see that if two vectors have an inner product of 0 they must be perpendicular to one another in some sense. This is a highly useful concept and we say that two vectors x and y are *orthogonal according to an inner product* $(\cdot, \cdot)$ if $(x, y) = 0$. We observe that the vector $\mathbf{0}$ is, by this definition, orthogonal to everything. We also note with satisfaction that it can be proved (see Problem 4.8) that any set of nonzero vectors that are mutually orthogonal according to some inner product is also linearly independent. Thus, in an n-dimensional vector space any n-vectors mutually orthogonal according to any inner product form a basis.

A set of vectors is said to be orthonormal according to a given inner product if the vectors are orthogonal and each vector in the set has unit length according to the norm induced by the inner product. An orthonormal basis is a basis made up of a set of orthonormal vectors.

The usefulness of an orthonormal basis is that it is especially easy to find the coefficients of the basis elements when expanding a vector in terms of the basis vectors: if $\{b^i | i = 1, 2, \ldots, n\}$ is an orthonormal basis according to the inner product $(\cdot\,,\,\cdot)$, then if

$$x = \sum_{i=1}^{n} \alpha_i b^i \tag{23}$$

it follows that

$$(x, b^k) = \sum_{i=1}^{n} \alpha_i (b^i, b^k) = \alpha_k \tag{24}$$

since $(b^i, b^k) = 0$ if $i \neq k$ and 1 if $i = k$. Thus, the coefficient of b_k in expanding x in terms of the b^i can be found simply by taking the inner product of x with b^k.

4.2.2 LINEAR LEAST-SQUARES ALGORITHM

The linear least-squares algorithm (Section 3.1.2) can be described more compactly and generally in terms of vector spaces and inner products, illustrating some of the power of these concepts.

The approximation problem can be expressed simply as the problem, given a function f, of choosing from a vector space of functions G, often not containing f, that function $\hat{f} \in G$ which minimizes $\|\hat{f} - f\|$. In the context of Section 3.1.2, the value of the function f is assumed known for a set of points $x_1, x_2, \ldots, x_N$ in its domain, and the norm is chosen so that

$$\|\hat{f} - f\| = \left(\sum_{i=1}^{N} (\hat{f}(x_i) - f(x_i))^2 \right)^{1/2} \tag{25}$$

Note that this is clearly a norm, provided that there is no nonzero member of G whose values at the x_i are all zero (see Problem 4.10b). When the norm is so chosen, the approximation problem becomes a least-squares problem. When the vector space G is chosen to be the set of generalized polynomials,

$$\hat{f}(x) = \sum_{j=1}^{n} \hat{a}_j f^j(x) \tag{26}$$

where $\{f^1, f^2, \ldots, f^n\}$ is some prechosen *basis* for the vector space G, and the coefficients $\hat{a}_j$ are arbitrary scalar constants ranging over the field of real numbers, then the problem is a *linear* least-squares problem. Note that the set $\{f^1, f^2, ..., f^n\}$ is a *basis* of G *by construction,* provided that the set is *linearly independent* and that the definition above does indeed define a vector space (see Problem 4.10a).

We showed in Section 3.1.2 that the prescribed values for the $\hat{a}_j$ were those values which satisfied the equation

$$\sum_{j=1}^{n} F_{kj} \hat{a}_j = d_k \qquad (k = 1, 2, ..., n) \tag{27}$$

where the F_{kj} and d_k were given by

$$F_{kj} = \sum_{i=1}^{N} f^k(x_i) f^j(x_i) \qquad (j, k = 1, 2, ..., n) \tag{28}$$

and

$$d_k = \sum_{i=1}^{N} f^k(x_i) f(x_i) \qquad (k = 1, 2, ..., n) \tag{29}$$

These expressions can be written in terms of the inner product that induces the norm being minimized (see equation 25),

$$(f, g) \triangleq \sum_{i=1}^{N} f(x_i) g(x_i) \tag{30}$$

so that F_{kj} and d_k can also be written

$$F_{kj} = (f^k, f^j) \qquad (j, k = 1, 2, ..., n) \tag{31}$$

and

$$d_k = (f^k, f) \qquad (k = 1, 2, ..., n) \tag{32}$$

Since we know (Section 1.4.2) that linear equations such as equation 27 can sometimes cause numerical difficulty and since we have seen (Section 3.1.2) that the problem can, in any event, be solved more efficiently if $[F_{kj}]$ is a diagonal matrix, it is clear that (where there is a choice) we should choose the basis functions to be mutually *orthogonal* according to the inner product. Then

$$F_{kj} = \begin{cases} 0 & (k \neq j) \\ (f^j, f^j) & (k = j) \end{cases} \tag{33}$$

for $j, k = 1, 2, ..., n$, and the coefficients which solve the least-squares problem are given by

$$\hat{a}_j = \frac{(f^j, f)}{(f^j, f^j)} \qquad (j = 1, 2, ..., n) \tag{34}$$

These algorithms are quite general and apply to quite arbitrary vector spaces (since any vector space is definable in terms of arbitrary linear combinations of its basis vectors). They also apply for any choice of inner product, if the minimization is with respect to the norm induced by the inner product (see Problem 4.11).

4.3 Operators on Vector Spaces

We have seen that both the desired solutions to numerical problems and the algorithms which we use to approximate these solutions can be thought of as operators applied to members of sets which often have the properties that define a vector space. For example, the integration operator is approximated by the composite Simpson's rule operator, both of which are applied to members of the vector space of integrable functions over the interval of integration. Similarly, the equation solution operator applied to the vector space of continuous functions of one variable is approximated by the Newton's method operator applied to the vector space of triples made up of a continuous function and two real numbers, the starting value and a tolerance for stopping the iteration. We will see that the integration operator and its Simpson's rule approximation are usefully distinguished from the other two operators just mentioned in that the former pair are *linear*. Let us define this term.

An operator T on members of a vector space V is said to be linear if for all vectors u, v in V and scalars α, β,

$$T(\alpha u + \beta v) = \alpha T(u) + \beta T(v) \qquad [35]$$

Equivalently (see Problem 4.13), T is linear if for all vectors u, v in V and scalars α,

$$T(u + v) = T(u) + T(v) \qquad [36]$$

and

$$T(\alpha v) = \alpha T(v) \qquad [37]$$

Linear operators have many useful properties. As a result of these they are often used to approximate linear and nonlinear operators that we desire to apply but cannot apply directly. Let us list a few of these properties. First, the set of images of a linear operator T applied to all the members of a vector space (the inputs) can be shown to form a vector space; the dimension of this vector space can be shown to be no greater than that of the input vector space (see Problem 4.14a,b). Second, the zero vector of the image vector space can be shown to be $T(0)$, the image of the zero vector of the input vector space (see Problem 4.14c). Third, if every vector in the input vector space maps into a different image, the operator T is said to be *nonsingular,* and the inverse operator T^{-1} mapping images under T to their corresponding input vectors can be shown to be linear (see Problem 4.14d). Finally, if a linear operator T is applied to the vector $x^* = x + \varepsilon_x$ instead of the desired input x, the result will be

$$T(x + \varepsilon_x) = T(x) + T(\varepsilon_x) \qquad [38]$$

which is in error — from the desired answer of $T(x)$ — by $T(\varepsilon_x)$, the result of applying the operator to the error vector.

We will be particularly worried about operators for which small input errors are mapped into large output errors, and especially operators for which input errors that are small relative to the input are mapped into output errors that are large relative to the output. Of course, a norm would be used to define all of these sizes.

If T is nonsingular and maps small vectors to large ones, T^{-1} will map large vectors to small ones. That is, T^{-1} will be approximately an operator S which maps large vectors to zero. It would be helpful to understand the behavior of such linear operators.

It can be shown (see Problem 4.14e) that the subset of all vectors which map into 0 under a linear operator S is a vector space. This space is called the *null space* of S. It includes at least the zero vector of its domain, and if it has dimension greater than zero it includes other (non-zero) vectors as well. It can be shown (see, for example, Franklin, 1968) that the dimension of the null space of S is the difference between the dimension of the vector space which is the domain of S and the dimension of the vector space of images under S. Therefore, any linear operator mapping a vector space into a space of lower dimension must map nonzero vectors into zero. Furthermore, any linear operator mapping a vector space into a space of the same dimension cannot map a nonzero vector to zero.

If a linear operator S maps two vectors, u and v, onto the same vector, it must map their difference, which is nonzero, onto $S(u) - S(v) = 0$. That is, linear operators which map more than one vector onto a single vector are those with a null space of dimension greater than zero, and vice versa. These are the singular operators — the ones that have no inverse.

4.4 Summary and Complements

We have developed the foregoing mathematical generalizations because they will allow us to make statements about and improve methods for many numerical problems at once. Many of the operators that we will deal with are linear, and many of the domains and images of the operators are vector spaces, but not all of the operators and sets of interest meet these specifications. In particular, the floating-point numbers with their addition operator do not form a vector space, though they are approximated by the vector space of real numbers. Similarly, many of the nonlinear operators that we will deal with can be well approximated, at least for the operands of interest, by linear operators.

Many of the operators that we have discussed have been applied to functions of one variable. For example, we have discussed evaluation and integration of functions of one variable. Similarly, we have examined differential equations the solutions of which are functions of one variable. We will largely restrict our attention in this book to operators on functions of one variable because these will suffice to illustrate our central issues of computing strategy, accuracy, and efficiency. But the reader should not get the impression that operators on functions of (or to) many variables are unimportant or uncommon. The solution of many nonlinear equations in many unknowns, the evaluation of tabulated functions of many variables, the solution of partial differential equations, and the like are numerical problems which frequently arise.

For the purposes of this book it will suffice to note that most of these multidimensional problems are solved by methods that are extensions to many variables of those that apply to single variable problems — those that *are* covered in this book. Such extension has already been exemplified by the solution of equations. As other examples, partial differential equations are often solved by marching along a multidimensional grid; many nonlinear equations in many unknowns are often solved by iteratively approximating by linear surfaces the nonlinear surfaces represented by the equations and solving the corresponding linear equations; and evaluation of multivariable tabulated functions is accomplished by fitting a function of many variables that can approximate well and can be easily evaluated. Finally, those methods that are not multidimensional extensions of one-dimensional methods are still instances of the basic computing strategies covered in this book.

REFERENCES

Franklin, J. N., *Matrix Theory*. Prentice-Hall, Englewood Cliffs, N.J., 1968.

PROBLEMS

4.1

(a) Show that the set of polynomials with natural polynomial addition forms a vector space.

(b) Show that the set of all vectors which solve $Ax = 0$ forms a vector space.

(c) Why is the set of all vectors which solve $Ax = b$ for $b \neq 0$ not a vector space?

4.2 Show that $\lim_{p \to \infty} (\sum_{i=1}^{n} |x_i|^p)^{1/p} = \max_{1 \leq i \leq n} |x_i|$. Thus the ℓ_∞ (or Tchebycheff) norm of an n-vector x is simply the magnitude of the element of x with largest magnitude.

4.3 Let V be a vector space and $(\cdot, \cdot)$ be an inner product on V. Show that $T(x) = (x, a)$ is a linear operator, where a is a given vector in V.

4.4 Show that if $(\cdot, \cdot)$ is an inner product, for any vectors x, y, and complex scalar λ

(a) $(x, \lambda y) = \bar{\lambda}(x, y)$.

(b) $(x, x) = 0$ iff $x = 0$.

4.5 Let V be a vector space and $(\cdot, \cdot)$ be an inner product on V. Show that $(x, x)^{1/2}$ has the properties of a norm.

4.6 Starting with the relation $0 \leq (z, z)$, when $z = x - [(x, y)/(y, y)]y$, show that $|(x, y)| \leq \|x\| \|y\|$, where $\|\cdot\|$ is the norm induced by the inner product $(\cdot, \cdot)$.

4.7 Let K be any $n \times n$ nonsingular real matrix. Show $(x, y) = (Kx)^T Ky$ is an inner product on L_n.

4.8 Let $\{x^i \mid i = 1, 2, ..., n\}$ be a set of vectors that are mutually orthogonal according to some inner product $(\cdot, \cdot)$. Show that the x^i are linearly independent.

4.9 Let $\{b^i\}$ be a basis for L_n. Let $x = \sum_{i=1}^{n} \xi_i b^i$ and $y = \sum_{i=1}^{n} \eta_i b^i$ both be real vectors in L_n. Let $f(x, y) = \sum_{i=1}^{n} \xi_i \eta_i$. Show $f(x, y)$ is an inner product on L_n. Then show that b^i and b^j are orthogonal according to this inner product if $i \neq j$.

4.10 Assume that we have n distinct arguments x_i, for $i = 1, 2, ..., N$, where each $x_i \in [a, b]$. Assume that we have a set of functions f^i, for $i = 1, 2, ..., N$, each on $[a, b]$ with the property that the f^i, for $i = 1, 2, ..., N$, are linearly independent.

(a) Show that the set G of linear combinations of the f^i is a vector space.

(b) Assume that $f \in G$ and $f(x_i) = 0$ at all x_i, for $i = 1, 2, \ldots, N$, iff $f = 0$. Show that $(\sum_{i=1}^N (f(x_i))^2)^{1/2}$ is a norm on G.

4.11 Assume that the functions f^i, for $i = 1, 2, \ldots, N$, are the basis for a vector space G, that G is a subspace of a vector space H, that $(\cdot, \cdot)$ is a real inner product on H such that the f^i, for $i = 1, 2, \ldots, N$, are mutually orthogonal according to the inner product, and that $\|\cdot\|$ is the norm induced by that inner product. Let $f \in H$. Show that the function $\hat{f} \in G$ that minimizes $\|\hat{f} - f\|$ is given by $\hat{f} = \sum_{j=1}^N \hat{a}_j f^j$, where $\hat{a}_j = (f^j, f)/(f^j, f^j)$, for $j = 1, 2, \ldots, N$.

4.12

(a) Show that the operator defined by $(f^j, f^k) = \sum_{i=1}^N [f_i^j f_i^k (x_i - x_{c_1})^2 (x_i - x_{c_2})^2]$ is an inner product.

(b) Show that in the least-squares problem where the approximating function is constrained to pass through (x_{c_1}, y_{c_1}) and (x_{c_2}, y_{c_2}), in order for the normal equations to be uncoupled, the basis functions should be orthogonal according to the inner product defined above.

4.13 Let V be a vector space, and let T be an operator on vectors in V. Show that $T(\alpha u + \beta v) = \alpha T(u) + \beta T(v)$ for all scalars α and β and all vectors u and v in V iff, for all such vectors and all scalars α, $T(u + v) = T(u) + T(v)$ and $T(\alpha v) = \alpha T(v)$.

4.14 Let V be a vector space, and let T be a linear operator on vectors in V.

(a) Show that $W = \{T(v) \mid v \in V\}$ is a vector space.

(b) Show that the dimension of W is no greater than the dimension of V.

(c) Show that the zero of W is $T(0)$.

(d) Show that, if T is nonsingular, T^{-1} is a linear operator from W to T.

(e) Show that $\{v \mid T(v) = 0\}$ is a vector space.

CHAPTER FIVE

Computer Representation of Mathematical Objects: Generated Error

5.1 Basic Concepts

When one is applying an operator T to inputs $\mathbf{x}$ to compute an output y, there are two basic sources of error in the output. First, there is the error due to applying only an approximation of the operator. This error is called the *generated* error due to applying T^* instead of T, though we will often sloppily call it the error generated by T. Here T^* is the computer operator used to approximate the desired mathematical operator. For example, such error may be made because floating-point arithmetic operations are used in place of the desired real arithmetic operations or because functional approximation is part of the process, for example, in numerical integration, when commonly an approximating polynomial is integrated rather than the underlying function.

Second, error arises due to error in the inputs, $\mathbf{x}$. This error is called the *propagated* error when applying T. Inputs will have error because they result from measurements or previous computation or because computer representation is used for real (mathematical) numbers.

It is important to specify the operator or pair of operators in question when categorizing error as propagated or generated, for two reasons. First, error that is generated by one operator may be propagated by a following operator. Second, error that is propagated by an operator S may be considered generated error with respect to an operator $U = ST$ (T followed by S) if the error in question was generated by the application of T.

We will throughout this book define error as the difference between the computed answer and the correct answer:

$$\varepsilon_y = y^* - y \qquad [1]$$

where ε_y is our notation for the error in y^* and y^* is the computed value of y. With

$$y = T(\mathbf{x}) \qquad [2]$$

we have

$$\varepsilon_y = T^*(\mathbf{x}^*) - T(\mathbf{x}) \qquad [3]$$

the difference between applying the approximate operator to the approximate input and applying the correct operator to the correct input. We will call this the *overall* error in applying T to $\mathbf{x}$ (or more correctly, in applying T^* to $\mathbf{x}^*$).

Subtracting and adding $T(\mathbf{x}^*)$ in equation 3, we obtain

$$\varepsilon_y = (T^*(\mathbf{x}^*) - T(\mathbf{x}^*)) + (T(\mathbf{x}^*) - T(\mathbf{x})) \qquad [4]$$

The first term in equation 4 is the error generated by using the operator T^* instead of T with the approximate input. The second term is the error propagated by T due to using input $\mathbf{x}^*$ in place of $\mathbf{x}$. Thus, *the overall error is equal to the sum of the generated error and the propagated error:*

$$\varepsilon_y = \varepsilon_y^{\text{gen}} + \varepsilon_y^{\text{prop}} \qquad [5]$$

An alternative expression for the overall error can be obtained by subtracting and adding $T^*(\mathbf{x})$ in equation 3. The result is

$$\varepsilon_y = (T^*(\mathbf{x}^*) - T^*(\mathbf{x})) + (T^*(\mathbf{x}) - T(\mathbf{x})) \qquad [6]$$

As with equation 4, this expression states that the overall error is equal to the sum of the generated error and the propagated error, but here the propagated error appears first and is the error propagated by the approximate operator rather than the correct operator, and the generated error appears second and is calculated with the correct input rather than the approximate input. In most cases it will make little difference whether we compute the propagated error using the correct or approximate operator or compute the generated error using the correct or approximate input, so we will freely use whichever is more convenient in each case.

Note that the generated error is analyzed as if the inputs had no error and the propagated error is analyzed as if the operator being applied was the desired operator. Thus, in Sections 5.3–5.5, where error generation is analyzed, all inputs will be assumed correct, and in Chapter

6, where error propagation is analyzed, all operators will be assumed correct.

5.2   Measures of Error

In general, we will not be able to compute the error in a result, for if we were able to do so we would simply subtract that error from the computed result to get the correct result. But we will be able to compute some measure of the error, commonly bounds on its value, a bound on its magnitude, or its standard deviation when the error-generation process is modeled probabilistically. In this book we will restrict ourselves to bounds and almost entirely to bounds on the magnitude of the error.

In some cases we will wish to talk about the error in a vector. This error, for instance ε_x, is the difference between the vector x^* and the vector x and is thus a vector in the same vector space as x^* and x. To produce a bound on such an error we need a concept of the size of this error vector, and this concept is realized by taking a norm of the error vector. Which norm we use will vary with the application. An error bound for a vector error will then be a bound on the norm of the error.

An error bound is often not very useful by itself. For example, knowing that the error is bounded by 0.01 does not give us an idea of the importance of the error. This error in a number like 1,000,000 will usually be negligible, but in a number like 0.001 it will be a disaster. To capture the importance of an error, we define the *relative error* in the number y as ε_y/y. The relative error tells the fraction of the correct value by which the computed value is off. The relative error, 10^{-8} in the first example but 10^1 in the second example, is a good indicator of the negligibility of the first error and the seriousness of the second one. To distinguish it from the relative error, the value ε_y is called the *absolute error* in y. Notice that the absolute error is not an absolute value, that is, it may have a negative value. We will use the notation b_y to refer to a bound on the magnitude of the absolute error in y and r_y to refer to a bound on the magnitude of the relative error in y. Note that

$$r_y = \frac{b_y}{|y|} \qquad [7]$$

We need a notion of the relative error in a vector. What we mean by this is the size of the error in x^* relative to the size of the vector x. Thus we will define the relative error in x^* as $\|\varepsilon_x\|/\|x\|$, where the norm used will vary from problem to problem but the same norm must be used in both the numerator and the denominator. Note that the relative error in a vector will always be non-negative.

We might have defined the relative error in an n-vector as the norm of the vector with elements ε_{x_i}/x_i. However, not only would this be ungeneralizable to vectors other than n-vectors, but it would measure the relative size of the element errors rather than the relative size of the error vector. For example, an error of $[1\ 1\ 1]^T$ in the vector $[1\ 1000\ 1000]^T$ is a small error in the whole vector despite the fact that one of the elements has 100 percent error.

The definition of relative error as ε_y/y is useful for mathematical analysis of error but is not as useful for computational analysis of error because in the latter case we have y^* rather than y. However, note that

$$\frac{\varepsilon_y}{y^*} = \frac{\varepsilon_y}{y + \varepsilon_y} = \frac{\varepsilon_y}{y}\left(\frac{y}{y + \varepsilon_y}\right)$$

$$= \frac{\varepsilon_y}{y}\left(1 - \frac{\varepsilon_y}{y} + \left(\frac{\varepsilon_y}{y}\right)^2 - \cdots\right) = \frac{\varepsilon_y}{y} + O\left(\frac{\varepsilon_y}{y}\right)^2 \quad \text{[8]†}$$

Thus, when the relative error is small, it is of little concern whether we use the definition ε_y/y or ε_y/y^*. When the relative error is large, we need only conclude that the process being analyzed must be rejected and we are not concerned with precisely how large the error is. Therefore we will use the two definitions of relative error interchangeably.

Let us develop expressions for the bounds on absolute and relative overall error in terms of the corresponding bounds on generated and propagated error. From equation 5, $\varepsilon_y = \varepsilon_y^{\text{gen}} + \varepsilon_y^{\text{prop}}$, it follows that

$$|\varepsilon_y| \leq |\varepsilon_y^{\text{gen}}| + |\varepsilon_y^{\text{prop}}| \quad \text{[9]}$$

and equality is achieved if $\varepsilon_y^{\text{gen}}$ and $\varepsilon_y^{\text{prop}}$ have the same sign. Thus,

$$b_y = b_y^{\text{gen}} + b_y^{\text{prop}} \quad \text{[10]}$$

Dividing both sides of equation 10 by $|y|$, we see

$$r_y = r_y^{\text{gen}} + r_y^{\text{prop}} \quad \text{[11]}$$

5.3 Arithmetic and Representation Error

The most basic operators for which we must discuss error generation are representation of input numbers in the computer and the arithmetic operators: addition, subtraction, multiplication, and division. We will

---

† In this chapter $O(g(u))$ means terms which go to zero as fast as $g(u)$ when $u \to 0$, that is, any function $f(u)$ such that

$$\limsup_{u \to 0} \left|\frac{f(u)}{g(u)}\right|$$

is bounded. This definition is the same as for $O(g(N))$ as used previously except that the limit is taken for $u \to 0$ instead of for $N \to \infty$.

call the error due to representing inputs *representation error,* that due
to the arithmetic operators *arithmetic error,* and that due to both combined *computation error.* In all cases we will assume the computer's
arithmetic operators are for base-B floating-point numbers with an n-digit
normalized fraction (high-order digit not equal to 0), as discussed in
Chapter 1. Examples of B and n used in various computers are $B = 16$,
$n = 6$; $B = 16$, $n = 14$; $B = 8$, $n = 13$; $B = 2$, $n = 24$. Remember
that we are assuming correct inputs.

For computer representation, the first n significant digits are represented exactly and all further digits are lost.† Thus, the worst loss is
of $B - 1$ in each of the places beyond the nth. That is, the absolute error
generated by computer representation is bounded in magnitude by
$B^{-n}B^{exp}$, where *exp* is the exponent in the floating-point representation
of the number in question. The relative error is worst when the fraction
part of the floating-point number is least in magnitude, that is, when the
floating-point number is $0.1_B \times B^{exp}$. Thus the relative error in computer
representation is bounded in magnitude by

$$R = \frac{B^{exp-n}}{(1/B)B^{exp}} = \frac{1}{B^{n-1}} \qquad [12]$$

In floating-point addition we assume a guard digit. That is, the addition process involves shifting the fraction of the summand with the
smaller exponent to the right, with the loss of all but the final digit shifted
out, and correspondingly increasing the exponent until it agrees with that
of the summand of greater magnitude. Then the two summands are added
together as $n + 1$-digit fractions, and the result is normalized and truncated
to n digits. There are three possible results of the addition of the two
fractions. The magnitude of the sum can be greater than or equal to 1
but less than 10_B (for example, $0.8_{10} + 0.3_{10} = 1.1_{10}$), in which case it
will be shifted one place right and the result truncated to n digits. It can
be less than 1 but have a nonzero digit in the first place to the right of
the radix point (for example, $0.3_{10} + 0.2_{10} = 0.5_{10}$), in which case the
digits beyond the nth will be truncated. Or it can be less than 1 but have
a zero in the digit immediately to the right of the radix point (this can
occur only if the operators are of opposite sign (for example, $0.34_{10} +
(-0.33_{10}) = 0.01_{10}$), in which case normalization will involve a left shift,
so the $n + 1$th digit, which was involved in the addition, and possibly
some following zeros will be shifted into the normalized result. Two
examples, for $n = 3$ and $B = 10$ are as follows: $-0.100 \times 10^0 +
0.999 \times 10^{-1} = -0.100 \times 10^0 + 0.0999 \times 10^0$ (the final 9 is not lost
because of the guard digit) $= -0.0001 \times 10^0 = -0.1000 \times 10^{-3}$ after
normalization; and $-0.100 \times 10^0 + 0.999 \times 10^{-2} = -0.100 \times 10^0 +$

† Some computers round rather than truncate. We will assume truncation in this book.

0.0099×10^0 (there is only one guard digit, so the third 9 is lost) $=$ $-0.0901 \times 10^0 = -0.901 \times 10^{-1}$ after normalization.

Let us begin by analyzing the first two cases. Here the error is due to truncation of digits beyond the nth in a normalized fraction. This truncation is of precisely the same form as that analyzed for computer representation. That is, if the sum of the two fractions is not a fraction beginning with the digit 0, the relative error generated by the addition is bounded by $1/B^{n-1}$.

We must now analyze the case where the guard digit [the $(n+1)$th] is shifted into the answer when the fraction part of the sum is normalized. If this happens when the smaller summand in magnitude has been shifted either zero or one place, no error is made, since all arithmetic is done to $n+1$ places and by assumption the $(n+1)$th digit is shifted into the answer during normalization. If the smaller summand has been shifted two or more places, the fraction part of the result of the addition (to $n + 1$ places) before normalization can be in magnitude no less than the fraction with 0 in the first place after the radix point and $B-1$ in the second $((B - 1) \times B^{-2})$, since the fraction part of the unshifted summand can be no less in magnitude than 0.1_B and the fraction part of the other summand after shifting is by assumption less in magnitude than 0.01_B. Since the fraction part of the result before normalization has exactly one high-order zero, there will be left shifting of one place in normalization. Since the arithmetic was done to $n + 1$ places, the normalized fraction is correct to n places, that is, the error is bounded in magnitude by $1/B^{n-1}$.

Note that had there been no guard digit a very serious error might have been made in the case where the guard digit was shifted into the result during normalization, in particular when the two summands differed in exponent by 1. For example, for $n = 3$ and $B = 10$ with no guard digit, the number $-0.100 \times 10^0 + 0.999 \times 10^{-1}$ is computed to be $-0.100 \times 10^0 + 0.099 \times 10^0 = -0.001 \times 10^0 = -0.1 \times 10^{-2}$, whereas the correct answer is -0.1×10^{-3}; a relative error of magnitude 900% has been made. It is to avoid this disastrous error that a guard digit is kept.

To summarize, if addition is done with a guard digit, the generated relative error is bounded in magnitude by $1/B^{n-1}$. Note the form of our analysis. Classes of operands for which the operation is similar are discerned, and for each class the worst-case relative error magnitude is determined. Then the worst case among the classes is determined by comparing the results.

Subtraction is just addition with the sign of one of the operands changed. Thus, the relative error generated in subtraction is also bounded in magnitude by $1/B^{n-1}$.

Floating-point multiplication operates by adding the exponents of the operands and multiplying the fractions. Since the fractions are assumed to be normalized, their product must be in magnitude less than 1 and greater than or equal to 0.01_B. Therefore, if $n + 1$ digits of the product are calculated beyond the radix point (in many machines all $2n$ digits are calculated), the normalization of the result by at most one place will result in a product fraction accurate to n places. That is, the error generated is that due to the truncation of digits beyond the nth, and we have seen that the magnitude of this relative error is bounded by $1/B^{n-1}$.

Floating-point division operates analogously. The exponents of the operands are subtracted, and the fractions are divided. Because each of the fractions is normalized, their quotient must be in magnitude less than 10_B and greater than 0.1_B. Therefore, if $n + 1$ digits beginning with the one immediately to the left of the radix point are calculated in the quotient, normalization, which will involve either no shifting or shifting one place to the right, followed by truncation will result in a generated error due to the loss of digits beyond the nth in a normalized n-digit number. That is, again the generated relative error will be bounded in magnitude by $1/B^{n-1}$.

We see therefore that in a computer which does its floating-point operations correctly to $n + 1$ digits and then normalizes and truncates to n digits the relative error generated by replacing real numbers by their floating-point representations or by replacing any real arithmetic operation by a floating-point operation is bounded in magnitude by $R = 1/B^{n-1}$. Remember that this bound must be added to a bound on the error due to errors in the inputs (the propagated error) to produce an overall error bound.

Usually an arithmetic expression involves more than one arithmetic operation or input representation. In the computation these are applied in sequence, and the error generated by any given operation is propagated by the operations which follow it. The arithmetic error generated in computing the entire expression is thus produced by a combination of the generation by the individual operators and propagation. Since error propagation is not covered in this text until Chapter 6, we can not conclude the analysis of arithmetic error until Section 7.2. For now, we will assume that we can compute the bounds b_y^{comp} and r_y^{comp}.

5.4   Approximation Error

Any error generated due to approximation of one function by another will be called *approximation error*. Such an error is generated whenever the computing strategy of approximate and operate is used. Let us use the notation $\hat{T}(\mathbf{x})$ to refer to the result of applying the operator involving functional approximation to the input $\mathbf{x}$ but making no arithmetic or representation error. We will let $\hat{T}^*(\mathbf{x})$ refer to the same result except that in this case we will assume that errors are made both in representing **x** and in arithmetic. The overall generated error is given by

$$\varepsilon_y^{gen} = \hat{T}^*(\mathbf{x}) - T(\mathbf{x}) \qquad [13]$$

As with the analysis in Section 5.1 when separating generated from propagated error, we subtract and add $\hat{T}(\mathbf{x})$ in equation 13, producing

$$\varepsilon_y^{gen} = (\hat{T}^*(\mathbf{x}) - \hat{T}(\mathbf{x})) + (\hat{T}(\mathbf{x}) - T(\mathbf{x})) = \varepsilon_y^{comp} + \varepsilon_y^{approx} \qquad [14]$$

where ε_y^{comp} is the arithmetic and representation error made in applying the operator involving functional approximation and ε_y^{approx} is the approximation error assuming all arithmetic operations are done correctly. From this result we can produce the generated error bounds,

$$b_y^{gen} = b_y^{comp} + b_y^{approx} \qquad [15]$$

and

$$r_y^{gen} = r_y^{comp} + r_y^{approx} \qquad [16]$$

In the following we will treat only the approximation error, ε_y^{approx}. That is, we will assume both the inputs and the arithmetic are exact. We will furthermore omit the superscript "approx" in this section.

5.4.1 EVALUATION OF FUNCTIONS

Let us first analyze the approximation error in the value of a function at a given argument. We will analyze here the error in simple polynomial exact-matching approximation because it is so common, with the understanding that similar analysis can be done for other approximations such as splines and least-squares approximations. This error is sometimes called *truncation error*, as it can be viewed as resulting from the truncation of an infinite series.

Point-in-Interval Approach

Simple Exact-Matching Approximation by a Single Polynomial. Assume we are approximating the underlying function $f(x)$ by the polynomial $\hat{p}(x)$ which exact-matches at the tabular arguments $x_0, x_1, ..., x_N$. Thus,

$$\varepsilon_{f(x)} = \hat{p}(x) - f(x) \qquad [17]$$

We know the error is zero at $x_0, x_1, ..., x_N$. Thus, we are motivated to write

$$\varepsilon_{f(x)} \overset{\Delta}{=} \hat{p}(x) - f(x) = K(x) \prod_{i=0}^{N} (x - x_i) \qquad [18]$$

where $K(x)$ is a function to be determined. Note that, as far as we know, the root of $\hat{p}(x) - f(x)$ at x_i could be of multiplicity† less than 1, in which case $K(x_i)$ would be infinite.

Let us assume f is $N + 1$ times differentiable. If K were a constant, we could solve for K by differentiating equation 18 $N + 1$ times with respect to x, producing

$$-f^{(N+1)}(x) = (N + 1)!K \qquad [19]$$

Since K is not a constant, differentiating equation 18 $N + 1$ times produces an equation involving derivatives of $K(x)$ as well as $K(x)$ itself, so this procedure is not directly helpful. However, if we choose z as any argument value not equal to a tabular argument value, we have

$$\hat{p}(x) - f(x) = K(z) \prod_{i=0}^{N} (x - x_i) \qquad \text{at} \quad x = z \qquad [20]$$

or equivalently,

$$\phi(x) \overset{\Delta}{=} \hat{p}(x) - f(x) - K(z) \prod_{i=0}^{N} (x - x_i) = 0 \qquad \text{at} \quad x = z \qquad [21]$$

Now $K(z)$ is a constant with respect to x, so differentiating $\phi(x)$ $N+1$ times produces

$$\phi^{(N+1)}(x) = -f^{(N+1)}(x) - (N + 1)!K(z) \qquad [22]$$

Now we know that $\phi(x)$ has $N + 2$ zeros: at each x_i and at z. We apply Rolle's theorem, which states that a differentiable function must have a relative maximum or relative minimum between two successive zeros, that is, if the function leaves the x axis and then produces another zero, it must have turned around somewhere to come back. Since we are assuming that ϕ is differentiable, we conclude that in the interval bounded by the x_i and z (which we write $[x_0, x_1, ..., x_N, z]$), $\phi'(x)$ has at least $N + 1$ zeros. Reapplying Rolle's theorem, we see $\phi''(x)$ must have N zeros in that interval, ..., so $\phi^{(N+1)}(x)$ must have at least one zero in that interval. Let the value of x where that zero occurs be ξ. Then

$$0 = \phi^{(N+1)}(\xi) = -f^{(N+1)}(\xi) - (N + 1)!K(z) \qquad [23]$$

so $K(z) = -f^{(N+1)}(\xi)/(N + 1)!$, for some $\xi \in [x_0, x_1, ..., x_N, z]$ [24]

Note that ξ is a function of z.

Equation 24 is true for any z not equal to a tabular argument, in particular for z equal to the evaluation argument x. Thus, substituting

† See Section 3.4.1 (page 71) for the definition of the multiplicity of a root.

x for z in equation 24 and the result in equation 18, we have for x not equal to any x_i,

$$\varepsilon_{f(x)} \overset{\Delta}{=} \hat{p}(x) - f(x) = -\prod_{i=0}^{N} (x - x_i) \frac{f^{(N+1)}(\xi(x))}{(N+1)!} \tag{25}$$

for some $\xi(x) \in [x_0, x_1, ..., x_N, x]$. We have assumed $f^{(N+1)}$ exists for all values in the interval in question. Therefore, equation 25 holds for $x = x_i$, since there the error is 0. Thus, the restriction $x \neq x_i$ may be dropped, and equation 25 holds for all x.

It is not particularly important to learn the above development, but it is very important to learn and understand the result given by equation 25. It states that the approximation error is given by the product of a polynomial in x, involving the tabular arguments, and a derivative of the underlying function evaluated at some unknown point in the interval defined by the tabular arguments and the evaluation argument.

We will give a name to the product polynomial in equation 25:

$$\psi(x) \overset{\Delta}{=} \prod_{i=0}^{N} (x - x_i) \tag{26}$$

It will also be helpful to give the name I to the interval for ξ in equation 25, namely, $[x_0, x_1, x_2, ..., x_N, x]$.

For a particular function f, set of tabular arguments $x_0, x_1, ..., x_N$, and evaluation argument x, we can bound the approximation error by using the largest value of $|f^{(N+1)}|$ over I, since we do not know what ξ is:

$$|\varepsilon_{f(x)}| \leq \frac{|\psi(x)|}{(N+1)!} \max_{\xi \in I} |f^{(N+1)}(\xi)| = b_{f(x)} \tag{27}$$

For example, consider approximating $\sin(\pi/4)$ by fitting a polynomial $\hat{p}$ through the tabular points $0, \pi/6, \pi/3, \pi/2$. Then

$$\varepsilon_{\sin(\pi/4)} = -\frac{(\pi/4 - 0)(\pi/4 - \pi/6)(\pi/4 - \pi/3)(\pi/4 - \pi/2)}{4!}$$
$$\times \left. \left(\frac{d^4}{dx^4} \sin(x) \right) \right|_{x=\xi} \tag{28}$$

where $\xi \in [0, \pi/2]$. Thus,

$$b_{\sin(\pi/4)} = \left(\frac{\pi}{4} \right)^4 \frac{1}{216} \max_{\xi \in [0, \pi/2]} |\sin(\xi)| = \left(\frac{\pi}{4} \right)^4 \frac{1}{216} \approx 1.8 \times 10^{-3} \tag{29}$$

Actually fitting the polynomial produces $\hat{p}(\pi/4) = 0.7059$ as compared to $\sin(\pi/4) = 0.7071$, so the error is -1.2×10^{-3}; our bound is reasonable. Note that in this case the sign of the fourth derivative is constant over. I, so we know that the error will be negative: $-1.8 \times 10^{-3} \leq \varepsilon_{\sin(\pi/4)} \leq 0$. We could reduce the error bound by subtracting half of its value from the polynomial approximation, that is, by subtracting -0.0009 from 0.7059, producing 0.7068 ± 0.0009.

The approximation error formula, equation 25, can be used more generally to evaluate approximation strategies. First, it tells us that the farther the evaluation argument x is from the tabular argument values, the larger $\psi(x)$ will be and thus the larger the anticipated error. We can conclude that (1) the $N + 1$ data points used for a particular approximation should ordinarily be those in the table whose argument values are closest to the argument values at which we are evaluating; and (2) *extrapolation,* which is approximation at a value x not in the interval bounded by the tabular arguments used, is a very error-prone procedure. Compared with interpolation, the error in extrapolation may be larger not only because of the large value of $\psi(x)$ but also because of the increased interval for ξ, which produces the potential for a larger value for the derivative factor in the error (equation 25). Extrapolation at any appreciable distance from our tabular arguments is very dangerous unless we know in some detail the form of the function being approximated. If we do not have such detailed knowledge, the approximating function may go in one direction and the function being approximated may go in the opposite direction [as do $\hat{p}(x)$ and $f(x)$, respectively, in Figure 5–1].

Another issue that equation 25 allows us to deal with is the determination of the optimum value for $N + 1$, the number of data points that should be used in a particular approximation. The denominator of equation 25 would lead us to believe that we should use as many points as possible if arithmetic error considerations do not make this approach unsatisfactory (but they do). However, in many cases, approximation error considerations alone indicate that increasing the number of points is not desirable. Introducing a tabular point, x_{N+1}, has three effects on the approximation error (equation 25).

1. It is multiplied by $x - x_{N+1}$ due to the additional term in the product $\psi(x)$.
2. It is divided by $N + 2$ due to the additional term in the factorial in the denominator.
3. The $(N+1)$th derivative is replaced by an $(N+2)$th derivative, and the interval in which the argument of the derivative may fall is slightly enlarged.

With regard to effect 1, we assume that the best tabular arguments of those available are used, that is, x_i is the $(i+1)$th closest tabular argument to x (see Figure 5–2). If the tabular arguments are approximately equally

FIGURE 5–1
A Danger of Extrapolation

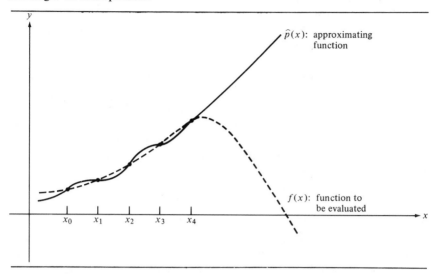

spaced with average spacing h, the x_i will be alternately on opposite sides of x and

$$x_{N+1} \approx x \pm \left(\frac{N+1}{2}\right) h \qquad [30]$$

Thus

$$|x_{N+1} - x| \approx \left(\frac{N+1}{2}\right) h \qquad [31]$$

so the combined result of effects 1 and 2 will be a multiplication in magnitude by $[(N+1)/(N+2)](h/2) \approx h/2$.

With regard to effect 3, many smooth functions have the property that $f^{(N+1)}$ decreases as N increases for a while but then increases with N. For example, consider the function $\ln(x)$. Its $(N+1)$th derivative is

FIGURE 5–2
Selecting the Closest of Equally Spaced Tabular Arguments to the Evaluation Argument. The symbol · indicates a tabular argument; subscripts are in order of closeness to x.

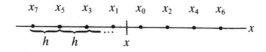

$(-1)^N N!/x^{N+1}$, which has the aforementioned property. If one is approximating a function f that behaves in this way, it is likely that the approximation error, which at each step reflects a multiplication by approximately $h/2$ and the behavior of the derivative as N increases, will have a minimum at some value of N, so further tabular points should not be used (see Figure 5–3).

Even if the above approximation error considerations do not lead to a limit on the number of tabular points, arithmetic error considerations do. Arithmetic error usually increases with the number of tabular points used (computations required). Therefore, it is almost always the case that the overall error reaches a minimum for some relatively small N, so further tabular points should not be used. We shall discuss this matter in more detail in Section 6.2.

We wish to bound the approximation error not only for a particular evaluation argument x but for a range of possible evaluation arguments, since when we construct an approximation subroutine we do not know exactly what the evaluation argument will be. Let us assume that the tabular arguments to be used are fixed and that the range of possible evaluation arguments J is included in the range of these tabular arguments I. Then the bound on the error over all $x \in J$ is

$$b_f = \max_{x \in J} b_{f(x)} = \max_{x \in J} \left[\frac{|\psi(x)|}{(N+1)!} \max_{\xi \in I} |f^{(N+1)}(\xi)| \right] \qquad [32]$$

Notice that since $x \in [x_0, x_1, ..., x_N]$, the interval I does not change with

FIGURE 5–3
Common Approximation Error Behavior

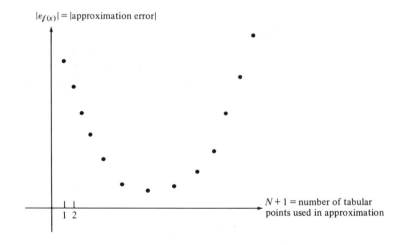

$|\varepsilon_{f(x)}| = |$approximation error$|$

$N + 1 =$ number of tabular
points used in approximation

1 2

x. Therefore,

$$b_f = \frac{1}{(N + 1)!} \left(\max_{x \in J} |\psi(x)| \right) \left(\max_{\xi \in I} |f^{(N+1)}(\xi)| \right) \quad [33]$$

Equation 33 states that for the case where the set of tabular arguments used does not change with the evaluation argument and the evaluation argument is always internal to the range of tabular arguments, the maximum of $|\psi(x)|$ and of $|f^{(N+1)}(\xi)|$ can be taken independently. Let us take as an example the problem of evaluating $\sin(x)$ using the tabular points 0, $\pi/6$, $\pi/3$, $\pi/2$, as discussed above, but here let x be only restricted to the interval $J = [\pi/6, \pi/3]$. Then as above, $I = [0, \pi/2]$, so

$$\max_{\xi \in I} |f^{(N+1)}(\xi)| = \max_{\xi \in [0, \pi/2]} |\sin(\xi)| = 1 \quad [34]$$

We can also compute

$$\max_{x \in J} |\psi(x)| = \max_{x \in [\pi/6, \pi/3]} \left| x \left(x - \frac{\pi}{6} \right) \left(x - \frac{\pi}{3} \right) \left(x - \frac{\pi}{2} \right) \right| \quad [35]$$

by taking the derivative of $\psi(x)$, setting it to 0, solving for x, and comparing the value of ψ at the results to the value of ψ at the boundaries of J. In this case the maximum is taken at $x = \pi/4$, where the value is 0.042. Thus $b_{\sin} = (1/4!)(0.042)(1) = 1.8 \times 10^{-3}$. Note that because the maximum for $\psi(x)$ occurred at $\pi/4$, the value specified for the evaluation argument in the earlier example, the bound is the same here as there.

Simple Piecewise Polynomial Exact-Matching Approximation. As discussed in Chapter 3, more common than having a fixed set of tabular arguments of which all are used in an approximation is to have a table, most often with equal intervals, in which piecewise simple exact matching is to be done; that is, simple exact matching is to be done through the $N + 1$ points in the table which are closest to the evaluation argument x, for some fixed number $N + 1$. In this case the range J of possible values for the evaluation argument x can be divided into intervals J_i such that for all $x \in J_i$ the same tabular arguments are the $N + 1$ closest. For example, let there be $n + 1$ tabular arguments equally spaced on $[0, 1]$ with interval $h = 1/n$, let the range of possible evaluation arguments be $[0, 1]$, and let $N + 1 = 2$; that is, piecewise linear exact matching is to be done. Then, for $x \in [0, h]$ the tabular points at 0 and at h will be used, for $x \in [h, 2h]$ the tabular arguments at h and $2h$ will be used, etc. (see Figure 5–4a). Thus, $J_1 = [0, h]$, $J_2 = [h, 2h]$, ..., $J_n = [(n - 1)h, 1]$. As another example, let J be as before, but let the tabular arguments be at $-h/2$, $h/2$, $3h/2$, ..., $1 - h/2$, $1 + h/2$, and let

FIGURE 5–4

Intervals J_i and I_i for (a) Linear and (b) Quadratic Equal-Interval Exact-Matching Approximation in [0, 1]. The tabular points in I_i are the closest $N + 1$ points to x if $x \in J_i$.

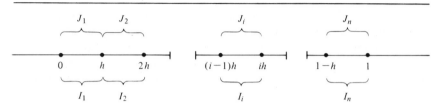

(a) $N+1=2$

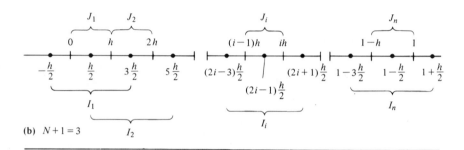

(b) $N+1=3$

$N + 1 = 3$; that is, piecewise simple quadratic exact matching is to be done. In this case we see that for $x \in [0, h]$ the tabular arguments at $-h/2, h/2$, and $3h/2$ will be used; for $x \in [h, 2h]$ the tabular arguments at $h/2, 3h/2$, and $5h/2$ will be used; and so on (see Figure 5–4b). Thus, in this example the J_i are the same as in the previous example, but here the J_i boundaries are halfway between tabular arguments rather than at tabular arguments.

Associated with each J_i is an interval I_i which is the union of J_i with the interval bounded by the tabular arguments used when $x \in J_i$. Unless there is extrapolation, the union with J_i is irrelevant. Thus in the linear interpolation example above,

$$I_1 = [0, h], \qquad I_2 = [h, 2h], \qquad \ldots, \qquad I_n = [(n - 1)h, 1]$$

In the quadratic interpolation example,

$$I_1 = \left[-\frac{h}{2}, \frac{3h}{2} \right], \qquad I_2 = \left[\frac{h}{2}, \frac{5h}{2} \right], \qquad \ldots, \qquad I_n = \left[1 - \frac{3h}{2}, 1 + \frac{h}{2} \right]$$

Also associated with each J_i is the polynomial

$$\psi_i(x) = \prod_{x_j \in I_i} (x - x_j) \tag{36}$$

which is used in the error formula for $x \in J_i$.

For $x \in J_i$, the approximation error is bounded by equation 33 with ψ_i in place of ψ, J_i in place of J, and I_i in place of I. Thus the error for all possible evaluation arguments $x \in J$ is

$$b_f = \max_i \frac{1}{(N+1)!} \left(\max_{x \in J_i} |\psi_i(x)| \right) \left(\max_{\xi \in I_i} |f^{(N+1)}(\xi)| \right) \tag{37}$$

For the case of equal intervals we can carry our analysis further. Consider that set of i for which the relation between I_i and J_i is the same. For example, in the case above with piecewise linear interpolation $I_i = J_i$ for $1 \leq i \leq n$, and in the case with piecewise quadratic interpolation I_i is J_i expanded by $h/2$ on each end for $1 \leq i \leq n$. Let the tabular arguments be numbered in increasing order such that x_1 is the leftmost tabular argument in I_1. Then

$$\begin{aligned}
\psi_i(x) &= (x - x_i)(x - x_{i+1}) \cdots (x - x_{i+N}) \\
&= (x - (x_0 + ih))(x - (x_0 + (i+1)h)) \cdots (x - (x_0 + (i+N)h)) \\
&= (x - x_0 - ih)((x - x_0 - ih) - h)((x - x_0 - ih) - 2h) \\
&\quad \cdots ((x - x_0 - ih) - Nh)
\end{aligned} \tag{38}$$

Note that

$$\begin{aligned}
J_i &= \left[\frac{x_i + x_{i+N}}{2} - \frac{h}{2}, \frac{x_i + x_{i+N}}{2} + \frac{h}{2} \right] \\[2mm]
&= \left[x_0 + \left(i + \frac{N-1}{2} \right) h, x_0 + \left(i + \frac{N+1}{2} \right) h \right]
\end{aligned} \tag{39}$$

With the change of variables

$$z = x - x_0 - \left(i + \frac{N}{2} \right) h \tag{40}$$

$$\begin{aligned}
\max_{x \in J_i} |\psi_i(x)| &= \max_{z \in [-h/2, h/2]} \left| \left(z + \frac{N}{2} h \right) \left(z + \left(\frac{N}{2} - 1 \right) h \right) \right. \\
&\quad \times \left(z + \left(\frac{N}{2} - 2 \right) h \right) \cdots \left(z - \left(\frac{N}{2} - 1 \right) h \right) \\
&\quad \left. \times \left(z - \frac{N}{2} h \right) \right|
\end{aligned} \tag{41}$$

For N odd the maximum occurs at $z = 0$ and has the value

$$h^{N+1} \prod_{i=1}^{(N+1)/2} \left(i - \frac{1}{2} \right)^2$$

For N even it can be shown that the derivative of

$$\left(z + \frac{N}{2}h \right)\left(z + \left(\frac{N}{2} - 1 \right)h \right)\left(z + \left(\frac{N}{2} - 2 \right)h \right)$$

$$\cdots \left(z - \left(\frac{N}{2} - 1 \right)h \right)\left(z - \frac{N}{2}h \right)$$

is not 0 in $[-h/2, h/2]$, so the maximum over the interval must be taken at $-h/2$ and $+h/2$ (see Problem 5.3). Thus the maximum value desired is the expression given by equation 41 evaluated at $h/2$:

$$\frac{h^{N+1}}{2} \prod_{i=1}^{N/2} \left(i^2 - \frac{1}{4} \right)$$

So we see that

$$\max_{x \in J_i} |\psi_i(x)| = K_N h^{N+1} \tag{42}$$

with

$$K_N = \begin{cases} \displaystyle\prod_{i=1}^{(N+1)/2} \left(i - \frac{1}{2} \right)^2 & \text{if } N \text{ odd} \\[2em] \displaystyle\frac{1}{2}\prod_{i=1}^{N/2} \left(i^2 - \frac{1}{4} \right) & \text{if } N \text{ even} \end{cases} \tag{43}$$

Note that $\max_{x \in J_i} |\psi_i(x)|$ is independent of i. Therefore it can be taken out of the maximum over i in equation 37. Thus

$$b_f = K_N \frac{h^{N+1}}{(N+1)!} \max_i \max_{\xi \in I_i} |f^{(N+1)}(\xi)| \tag{44}$$

But

$$\max_i \max_{\xi \in I_i} |f^{(N+1)}(\xi)| = \max_{\xi \in \cup_i I_i} |f^{(N+1)}(\xi)| = \max_{\xi \in I} |f^{(N+1)}(\xi)| \tag{45}$$

where I is the interval enclosed by all of the tabular arguments which might be used for $x \in J$. So we have the final result that in the case where the tabular arguments are equally spaced, with enough tabular arguments outside of J so that all intervals are standard, piecewise simple polynomial exact matching produces an approximation error bound of

$$b_f = K_N \frac{h^{N+1}}{(N+1)!} \max_{\xi \in I} |f^{(N+1)}(\xi)| \tag{46}$$

If there are not enough tabular arguments outside of J, the leftmost $N + 1$ tabular arguments will be used for an interval wider than width h, and likewise for the rightmost $N + 1$ tabular arguments. For example, if in Figure 5–4b $J = [-h/2, 1 + h/2]$ instead of $[0, 1]$, then $J_1 = [-h/2, h]$ and $J_n = [1 - h, 1 + h/2]$, and $I_1 = [-h/2, 3h/2]$ is not J_1 extended by $h/2$ on each end and similarly for I_n and J_n, unlike all the rest of the I_i, J_i pairs. In this case the bounds for the two intervals, J_1 and J_n, must be treated separately, and the maximum for the standard central intervals will be given by equation 46 but with the maximum only over $\cup_{i=2}^{n-1} I_i$, the range of tabular arguments used for $x \in \cup_{i=2}^{n-1} J_i$. Then the overall bound desired is the maximum of the bounds for the two edge intervals and that for the central intervals.

The result given by the foregoing discussion can be used not only to bound the approximation error for a given tabular interval h but also to determine h if a function is to be tabulated at equal intervals so as to assure that the approximation error will be less than some tolerance, assuming a given value of N is used. In equation 46 the only term involving h is h^{N+1}. All the rest can be determined if only J and N are known. With these values we can find the value of h such that b_f is equal to the tolerance, and we must tabulate with an interval width no greater than this. For example, assume that we wish to approximate $\sin(x)$ by a piecewise simple quadratic exact-matching polynomial to within 10^{-6} over $[0, \pi/2]$ assuming tabular points are provided in steps of h from $-h/2$ to $\pi/2 + h/2$. Here $N = 2$, so $b_{\sin} = (3/8)(h^3/3!)(1) = h^3/16$. Thus, we require that $h^3/16 \leq 10^{-6}$, that is, $h \leq 16^{1/3} \times 10^{-2} = 0.0252$.

Error Estimation Using the Difference Formula. All of the above is based on the assumption that the function f is analytically available so that we can take its derivative. This is often not the case. When we can not compute $f^{(N+1)}$, we can not obtain a bound on the approximation error but we can obtain an estimate of the error as follows.

Let us examine equation 25 of Chapter 3, produced by us in the course of development of the Newton divided-difference technique:

$$f(x) = f(x_0) + (x - x_0)f[x_0, x_1] + (x - x_0)(x - x_1)f[x_0, x_1, x_2]$$
$$+ \cdots + (x - x_0)(x - x_1) \cdots (x - x_{N-1})f[x_0, x_1, ..., x_N] \quad [47]$$
$$+ (x - x_0)(x - x_1) \cdots (x - x_N)f[x_0, x_1, ..., x_N, x]$$

We also showed in Chapter 3 that the exact matching polynomial through $x_0, x_1, ..., x_N$ is given by all but the last term of equation 47. Thus we have

$$f(x) = \hat{p}(x) + \psi(x)f[x_0, x_1, ..., x_N, x] \quad [48]$$

that is,

$$\varepsilon_{f(x)} = \hat{p}(x) - f(x) = -\psi(x)f[x_0, x_1, ..., x_N, x] \quad [49]$$

Comparing this equation with equation 25, we see that

$$f[x_0, x_1, ..., x_N, x] = \frac{f^{(N+1)}(\xi(x))}{(N+1)!} \qquad [50]$$

for some $\xi \in [x_0, x_1, ..., x_N, x]$.

We can not evaluate the divided difference $f[x_0, x_1, ..., x_N, x]$ because its evaluation requires knowledge of $f(x)$, the value we are looking for. But if we evaluate equation 50 at $x = x_{N+1}$, we obtain

$$f[x_0, x_1, x_2, ..., x_{N+1}] = \frac{f^{(N+1)}(\xi(x_{N+1}))}{(N+1)!} \qquad [51]$$

where $\xi \in [x_0, x_1, x_2, ..., x_{N+1}]$. If x_{N+1} is the tabular argument closest to x except for $x_0, x_1, ..., x_N$, then the interval for $\xi(x_{N+1})$ in equation 51 is close to the interval for $\xi(x)$ in equation 50. Therefore, if $f^{(N+1)}$ does not change value a great deal over that interval,

$$f^{(N+1)}(\xi(x)) \approx f^{(N+1)}(\xi(x_{N+1})) \qquad [52]$$

so

$$f[x_0, x_1, x_2, ..., x_{N+1}] \approx f[x_0, x_1, ..., x_N, x] \qquad [53]$$

Even if $f^{(N+1)}$ varies a great deal over the interval, it is possible that $\xi(x) \approx \xi(x_{N+1})$ and therefore approximation 53 holds. In either case,

$$\varepsilon_{f(x)} \approx -\psi(x)f[x_0, x_1, x_2, ..., x_{N+1}] \qquad [54]$$

If we had a computed value and knew the error, we would subtract the error from the computed value to get the correct value. In the same way, we can improve the approximation obtained from the first $N + 1$ terms of the divided-difference formula by adding the negative of the right side of approximation 54. But this value is precisely the next term that would be added in the normal Newton divided-difference series. In other words, the error at any step is approximately equal in magnitude to the next term to be added in the series.

We can base our choice of $N + 1$, the number of tabular points to exact-match, on the above results when we can not compute the error bound analytically. Given a tolerance on the approximation error, we should stop adding terms in the Newton divided-difference series when the error estimate is less in magnitude than the tolerance. That is, we should stop adding terms when the terms added are less than the tolerance if this occurs. We should stop earlier if the terms added, the error estimates, begin to increase. In this latter case we should stop just before we have added in the least term in magnitude. We will say more on the subject of choosing N in Section 6.2 when we have covered propagated error.

Taylor Series Approach

The approximation error formulas given thus far are based on equation 25. Like equation 25, many approximation error formulas are developed using Rolle's theorem, the mean value theorem, or the integral mean value theorem and involve a parameter whose value is unknown but which is restricted to some interval. Another very common approach to developing approximation error formulas is using Taylor series. This approach is especially useful when equal interval approximation is used. The idea is to express everything in sight in both the approximate and desired result as Taylor series in h about an appropriate point. Let us demonstrate this approach with the modified Euler predictor for solving differential equations.

The modified Euler predictor is given by

$$\hat{y}_{i+1} = y_{i-1} + 2hy_i' \qquad [55]$$

The approximation error is given by

$$\varepsilon_{y_{i+1}} = \hat{y}_{i+1} - y_{i+1} = y_{i-1} + 2hy_i' - y_{i+1} \qquad [56]$$

(assuming that the inputs y_{i-1} and y_i' are correct and the arithmetic is done accurately). Remember that $y_{i+1} = y(x_i + h)$ and $y_{i-1} = y(x_i - h)$. Expanding everything in the expression for the error in a Taylor series about x_i produces

$$\varepsilon_{y_{i+1}} = y_i - hy_i' + \frac{h^2}{2}y_i'' - \frac{h^3}{6}y_i''' + O(h^4)$$

$$+ 2hy_i' - \left(y_i + hy_i' + \frac{h^2}{2}y_i'' + \frac{h^3}{6}y_i''' + O(h^4) \right) \qquad [57]$$

$$= -\frac{h^3}{3}y_i''' + O(h^4)$$

By the same approach the error generated by Heun's method,

$$\hat{y}_{i+1} = y_i + \frac{h}{2}(y_i' + y_{i+1}') \qquad [58]$$

(assuming no error in the arithmetic or the input, including the predicted value of y_{i+1} needed to compute y_{i+1}') is

$$\varepsilon_{y_{i+1}} = \hat{y}_{i+1} - y_{i+1} = y_i + \frac{h}{2}(y_i' + y_{i+1}') - y_{i+1} \qquad [59]$$

which, when everything is expanded in Taylor series about x_i, gives

$$\varepsilon_{y_{i+1}} = y_i + \frac{h}{2}\left(y_i' + y_i' + hy_i'' + \frac{h^2}{2}y_i''' + O(h^3)\right)$$

$$- \left(y_i + hy_i' + \frac{h^2}{2}y_i'' + \frac{h^3}{3!}y_i''' + O(h^4)\right) \qquad [60]$$

$$= \frac{h^3}{12}y_i''' + O(h^4)$$

We note in passing that the errors in the modified Euler predictor and the Heun corrector have the same order in h and are of opposite sign. This behavior is very common in predictor–corrector methods and will be made use of in Section 7.4.

By similar means we can show that the error in the numerical differentiation formula,

$$\hat{y}_i' = \frac{1}{2h}(y_{i+1} - y_{i-1}) \qquad [61]$$

can be obtained by expanding everything in $\hat{y}_i' - y_i'$ in a Taylor series about x_i to produce

$$\varepsilon_{y_i'} = \frac{h^2}{6}y_i''' + O(h^3) \qquad [62]$$

The same idea can be applied to numerical integration formulas. For example, for Simpson's rule with integration interval of width $2h$ centered at x_i

$$I_i \triangleq \int_{x_{i-1}}^{x_{i+1}} y(x)\, dx \qquad [63]$$

and

$$\hat{I}_i = \frac{h}{3}(y_{i-1} + 4y_i + y_{i+1}) \qquad [64]$$

so

$$\varepsilon_{I_i} = \hat{I}_i - I_i = \frac{h}{3}\left(y_i - hy_i' + \frac{h^2}{2}y_i'' - \frac{h^3}{6}y_i''' + \frac{h^4}{24}y_i'''' + O(h^5)\right)$$

$$+ 4y_i$$

$$+ y_i + hy_i' + \frac{h^2}{2}y_i'' + \frac{h^3}{6}y_i''' + \frac{h^4}{24}y_i'''' + O(h^5)\Bigg)$$

$$-\int_{x_{i-1}}^{x_{i+1}} \left(y_i + (x - x_i)y_i' + \frac{(x - x_i)^2}{2} y_i'' \right.$$

$$\left. + \frac{(x - x_i)^3}{6} y_i''' + \frac{(x - x_i)^4}{24} y_i'''' + \cdots \right) dx$$

$$= 2hy_i + \frac{h^3}{3} y_i'' + \frac{h^5}{36} y_i'''' + O(h^6)$$

$$-\left(xy_i + \frac{(x - x_i)^2}{2} y_i' + \frac{(x - x_i)^3}{6} y_i'' \right.$$

$$\left. + \frac{(x - x_i)^4}{24} y_i''' + \frac{(x - x_i)^5}{120} y_i'''' + \cdots \right)\Bigg|_{x_{i-1}}^{x_{i+1}} \qquad [65]$$

$$= 2hy_i + \frac{h^3}{3} y_i'' + \frac{h^5}{36} y_i'''' + O(h^6)$$

$$-\left(2hy_i + \frac{h^3}{3} y_i'' + \frac{h^5}{60} y_i'''' + O(h^6) \right)$$

$$= \frac{h^5}{90} y_i'''' + O(h^6)$$

For all of the error formulas produced by the Taylor method there is a corresponding error formula, produced by methods using mean value theorems, etc., which is of precisely the same form except that instead of having a derivative evaluated at a tabular argument and adding a term of higher order in h, the extra term is deleted and the derivative is evaluated at some unknown point in the interval bounded by the extreme tabular arguments used in either the approximation or the argument of the result. For example, with the modified Euler predictor the extreme points are x_{i-1} (used in the formula) and x_{i+1} (the argument of the result), so it can be shown that

$$\varepsilon_{y_{i+1}} = -\frac{h^3}{3} y''' (\xi) \qquad \text{for some} \quad \xi \in [x_{i-1}, x_{i+1}] \qquad [66]$$

(cf. equation 57). Similarly, by examining equation 60 we know that it can be shown that for Heun's formula

$$\varepsilon_{y_{i+1}} = \frac{h^3}{12} y''' (\xi) \qquad \text{for some} \quad \xi \in [x_i, x_{i+1}]; \qquad [67]$$

by examining equation 62 that the error in the derivative formula given by equation 61 can be found to be

$$\varepsilon_{y_i'} = \frac{h^2}{6} y''' (\xi) \qquad \text{for some} \quad \xi \in [x_{i-1}, x_{i+1}]; \qquad [68]$$

and by examining equation 65 that the error in the simple Simpson's rule (equation 64) can be found to be

$$\varepsilon_{I_i} = \frac{h^5}{90} y'''' (\xi) \qquad \text{for some} \quad \xi \in [x_{i-1}, x_{i+1}] \qquad [69]$$

The Taylor series approach is not only one of the most common for finding an expression for the approximation error but it can also be used to develop approximation formulas. For example, say we wish to develop a method to be used in the solution of differential equations which predicts y_{i+1} given y_{i-1}, y_i, and y_i' by

$$\hat{y}_{i+1} = Ay_{i-1} + By_i + Cy_i' \qquad [70]$$

We can write the error in $\hat{y}_{i+1}$ by expanding everything in Taylor series. We can then set the three parameters, A, B, and C, so that the terms in y_i, y_i', and y_i'' in this error will be zero. Since these will be terms in h^0, h^1, and h^2, we will expect the error to be of order h^3. We write

$$\varepsilon_{y_{i+1}} = \hat{y}_{i+1} - y_{i+1} = A\left(y_i - hy_i' + \frac{h^2}{2} y_i'' - \frac{h^3}{6} y_i''' + O(h^4)\right)$$

$$+ By_i + Cy_i' \qquad [71]$$

$$-\left(y_i + hy_i' + \frac{h^2}{2} y_i'' + \frac{h^3}{6} y_i''' + O(h^4)\right)$$

Setting the term in y_i to 0 produces

$$A + B = 1 \qquad [72]$$

Setting the term in y_i' to 0 produces

$$-Ah + C = h \qquad [73]$$

And setting the term in y_i'' to 0 produces

$$A\frac{h^2}{2} = \frac{h^2}{2} \qquad [74]$$

From these three equations we see $A = 1$, $B = 0$, and $C = 2h$. Note that the resulting formula is precisely the modified Euler formula. Furthermore, the error can be written as

$$\left(-A\frac{h^3}{6} - \frac{h^3}{6} \right) y_i''' + O(h^4) = -\frac{h^3}{3} y_i''' + O(h^4) \qquad [75]$$

precisely the error term developed above for the modified Euler formula.

The same approach based on Taylor series can be used to develop equal interval approximation formulas for many different purposes.

5.4.2 OPERATORS ON FUNCTIONS

We have seen that a common computing strategy is approximate and operate. In this strategy, to apply an operator to a function represented by given data, we find a function approximating the data, and carry out the operator on that function. That is,

$$\hat{T}(f(x)) = T(\hat{f}(x)) \qquad [76]$$

The approximation error in this operation is

$$\varepsilon_y = \hat{T}(f(x)) - T(f(x)) = T(\hat{f}(x)) - T(f(x)) \qquad [77]$$

If T is a linear operator, it follows from equation 77 that

$$\varepsilon_y = T(\hat{f}(x) - f(x)) = T(\varepsilon_{f(x)}) \qquad [78]$$

We will see equations 77 and 78 again when we discuss error propagation, for they state that the error in $T(\hat{f}(x))$ is the result of propagating the error in $\hat{f}$ with the operator T. Here let us apply equation 78 to a particular approximation problem.

Equation 78 states that to find the approximation error in approximating and operating with a linear operator we apply the operator to the error in the approximation. Let us take as an example numerical integration using the simple trapezoidal rule, for which

$$T(f(x)) = \int_{x_0}^{x_1} f(x)\, dx \qquad [79]$$

and

$$\hat{T}(f(x)) = \frac{h}{2}(f(x_0) + f(x_1)) \qquad [80]$$

The trapezoidal rule was created by integrating the simple exact-matching polynomial at arguments x_0 and x_1. The error in this approximation is known from equation 25 to be

$$\varepsilon_{f(x)} = -\frac{(x - x_0)(x - x_1)}{2!} f''(\xi(x)) \qquad \text{for some} \quad \xi(x) \in [x_0, x_1, x] \qquad [81]$$

Thus, the error in the trapezoidal rule is obtained by applying the operator, integration, to this error:

$$\varepsilon_I = \int_{x_0}^{x_1} \varepsilon_{f(x)} \, dx = -\frac{1}{2} \int_{x_0}^{x_1} (x - x_0)(x - x_1) f''(\xi(x)) \, dx \qquad [82]$$

The integrand on the right side of equation 82 is the product of $(x - x_0)(x - x_1)$, which is nonpositive on the integration interval $[x_0, x_1]$, and $f''(\xi(x))$. Thus, by the integral mean value theorem

$$\varepsilon_I = -\frac{1}{2} f''(\xi(\zeta)) \int_{x_0}^{x_1} (x - x_0)(x - x_1) \, dx = \frac{h^3}{12} f''(\xi(\zeta)) \qquad [83]$$

for some ζ in $[x_0, x_1]$. Since for all x in the integration interval $[x_0, x_1, x]$ = $[x_0, x_1]$, $\xi(x) \in [x_0, x_1]$ for all $x \in [x_0, x_1]$, and therefore $\eta \triangleq \xi(\zeta)$ is a value in $[x_0, x_1]$, so the error in simple trapezoidal rule integration is

$$\varepsilon_I = \frac{h^3}{12} f''(\eta) \qquad \text{for some} \quad \eta \in [x_0, x_1] \qquad [84]$$

The trapezoidal rule is normally used not in its simple form but in its composite form. Here

$$T(f(x)) = \int_{x_0}^{x_N} f(x) \, dx = \sum_{i=1}^{N} \int_{x_{i-1}}^{x_i} f(x) \, dx \qquad [85]$$

and

$$\hat{T}(f(x)) = \frac{h}{2}\left(f(x_0) + 2\sum_{i=1}^{N-1} f(x_i) + f(x_N) \right) = \sum_{i=1}^{N} \int_{x_{i-1}}^{x_i} \hat{f}_i(x) \, dx \qquad [86]$$

where $\hat{f}_i(x)$ is the simple exact-matching polynomial approximation using the tabular arguments x_{i-1} and x_i. Since $\sum_{i=1}^{N} \int_{x_{i-1}}^{x_i}$ is a linear operator, the error in the composite rule is given by

$$\varepsilon_I^{\text{composite}} = \sum_{i=1}^{N} \int_{x_{i-1}}^{x_i} \varepsilon_{f(x)} \, dx \qquad [87]$$

But from our discussion of the simple trapezoidal rule we know that

$$\int_{x_{i-1}}^{x_i} \varepsilon_{f(x)} \, dx = \frac{h^3}{12} f''(\eta_i) \qquad \text{for some} \quad \eta_i \in [x_{i-1}, x_i] \qquad [88]$$

so

$$\varepsilon_I^{\text{composite}} = \sum_{i=1}^{N} \frac{h^3}{12} f''(\eta_i) = \frac{h^3}{12} \sum_{i=1}^{N} f''(\eta_i) \qquad [89]$$

The sum in this equation can be written as the product of 1, which is non-negative for all i, and $f''(\eta_i)$, so the discrete form of the integral mean value theorem can be applied, producing

$$\varepsilon_I^{\text{composite}} = \frac{h^3}{12} v \sum_{i=1}^{N} 1 = \frac{Nh^3}{12} v \qquad\qquad [90]$$

where v is some number between the maximum and minimum values over i of $f''(\eta_i)$. On the assumption that f'' is continuous, f'' takes this value v somewhere in the range of the η_i, which is included in $[x_0, x_N]$. Therefore, we have

$$\varepsilon_I^{\text{composite}} = \frac{Nh^3}{12} f''(\phi) \qquad \text{for some} \quad \phi \in [x_0, x_N] \qquad [91]$$

Finally we note that $Nh = x_N - x_0$, so we reach the most useful form of the error in the composite trapezoidal rule:

$$\varepsilon_I^{\text{composite}} = \frac{(x_N - x_0)h^2}{12} f''(\phi) \qquad \text{for some} \quad \phi \in [x_0, x_N] \qquad [92]$$

This error can, of course, be bounded by finding the maximum magnitude of f'' over $[x_0, x_N]$.

The approximation error in Simpson's rule integration can be obtained in a similar fashion, namely, by integrating the error in the quadratic approximation on which it is based. The development is somewhat more complicated (see Pizer, 1975, pp. 324–329), and it will not be given here. However, from the Taylor's approach development given by equation 65 we see that the error in the simple Simpson's rule is $O(h^5)$. This is an unexpected result since $\varepsilon_{f(x)}$ has one more factor, $x - x_i$, than with the simple trapezoidal rule and thus would be expected to have one higher power of h than the simple trapezoidal rule error, which is $O(h^3)$. The extra power of h comes from the fact that the polynomial $\psi(x)$ in $\varepsilon_{f(x)}$ has odd symmetry in the integration interval and thus integrates to 0. The result is that the next-higher-order term in the error is the one whose integration produces the low-order error term. We can thus see that it is reasonable that for simple Simpson's rule integration over $[x_0, x_2]$

$$\varepsilon_I = \frac{h^5}{90} f^{(4)}(\eta) \qquad \text{for some} \quad \eta \in [x_0, x_2] \qquad [93]$$

By the same reasoning as with the trapezoidal rule it follows that for Simpson's rule

$$\varepsilon_I^{\text{composite}} = \frac{(x_N - x_0)h^4}{180} f^{(4)}(\phi) \qquad \text{for some} \quad \phi \in [x_0, x_N] \qquad [94]$$

This behavior — that the error is of one higher order in h than expected — is an advantage for small h. The behavior holds for all the Newton-Cotes integration rules based on approximation by polynomials of even degree. We thus find that for composite rules both the midinterval rectangular rule and the trapezoidal rule have approximation error of order h^2, the quadratic (Simpson's) rule and the cubic rule both have error of order h^4, etc. Moreover, for $n > 0$, the constant multiplying $h^{2n+2} f^{(2n+2)}(\eta)$ is less in magnitude for the rule based on a $2n$th-degree polynomial than on a $(2n+1)$th-degree polynomial. Therefore, except for the trapezoidal rule, only even-order Newton-Cotes rules are used.

The approximation error in numerical differentiation, or any other method based on the strategy of approximate and operate with a linear operator, can be obtained, as above, by applying the operator to the error in the approximation function. This holds even when the approximation is not by simple polynomial exact matching.

5.5 Summary and Complements

We have seen that overall error is the sum of propagated and generated error and that generated error is the sum of arithmetic error and approximation error (this is also true for bounds). Thus overall error is the sum of propagated, computation, and approximation errors:

$$\varepsilon_y = \varepsilon_y^{\text{prop}} + \varepsilon_y^{\text{comp}} + \varepsilon_y^{\text{approx}} \qquad [95]$$

and

$$b_y = b_y^{\text{prop}} + b_y^{\text{comp}} + b_y^{\text{approx}} \qquad [96]$$

Computation error generated by representation or any arithmetic operation is bounded in magnitude by R times the result, where $R = 1/B^{n-1}$ for a base-B, n-digit floating-point fraction and arithmetic accurate to $n + 1$ digits. All that remains to be covered on this matter is the error generated by an expression with more than one operation.

We have developed in some detail bounds on the approximation error generated by simple polynomial exact matching. We have produced formulas —

1. to bound error at a particular evaluation argument (equation 27);
2. to bound the error over an evaluation argument range in which the set of tabular arguments producing the approximation is fixed (equation 33); and
3. to bound the error over an evaluation argument range such that different pieces of a piecewise fit are used (equation 46).

We have seen two analytic approaches to producing such error bounds: the point-in-interval approach and the Taylor series approach. The approximation error due to spline approximation and other exact-matching approximations can also be bounded by analytic approaches. All of these methods are based on the assumption that the data are exact.

In contrast, approximation methods such as least-squares methods which involve norm minimization are based on the assumption that there is error in the data and that the error is characterized by being a sample from a probability distribution. Often these probability distributions are nonzero for all values of error though the probability of large error is very tiny. In these cases it is not possible to bound the error in an approximation. Even in cases where the probability distribution has nonzero values only for bounded errors, it is more appropriate to measure errors by standard deviations of their probability distributions. Thus, with norm-minimization approximation methods one more often develops methods to determine the standard deviation of the approximation error than the bound on the approximation error. This matter is discussed in Section 4.3 of Pizer (1975).

Computation error can also be modeled probabilistically, as can error propagation. This approach is covered in Section 1.4 of Pizer (1975).

However approximation error is modeled, when the strategy of approximate and operate leads to a linear operator being applied to the approximation, the error in the result can be found by applying the operator to the expression for approximation error (equation 78). We saw such an application only in the exact-matching (nonprobabilistic) case, namely, for Newton-Cotes numerical integration.

REFERENCES

Pizer, S. M., *Numerical Computing and Mathematical Analysis*. Science Research Associates, Chicago, 1975.

PROBLEMS

5.1 Determine a bound on the relative error generated by addition of normalized floating-point numbers in base 16 with a d-digit fraction, using a single guard bit (not, say, a base 16 digit), and chopping (truncating) of the result. Use ε_x/x as the definition of relative error.

5.2 Assume we are given $\{(x_i, y_i, y_i') \mid i = 0, 1, ..., N\}$, a set of argument values and for each argument the value of a function f and a value for its derivative at that argument. In Problem 3.2 you found the Hermite interpolating polynomial $\hat{p}(x)$ such that $\hat{p}(x_i) = y_i$ and $\hat{p}'(x_i) = y_i'$ for $i = 0, 1, 2, ..., N$.

(a) Find the error in this approximation at an arbitrary value of x.
(b) For what values of x would you expect Hermite interpolation using the tabular arguments $x_1, x_3, x_5, ..., x_{2N+1}$ to be superior to Lagrange interpolation using the tabular arguments $x_0, x_1, x_2, ..., x_{2N+1}$?

5.3 Consider the form

$$f(z) = \left[z - \frac{N}{2}h \right]\left[z - \left(\frac{N}{2} - 1\right)h \right]\left[z - \left(\frac{N}{2} - 2\right)h \right]$$
$$\cdots \left[z + \left(\frac{N}{2} - 1\right)h \right]\left[z + \frac{N}{2}h \right]$$

(a) For N even $f(z) = z\, \Pi_{k=1}^{N/2}[z^2 - (kh)^2]$. Show that $f'(z) \neq 0$ for $z \in [-h/2, h/2]$. You may use the fact that

$$\sum_{k=1}^{N/2} \frac{1}{4k^2 - 1} = \frac{1}{2}\left(1 - \frac{1}{N + 1}\right)$$

(b) For N odd $f(z) = \Pi_{k=1}^{(N+1)/2}[z^2 - ((k - 1/2)h)^2]$. Show that $f'(0) = 0$.

5.4 Assume the function $f(x) = e^{-x^2/4}$ is tabulated with equal spacing $h = 0.2$ on the interval $[-0.2, 2.2]$. Bound the maximum approximation error magnitude over all approximation arguments in the interval $[0, 2]$ if quadratic exact-matching interpolation is used.

5.5 Assume that the function $f(x) = (2/\sqrt{\pi}) \int_0^x u^{1/2} e^{-u}\, du$ is to be tabulated at points equally spaced in the interval given below in order to use linear exact-matching interpolation to approximate $f(x)$ for any x in the interval with an approximation error of no more than 10^{-4}. For the interval $[1, 3]$, how many tabular points need there be?

5.6   Let $f(x) = x^{2/3}$ be tabulated on [1, 2] with an interargument interval of 0.01. How many points must we use to interpolate $f(x)$, $1 \leqslant x \leqslant 2$, if the approximation error magnitude in an exact-matching polynomial is to be no more than 10^{-6}?

5.7   Consider the following tabulated function.

x	$f(x)$
0.5	-4.125
1.0	-4.000
1.5	-1.875
2.0	3.000

(a)   Using the divided difference method, find the best approximation for the root of f between 1.5 and 2.0.

(b)   If you are told that $f(x) = (x^2 - 3)(x + 1)$ [the root is therefore at $x = 1.732$], can you explain the behavior?

5.8   Explain the behavior observed in polynomial exact-matching approximation to $\tan^{-1}(x)$ at -10, -8, -6, ..., 10 in Figure 3–9.

5.9   Given the numerical differentiation formula $y'(x_0) \approx (y_1 - y_{-1})/2h$, in terms of h and the derivatives of f, what is the approximation error of this approximation? Give the answer both with the derivative evaluated at x_0 and as a factor times a power of h.

5.10   In Problem 7.25 it is shown that $\delta^2 y_0/h^2$ produces an approximation to y_0'' derived from exact matching. Show that the error in this approximation is $(h^2/12)f^{(4)}(\xi)$, where $\xi \in [x_{-1}, x_1]$.

5.11   Given the differential equation $y' = f(x, y)$, $y(x_0) = y_0$, consider a predictor of the form $y_{i+1}^{(p)} = Ay_i + By_{i-1} + h(Cy_i' + Dy_{i-1}')$ with approximation error $O(h^4)$.

(a)   Find A, B, C, and D.

(b)   Find the approximation error of this predictor.

5.12   Given the differential equation $y' = f(x, y)$, $y(x_0) = y_0$, consider a corrector of the form $y_{i+1}^{(c)} = Fy_i + Gy_{i-1} + h(Hy_i' + Iy_{i+1}')$ with approximation error $O(h^4)$.

(a)   Find F, G, H, and I.

(b)   Find the approximation error of this corrector (assuming no error in y_{i+1}').

5.13   Given the differential equation $y' = f(x, y)$, $y(x_0) = y_0$,:

(a)   Consider a predictor of the form

$$\hat{y}_{i+1} = A_0 y_i + A_1 y_{i-1} + h(B_0 y_i' + B_1 y_{i-1}' + B_2 y_{i-2}') + \frac{Eh^4 y^{(4)}(\theta)}{4!}$$

Let A_1 be a parameter and find A_0, B_0, B_1, B_2, and E in terms of A_1.
(b)   Consider a corrector of the form

$$\hat{y}_{i+1} = C_0 y_i + C_1 y_{i-1} + h(D_{-1} y_{i+1}' + D_0 y_i' + D_1 y_{i-1}') + \frac{Fh^4 y^{(4)}(\theta)}{4!}$$

Let C_1 be a parameter and find C_0, D_{-1}, D_0, D_1, and F in terms of C_1.

5.14   Consider the integral $\int_0^1 e^{-x} \, dx$. Say we wish to integrate this numerically with an approximation error of magnitude less than 2×10^{-5}.

(a)   What interval width and thus how many data points will we need if we wish to use the trapezoidal rule?
(b)   — if we wish to use Simpson's rule?

5.15   Assume that you are given the value of a function f at two points b and a and that you are told that $f'(b) = 0$.

(a)   Give a formula of the form $\hat{f}[(a + b)/2] = A f(a) + B f(b)$ which estimates $f[(a + b)/2]$ as well as possible assuming that $|b - a|$ is small and derivatives of f tend to decrease in magnitude with the order of the derivative for orders between 2 and 6.
(b)   Give, in point-in-interval form, the approximation error in $\hat{f}[(a + b)/2]$.
(c)   If $f(x) = \cos(\frac{1}{2}(b - x))$ and $b - a = \pi/3$, give a bound on the magnitude of the absolute approximation error in $\hat{f}[(a + b)/2]$.

CHAPTER SIX

Propagation of Error

6.1 Basic Concepts

We have seen that we must analyze the error in a result due to errors in the inputs to the algorithm, that is, the propagated error. We are concerned with this analysis for all operators of interest, most of which consist of the compound or repetitive application of simple operators. But to do this analysis for complicated operators, we must first be able to analyze their component simple operators. In this chapter we will be treating simple operators. The error propagation due to operators that are repetitively or successively applied will be treated in Chapters 7 through 9.

We need to obtain a measure of the error in the result of a computation given measures of the errors in the inputs to the computation, under the assumption that no error is generated in the process. In this book the usual measure of error is an upper bound on its magnitude. Thus, for example, if $f(x, y, z) = x \sin(y^2) + z$, we must bound the error in f, given bounds on the errors in x, y, and z. In some cases the bounds will be on absolute errors, but in most cases they will be on relative errors, these being more informative.

The bounds on the inputs can be given in two ways: bounds on each of the errors (absolute or relative) in the individual input values or a bound on the norm of the vector of errors in the input values (or the corresponding relative error). In the above example, we might be given bounds on each of $|\varepsilon_x|$, $|\varepsilon_y|$, and $|\varepsilon_z|$ or on each of $|\varepsilon_x/x|$, $|\varepsilon_y/y|$, and $|\varepsilon_z/z|$, or we might be given a bound on $\|(\varepsilon_x, \varepsilon_y, \varepsilon_z)^T\|$ or a bound on $\|(\varepsilon_x, \varepsilon_y, \varepsilon_z)^T\|/\|(x, y, z)^T\|$. We will treat first the case where bounds on the individual input errors are given and then the case where the error vector bound is given.

6.2   Bounding Propagated Error from Individual Input Error Bounds

To start simply, let f be a function of one variable x. For example, $f(x) = e^x$. We wish to know the absolute propagated error†

$$\varepsilon_f \triangleq f(x^*) - f(x) \qquad [1]$$

where

$$x^* \triangleq x + \varepsilon_x \qquad [2]$$

From Taylor's theorem, if f is appropriately differentiable on $[x, x^*]$,

$$f(x + \varepsilon_x) = f(x) + \varepsilon_x f'(x) + \varepsilon_x^2 \frac{f''(\theta)}{2} \qquad \text{for some} \quad \theta \in (x, x^*) \qquad [3]$$

so

$$\varepsilon_f = f(x + \varepsilon_x) - f(x) = \varepsilon_x f'(x) + \varepsilon_x^2 \frac{f''(\theta)}{2} \qquad [4]$$

Assuming that ε_x is small enough so that

$$\left| \varepsilon_x \frac{f''(\theta)}{2} \right| \ll f'(x)$$

we have

$$\varepsilon_f \approx \varepsilon_x f'(x) \qquad [5]$$

and thus

$$\frac{\varepsilon_f}{f} \approx \left(\frac{\varepsilon_x}{x} \right) \left(\frac{x f'(x)}{f(x)} \right) \qquad [6]$$

From equations 5 and 6, respectively, we can conclude that if $|\varepsilon_x|$ is bounded by b_x, then

$$b_f \approx b_x |f'(x)| \qquad [7]$$

and if $|\varepsilon_x / x|$ is bounded by r_x, then

$$r_f \approx r_x \left| \frac{x f'(x)}{f(x)} \right| \qquad [8]$$

---

† We will write "ε_f" as a shorthand for "$\varepsilon_{f(x)}$," realizing that here we are not discussing the error in the function but the error in its value at an argument. Furthermore, throughout this chapter we will omit the superscript "prop," since almost all errors under discussion here are propagated errors.

As an example, assume we are interested in evaluating e^x at $x = 2$ and we are given that the bound on the magnitude of the absolute error in x is 10^{-4}. Then we can determine the bound on the absolute error in e^x as $b_{e^x} \approx 10^{-4}e^2 = 7.4 \times 10^{-4}$. If alternatively we are given $r_x = 5.0 \times 10^{-5}$, we can bound the relative error in e^x as $r_{e^x} \approx 5.0 \times 10^{-5}$ $(2e^2/e^2) = 1.0 \times 10^{-4}$.

Assuming that terms in ε_x^2 are negligible compared to terms in ε_x, we can obtain the following alternative results to equations 7 and 8. In the alternatives functions and derivatives are evaluated at x^* rather than x (see Problem 6.1):

$$b_f \approx b_x \, |f'(x^*)| \qquad [9]$$

and

$$r_f \approx r_x \left| \frac{x^* f'(x^*)}{f(x^*)} \right| \qquad [10]$$

For the case of functions of more than one variable, $f(x_1, x_2, ..., x_n)$, the same approach as the above is applicable. The n-dimensional Taylor series applied to

$$f(x_1^*, x_2^*, ..., x_n^*) \equiv f(x_1 + \varepsilon_{x_1}, \quad x_2 + \varepsilon_{x_2}, \quad ..., \quad x_n + \varepsilon_{x_n})$$

produces

$$\varepsilon_f \triangleq f(x_1^*, x_2^*, ..., x_n^*) - f(x_1, x_2, ..., x_n) \qquad [11]$$

$$= \sum_{i=1}^{n} \varepsilon_{x_i} \frac{\partial f}{\partial x_i} (x_1, x_2, ..., x_n) + \frac{1}{2} \sum_{i=1}^{n} \sum_{j=1}^{n} \varepsilon_{x_i} \varepsilon_{x_j} \frac{\partial^2 f}{\partial x_i \, \partial x_j} (\theta)$$

where θ is on the line between

$$(x_1, x_2, ..., x_n) \quad \text{and} \quad (x_1^*, x_2^*, ..., x_n^*)$$

Assuming that all the ε_{x_i} are small enough that the terms in $\varepsilon_{x_i} \varepsilon_{x_j}$ are negligible compared to the terms in ε_{x_i}, we have

$$\varepsilon_f \approx \sum_{i=1}^{n} \varepsilon_{x_i} \frac{\partial f}{\partial x_i} (x_1, x_2, ..., x_n) \qquad [12]$$

so

$$b_f \approx \sum_{i=1}^{n} b_{x_i} \left| \frac{\partial f}{\partial x_i} (x_1, x_2, ..., x_n) \right| \qquad [13]$$

or equivalently

$$r_f \approx \sum_{i=1}^{n} r_{x_i} \left| x_i \frac{\partial f}{\partial x_i} (x_1, x_2, ..., x_n) \right| \bigg/ \left| f(x_1, x_2, ..., x_n) \right| \qquad [14]$$

As with the one variable case, expressions that are the same as equations 13 and 14 but with x_i^* in place of each x_i are also valid approximations if terms in $\varepsilon_{x_i} \varepsilon_{x_j}$ are negligible compared to terms in ε_{x_i}.

Let us apply equation 13 to addition:

$$f(x, y) = x + y \qquad [15]$$

Then

$$b_{x+y} \approx b_x \cdot 1 + b_y \cdot 1 = b_x + b_y \qquad [16]$$

Thus, the bound on the absolute error in a sum is the sum of the bounds on the absolute errors in the summands. The same result could have been obtained directly from

$$\varepsilon_{x+y} = (x^* + y^*) - (x + y)$$
$$= (x + \varepsilon_x + y + \varepsilon_y) - (x + y) = \varepsilon_x + \varepsilon_y \qquad [17]$$

so

$$b_{x+y} = b_x + b_y \qquad [18]$$

If

$$f(x, y) = x - y$$

equation 13 again produces

$$b_{x-y} = b_x + b_y \qquad [19]$$

so *for addition and subtraction, absolute error bounds add.*

For multiplication, $f(x, y) = xy$, equation 14 produces

$$r_{xy} \approx \frac{r_x|xy| + r_y|yx|}{|xy|} = r_x + r_y \qquad [20]$$

Thus, the bound on the relative error in a product is the sum of the bounds on the relative errors in the factors. Similarly, it can be shown (Problem 6.2) that the same result holds for division:

$$r_{x/y} \approx r_x + r_y \qquad [21]$$

Thus, *for multiplication and division, relative error bounds add.*

The above rules on the error propagation by the simple arithmetic operators are so basic that they should be memorized. It will be necessary to use them as we analyze the error propagation of more complicated functions. With these rules plus the transformations between absolute and relative error by the formulas

$$r_z = \frac{b_z}{|z|} \approx \frac{b_z}{|z^*|} \qquad [22]$$

and

$$b_z = |z|\, r_z \approx |z^*|\, r_z \tag{23}$$

we can analyze most any computation, by analyzing each arithmetic operation in turn, in order of its application.

For example, take the function $x \sin(y^2) + z$. The first value that will be computed is y^2, so we must analyze its error first. The multiplication rule is applicable, giving

$$r_{y \cdot y} = r_y + r_y = 2r_y \tag{24}$$

The next computation, $\sin(u)$, is analyzed by equation 7 to produce

$$b_{\sin(u)} = b_u |\cos(u)| \tag{25}$$

so

$$b_{\sin(y^2)} = b_{y^2} |\cos(y^2)| \tag{26}$$

We do not have b_{y^2} but rather r_{y^2}, so we use the transformation rule given by equation 23 to produce

$$b_{y^2} = y^2 r_{y^2} \tag{27}$$

so from equations 24 and 26

$$b_{\sin(y^2)} = 2y^2 |\cos(y^2)| r_y \tag{28}$$

The next operation to be applied is the multiplication of x and $\sin(y^2)$. The multiplication rule states that

$$r_{x \sin(y^2)} = r_x + r_{\sin(y^2)} \tag{29}$$

We must compute $r_{\sin(y^2)}$ from $b_{\sin(y^2)}$ by equation 22 to produce

$$r_{\sin(y^2)} = \frac{b_{\sin(y^2)}}{|\sin(y^2)|} \tag{30}$$

so from equation 28

$$r_{\sin(y^2)} = 2y^2 |\cot(y^2)| r_y \tag{31}$$

and thus

$$r_{x \sin(y^2)} = r_x + 2y^2 |\cot(y^2)| r_y \tag{32}$$

The final operation to be applied is the addition. The addition rule gives

$$b_{x \sin(y^2) + z} = b_{x \sin(y^2)} + b_z \tag{33}$$

Using equation 23 to transform $r_{x \sin(y^2)}$ to $b_{x \sin(y^2)}$, we finally obtain

$$b_{x \sin(y^2) + z} = |x \sin(y^2)|r_x + 2y^2|x \cos(y^2)|r_y + b_z$$

$$= |\sin(y^2)|b_x + |2xy \cos(y^2)|b_y + b_z \qquad [34]$$

The result given by equation 34 could have been obtained more simply by direct application of equation 13:

$$b_{x \sin(y^2) + z} = \left|\frac{\partial}{\partial x}(x \sin(y^2) + z)\right|b_x + \left|\frac{\partial}{\partial y}(x \sin(y^2) + z)\right|b_y$$

$$+ \left|\frac{\partial}{\partial z}(x \sin(y^2) + z)\right|b_z \qquad [35]$$

$$= |\sin(y^2)|b_x + |2xy \cos(y^2)|b_y + b_z$$

In general, the result from the direct application of the partial derivative approach will differ, in where the absolute values signs appear, from the result produced by step-by-step application of the rules for simple arithmetic operators and functions. The direct application will give a tighter bound than the step-by-step application and is therefore to be preferred. For example, if the function is $x \sin(y^2) + yz$, the direct application of the partial derivative approach will give

$$b_{x \sin(y^2) + yz} = |\sin(y^2)|b_x + |2xy \cos(y^2) + z|b_y + |y|b_z \qquad [36]$$

whereas the step-by-step approach will give

$$b_{x \sin(y^2) + yz} = |\sin(y^2)|b_x + (|2xy \cos(y^2)| + |z|)b_y + |y|b_z \qquad [37]$$

From these examples the reader should be able to see how any computation can be analyzed. In most cases the partial derivative approach should be applied directly, but we will see cases in Section 7.3 where complications due to the combination of approximation, arithmetic, and generated errors make it desirable to analyze certain arithmetic operations separately.

As another example, consider the differences in an equal-interval difference table, computed from

$$\delta^{m+1}y_{i+1/2} = \delta^m y_{i+1} - \delta^m y_i \qquad [38]$$

If $b_{y_i} = b$, a constant for all i, then

$$b_{\delta y_i} = b + b = 2b \qquad [39]$$

Recurring this analysis,

$$b_{\delta^m y_i} = 2^m b \qquad [40]$$

This growth of the propagated error bound with the order of the difference is faster than that of a measure of the error which reflects the probable size rather than the bound, but both measures grow exponentially with m. This exponential growth of the absolute error, together with the fact that for a smooth underlying function the differences get smaller for the first few values of m, makes the relative error grow very quickly with m, becoming 100% for small m. Adding a term involving such a difference into a difference formula frequently decreases the approximation error, but it increases the propagated error by the propagated error of the term. We recall that the decrease in the approximation error is approximately equal in magnitude to the term to be added and that the overall error is the sum of the propagated and the approximation errors (since the arithmetic error is normally negligible here), and we note that the relative propagated error in the term to be added is approximately equal in magnitude to that of the difference it involves. Thus, adding a term involving a difference with 100% or more error may increase the propagated error by the magnitude of the term being added or more, that is, by at least as much as it decreases the approximation error. Therefore, this and succeeding terms should not be added, as they may degrade the overall accuracy of the value computed by the formula.

We have concluded that paths in the difference table should not be carried to the point in the table where relative errors in the differences are 100 percent or more. Such a point can be detected by error analysis of the type above (stop when $b_{\delta^m y_i}$ is greater than $|\delta^m y_i|$, where by equation 40 $b_{\delta^m y_i} = 2^m b$, with b normally being the bound on representation error) or, better, by error analysis using the probable error measure, not covered in this book (see Pizer, 1975). Alternatively, 100% error in the differences can be detected empirically using the fact that if errors are of the same magnitude as difference values, we can expect the errors to cause some positive values to become negative and vice versa, so the differences in a column with such errors will fluctuate in sign (more than once). Even if this sign fluctuation is due to accurate differences, differences in succeeding columns will increase because of subtraction of adjacent differences with opposite signs. When differences increase, the terms to be added and thus the error estimates normally increase, so we should not add further terms to the approximation. Therefore, whatever the cause, if fluctuations in sign are common in an area of the difference column along our path in the difference table, we should stop adding terms.

As an example, consider the difference table in Figure 6–1. The differences are not computed beyond the fifth difference because there is sign fluctuation in some areas of the δ^4 column and in all areas of the δ^5 column not adjacent to the areas of sign fluctuation in the δ^4 column, so no formula should use differences beyond the fourth. By our criteria,

FIGURE 6–1

An Equal-Interval Difference Table

x	y	δ	δ^2	δ^3	δ^4	δ^5
0.2	0.0034					
		103				
0.4	0.0137		− 10			
		93		2		
0.6	0.0230		− 8		9	
		85		11		1
0.8	0.0315		3		10	
		88		21		− 2
1.0	0.0403		24		8	
		112		29		− 2
1.2	0.0515		53		6	
		165		35		− 2
1.4	0.0680		88		4	
		253		39		− 2
1.6	0.0933		127		2	
		380		41		− 1
1.8	0.1313		168		1	
		548		42		− 2
2.0	0.1861		210		− 1	
		758		41		3
2.2	0.2619		251		2	
		1009		43		− 4
2.4	0.3628		294		− 2	
		1303		41		3
2.6	0.4931		335		1	
		1638		42		
2.8	0.6569		377			
		2015				
3.0	0.8584					

the fourth differences in the top part of the table may be used, but those in the bottom part fluctuate in sign so should not be used.

Calculating bounds on the error in the differences supports these decisions made on the basis of sign fluctuation. The error in the y_i is bounded by 5×10^{-5}. Thus, the propagated error in the values in the δ^3 column is bounded by $2^3 \times 5 \times 10^{-5} = 4 \times 10^{-4}$, which is less in magnitude than the values in that column. For the δ^4 column, the propagated error is bounded by $2^4 \times 5 \times 10^{-5} = 8 \times 10^{-4}$. The probable error magnitude is less than this, approximately 2.4×10^{-4}. Thus, the error in the values in the δ^4 column has approximately the same magnitude as the differences in the bottom part of that column but is smaller than the differences at the top. The error in the values in the δ^5 column, bounded by 1.6×10^{-3}, is larger than all values in that column.

We conclude that to compute $\hat{f}(2.46)$ with maximum accuracy we should stop at the third difference, whereas to compute $\hat{f}(0.82)$ with maximum accuracy we should stop at the fourth difference, at least as far as propagated error considerations are concerned.

We are finally ready to conclude our discussion of how to choose the number of points, $N + 1$, to use in a simple polynomial exact-matching approximation. We should stop adding terms in a difference series if the tolerance desired is obtained ($N = L$ below), or if the approximation error begins to increase as terms are added ($N = M$ below), or if the error in a term to be added is greater in magnitude than the term itself ($N = P$ below). Practically, we choose the stopping column N for a given evaluation argument x as follows:

Let $P + 1$ be the number of the earliest difference column such that along our path $|\varepsilon_{diff}/diff| \approx 1$, where the y_i values are in the 0th-difference column, ε_{diff} is the error in a given difference near our path, and $diff$ is the value of that difference. The value of $P + 1$ is either determined using numerical error estimates normally based on analysis of propagation of representation error or it is taken to be the number of the earliest difference column such that the differences in that column fluctuate in sign near the path.

1. If $f^{(N+1)}(x)$ can be computed analytically for each N, then choose $N = \min(L, M, P)$, where L is the smallest n such that

$$b_n(x) \triangleq \frac{\left| \prod_{i=0}^{n} (x - x_i) \right|}{(n + 1)!} \max_{\xi \in [x_0, x_1, \ldots, x_n, x]} |f^{(n+1)}(\xi)| < \beta \tag{41}$$

where β is a desired absolute error tolerance in the approximation to $f(x)$; and M is the value of n such that

$$b_M(x) = \min_n b_n(x) \tag{42}$$

2. If $f^{(N+1)}(x)$ can not be computed analytically, then choose $N = \min(L, M, P)$, where L is the smallest n such that

$$c_n(x) \triangleq \left| \prod_{i=0}^{n} (x - x_i) \right| |f[x_0, x_1, \ldots, x_{n+1}]| < \beta \tag{43}$$

(note that $c_n(x)$ without the absolute-value signs is the next term to be added to the approximation if the inequality is not satisfied); and M is the value of n such that

$$c_M(x) = \min_n c_n(x) \qquad [44]$$

The Newton divided-difference method, using criterion 2 above for stopping, and assuming P is an input parameter, is fully specified in Program 3–2 (p. 93).

6.3 Bounding Propagated Error from a Bound on the Norm of the Input Error

Let ε be a vector expressing the error in any input vector x. Often we are given a bound on $\|\varepsilon\|$ for some norm or alternatively a bound on the relative error, $\|\varepsilon\|/\|x\|$. We need to go from that bound to a bound on the output error. Unlike the approach in Section 6.2 this approach applies not only when x, and thus ε, is an n-vector but when x is any vector, for example, a function, a polynomial, a matrix, or an n-vector, and likewise when $T(x)$ is any vector.

The common approach is to determine the largest constant by which error may be magnified from input to output:

$$c = \max_{\text{input errors} \neq 0} \frac{\text{output error size}}{\text{input error size}} \qquad [45]$$

which we will call the maximum error magnification factor. The reason for this is that for linear operators, multiplying the input error by a number will multiply the output by the same number, so the ratio between the two errors is the relevant measure. Note that a large error magnification factor is not a guarantee that a given error will be magnified greatly but is only a worst case measure.

6.3.1 ABSOLUTE ERROR: LINEAR OPERATORS

In the following the effects of linear operators will be emphasized. Nonlinear operators will be discussed in passing. For linear operators, T, on a vector space, V, the propagated error is given by

$$T(x^*) - T(x) = T(x + \varepsilon) - T(x) = T(x) + T(\varepsilon) - T(x) = T(\varepsilon) \qquad [46]$$

Thus the absolute error magnification factor is

$$\beta_\varepsilon = \frac{\|T(\varepsilon)\|}{\|\varepsilon\|} \qquad [47]$$

for appropriate norms in the image space (the norm in the numerator) and in the input space (the norm in the denominator). The maximum of this absolute error magnification factor over all nonzero input errors is called the *norm of the operator T*:

$$\|T\| = \max_{\varepsilon \neq 0} \beta_\varepsilon = \max_{\varepsilon \neq 0} \frac{\|T(\varepsilon)\|}{\|\varepsilon\|} \tag{48}$$

Note that the definition of $\|T\|$ depends on two vector norms, which may be different. For example, let

$$x = \begin{bmatrix} u \\ v \end{bmatrix} \in L_2 \quad \text{and} \quad T(x) = u + v \in L_1 \tag{49}$$

where L_n is the space of all n-vectors of reals. Then using the ℓ_1 norm on the input space, L_2, and the absolute value norm on the output space, L_1, we have

$$\|T\| = \max_{(\varepsilon_u, \varepsilon_v)^T \neq 0} \frac{|T((\varepsilon_u, \varepsilon_v)^T)|}{\|(\varepsilon_u, \varepsilon_v)^T\|_1} = \max_{(\varepsilon_u, \varepsilon_v)^T \neq 0} \frac{|\varepsilon_u + \varepsilon_v|}{|\varepsilon_u| + |\varepsilon_v|} \tag{50}$$

$|\varepsilon_u + \varepsilon_v|/(|\varepsilon_u| + |\varepsilon_v|)$ takes its maximum value of 1 when ε_u and ε_v have the same sign, so for these norms $\|T\| = 1$.

If we had used the norm $\|x\| = 2|x|$ on L_1 and the ℓ_∞ norm on L_2, we would have obtained

$$\|T\| = \max_{(\varepsilon_u, \varepsilon_v)^T \neq 0} \frac{2|\varepsilon_u + \varepsilon_v|}{\max(|\varepsilon_u|, |\varepsilon_v|)} = 4 \tag{51}$$

the maximum being achieved when $\varepsilon_u = \varepsilon_v$.

Interpreting the two different results for $\|T\|$ in light of the definition of $\|T\|$ as the maximum absolute error magnification factor, we see that if the error, measured by the ℓ_1 norm, in the input (vector) to an addition is bounded by $b_{x,1}$, the output error, measured by the absolute value norm, will be bounded by $\|T\|b_{x,1} = 1 \cdot b_{x,1} = b_{x,1}$; whereas if the error, measured by the ℓ_∞ norm in the input to an addition is bounded by $b_{x,\infty}$, the output error, measured by the twice-the-absolute-value norm, will be bounded by $\|T\|b_{x,\infty} = 4b_{x,\infty}$.

Let us consider a similar linear operator:

$$T((u, v)^T) = u + 2v \in L_1 \tag{52}$$

In this case if $\|\varepsilon\|_1 \leq b_{x,1}$, $|\varepsilon_{T(x)}| \leq \|T\| b_{x,1}$, where

$$\|T\| = \max_{(\varepsilon_u, \varepsilon_v)^T \neq 0} \frac{|\varepsilon_u + 2\varepsilon_v|}{|\varepsilon_u| + |\varepsilon_v|} \tag{53}$$

This maximum is achieved when for a given value of $|\varepsilon_u| + |\varepsilon_v|$, $|\varepsilon_u + 2\varepsilon_v|$ is as large as possible, that is, when $|\varepsilon_v| = |\varepsilon_u| + |\varepsilon_v|$ so $\varepsilon_u = 0$. Thus $\|T\| = 2$, so $|\varepsilon_{T(x)}| \le 2b_{x,1}$.

Using our other norm pair, if $\|\varepsilon\|_\infty \le b_{x,\infty}$ and $\|\varepsilon_{T(x)}\| = 2|\varepsilon_{T(x)}|$, then $\|\varepsilon_{T(x)}\| \le \|T\| b_{x,\infty}$, where

$$\|T\| = \max_{(\varepsilon_u, \varepsilon_v)^T \ne 0} \frac{2|\varepsilon_u + 2\varepsilon_v|}{\max(|\varepsilon_u|, |\varepsilon_v|)} = 6 \tag{54}$$

the maximum being achieved when $\varepsilon_u = \varepsilon_v$. Thus $\|\varepsilon_{T(x)}\| \le 6b_{x,\infty}$.

An alternative expression for $\|T\|$ can be obtained. If λ is any scalar,

$$\frac{\|T(\lambda\varepsilon)\|}{\|\lambda\varepsilon\|} = \frac{\|\lambda T(\varepsilon)\|}{\|\lambda\varepsilon\|} = \frac{|\lambda| \, \|T(\varepsilon)\|}{|\lambda| \, \|\varepsilon\|} = \frac{\|T(\varepsilon)\|}{\|\varepsilon\|} \tag{55}$$

In particular, for $\lambda = 1/\|\varepsilon\|$

$$\frac{\|T(\varepsilon)\|}{\|\varepsilon\|} = \frac{\left\|T\left(\dfrac{\varepsilon}{\|\varepsilon\|}\right)\right\|}{\left\|\dfrac{\varepsilon}{\|\varepsilon\|}\right\|} = \left\|T\left(\frac{\varepsilon}{\|\varepsilon\|}\right)\right\| \tag{56}$$

since

$$\left\|\frac{\varepsilon}{\|\varepsilon\|}\right\| = 1 \tag{57}$$

Thus,

$$\|T\| \overset{\Delta}{=} \max_{\varepsilon \ne 0} \frac{\|T(\varepsilon)\|}{\|\varepsilon\|} = \max_{\varepsilon \ne 0} \left\|T\left(\frac{\varepsilon}{\|\varepsilon\|}\right)\right\| = \max_{\|\varepsilon\|=1} \|T(\varepsilon)\| \tag{58}$$

an alternative expression for $\|T\|$ applicable to linear operators. Thus, for example, equation 53 could be rewritten

$$\|T\| = \max_{|\varepsilon_u| + |\varepsilon_v| = 1} |\varepsilon_u + 2\varepsilon_v| = 2 \tag{59}$$

As another example, let

$$T(b) = x \tag{60}$$

where x is the solution of the set of linear equations, $Ax = b$, for a given $n \times n$ matrix A, that is,

$$T(b) = A^{-1}b \tag{61}$$

Since in this case the input vector, b, and the output vector, x, are both in the same space, it is common to choose the same norm in both the input and output spaces. Thus we can have a meaning, for example, for $\|T\|_\infty$:

$$\|T\|_\infty = \max_{\varepsilon \neq 0} \frac{\|T(\varepsilon)\|_\infty}{\|\varepsilon\|_\infty} \qquad [62]$$

or equivalently

$$\|T\|_\infty = \max_{\|\varepsilon\|_\infty = 1} \|T(\varepsilon)\|_\infty \qquad [63]$$

In our example

$$T(\varepsilon) = A^{-1}\varepsilon \qquad [64]$$

so

$$\|T\|_\infty = \max_{\|\varepsilon\|_\infty = 1} \|A^{-1}\varepsilon\|_\infty \qquad [65]$$

From equation 65 we can show that for this operation T, $\|T\|_\infty$ is the maximum row sum of magnitudes in the matrix A^{-1}:

$$\|T\|_\infty = \max_i \sum_{j=1}^n |A_{ij}^{-1}| \qquad [66]$$

THEOREM: For any $n \times n$ matrix, B,

$$\|B\|_\infty = \max_i \sum_{j=1}^n |B_{ij}| \qquad [67]$$

Proof:

$$\|B\|_\infty \overset{\Delta}{=} \max_{\|y\|_\infty = 1} \|By\|_\infty \qquad [68]$$

Since $\|y\|_\infty = \max_i |y_i| = 1$, $|y_j| \leq 1$ for all j. So for any y,

$$\|By\|_\infty = \max_i |(By)_i|$$
$$= \max_i \left| \sum_j B_{ij} y_j \right| \leq \max_i \sum_j |B_{ij}||y_j| \leq \max_i \sum_j |B_{ij}| \qquad [69]$$

Thus

$$\|B\|_\infty \leq \max_i \sum_j |B_{ij}| \qquad [70]$$

Let i_0 be the row such that $\Sigma_j |B_{ij}|$ is maximum. Let y^0 be a vector such that

$$y_j^0 = \begin{cases} +1 & \text{if } B_{i_0 j} \geq 0 \\ -1 & \text{if } B_{i_0 j} < 0 \end{cases}$$

Note $\|y^0\|_\infty = 1$. Then

$$\|By^0\|_\infty = (By^0)_{i_0} = \sum_j |B_{i_0 j}| = max_i \sum_j |B_{ij}| \qquad [71]$$

because for $i \neq i_0$,

$$|(By^0)_i| \leq \sum_j |B_{ij}| \leq \sum_j |B_{i_0 j}| \qquad [72]$$

Since for a particular vector y^0 such that $\|y^0\|_\infty = 1$, $\|By^0\|_\infty = max_i \sum_j |B_{ij}|$, it follows that

$$\|B\|_\infty = \max_{\|y\|_\infty = 1} \|By\|_\infty \geq \max_i \sum_j |B_{ij}| \qquad [73]$$

Combining equations 69 and 73, we obtain $\|B\|_\infty = max_i \sum_j |B_{ij}|$.

Q.E.D.

Concluding our example, we can use the result of the above theorem to show that for $x = T(b) = A^{-1}b$

$$\|\varepsilon_x\|_\infty \leq (\max_i \sum_j |A_{ij}^{-1}|) \|\varepsilon_b\|_\infty \qquad [74]$$

It can similarly be shown (Problem 6.8) for this T that

$$\|T\|_1 = \max_j \sum_i |A_{ij}^{-1}|$$

the maximum column sum of magnitudes in the matrix A^{-1}, so

$$\|\varepsilon_x\|_1 \leq (\max_j \sum_i |A_{ij}^{-1}|) \|\varepsilon_b\|_1 \qquad [75]$$

As a final example consider the error in integration of $\int_c^d f(u)\, du$ when the function $f(u)$ is approximated by a function $\hat{f}(u)$. In this case T is integration, the input vector space is the set of integrable functions on $[c, d]$, and the output vector space is L_1. If we use the $\mathcal{L}_\infty$ norm in the input space and the absolute value norm on L_1,

$$\|T\| = \max_{\|\varepsilon(x)\|_\infty = 1} \left| \int_c^d \varepsilon(x)\, dx \right| = \max_{\substack{\varepsilon(x) \ni:\ max\ |\varepsilon(x)| = 1 \\ [c,d]}} \left| \int_c^d \varepsilon(x)\, dx \right| = d - c \qquad [76]$$

Thus if

$$\|\hat{f}(x) - f(x)\|_\infty \triangleq \|\varepsilon(x)\|_\infty \triangleq \max_{x \in [c,d]} |\varepsilon(x)| \leq b_f$$

then the absolute error in $\int_c^d \hat{f}(x)\, dx$ will be bounded in magnitude by

$$b_{T(f)} = (d - c)b_f \qquad [77]$$

6.3.2 ABSOLUTE ERROR: NONLINEAR OPERATORS

For a nonlinear operator $T(x)$, $T(x^*) - T(x) \neq T(\varepsilon)$, so the equation for the absolute error magnification factor, β_ε, given by equation 47 ($\beta_\varepsilon = \|T(\varepsilon)\|/\|\varepsilon\|$) no longer holds. We must write

$$\beta_{x,\varepsilon} = \frac{\|T(x + \varepsilon) - T(x)\|}{\|\varepsilon\|} \qquad [78]$$

noting that the magnification factor depends on the correct input x as well as the input error ε. Furthermore, since multiplying ε by a constant λ will not necessarily multiply $\|T(x + \varepsilon) - T(x)\|$ by λ, $\beta_{x,\varepsilon}$ depends on the size of the error as well as its "direction."

Since we are normally interested in the magnification of error by T for small input errors, a measure of interest would be the biggest magnification factor at input x for very small (infinitesimal) errors. Mathematically, this factor is written

$$\|T|_x\| \triangleq \lim_{\delta \to 0} \max_{0 < \|\varepsilon\| < \delta} \beta_{x,\varepsilon} = \lim_{\delta \to 0} \max_{0 < \|\varepsilon\| < \delta} \frac{\|T(x + \varepsilon) - T(x)\|}{\|\varepsilon\|} \qquad [79]$$

for appropriate norms in the input and output spaces. Note that $\|T|_x\| = \|T\|$ if T is a linear operator.

If $T(x) \in L_1$ (is real), the absolute value norm is commonly used in the numerator of equation 79. If x and thus ε are n-vectors and ε is infinitesimal,

$$T(x + \varepsilon) = T(x) + \sum_{i=1}^{n} \varepsilon_{x_i} \frac{\partial T}{\partial x_i}(x) \qquad [80]$$

Thus,

$$T(x + \varepsilon) - T(x) = \sum_{i=1}^{n} \varepsilon_{x_i} \frac{\partial T(x)}{\partial x_i} \qquad [81]$$

which is maximized in magnitude when, for all i, ε_{x_i} has the same sign as $\partial T(x)/\partial x_i$. Therefore, when $T(x)$ is real and x is an n-vector

$$\|T|_x\| = \lim_{\delta \to 0} \max_{0 < \|\varepsilon\| < \delta} \frac{\sum_{i=1}^{n} |\varepsilon_{x_i}| \left| \dfrac{\partial T(x)}{\partial x_i} \right|}{\|(\varepsilon_{x_1}, \varepsilon_{x_2}, \dots, \varepsilon_{x_n})^T\|} \qquad [82]$$

If the ℓ_1 norm is used in the input space, the maximum is taken when all but one ε_{x_i} is zero, and we conclude

$$\|T|_x\| = \max_i \left| \frac{\partial T(x)}{\partial x_i} \right| \qquad [83]$$

If the ℓ_∞ norm is used in the input space, the maximum is taken when all the ε_{x_i} are equal in magnitude, and we conclude

$$\|T|_x\| = \sum_{i=1}^{n} \left| \frac{\partial T(x)}{\partial x_i} \right| \qquad [84]$$

For example, let

$$T((u, v, w)^T) = u \sin(v^2) + w \qquad [85]$$

the first example of this chapter. Then, using the ℓ_1 norm in the input space,

$$\|T|_x\| = \max(|\sin(v^2)|, |2uv \cos(v^2)|, 1) \qquad [86]$$

where $x = (u, v, w)^T$. Thus, for example, we could say that at $u = 2$, $v = \sqrt{\pi}$, $w = 10$, if $\|\varepsilon_x\|_1 \leq b_{x,1}$ and $b_{x,1}$ is small,

$$|\varepsilon_{2\sin(\pi)+10}| \leq \max(0, 4 \sqrt{\pi}, 1)b_{x,1} = 4 \sqrt{\pi} \, b_{x,1} \qquad [87]$$

Similar analysis can be done if $T(x)$ is itself an m-vector, but this will not be done here.

6.3.3 RELATIVE ERROR: LINEAR OPERATORS

Sensitivity

Quite often, we know a bound, r_x, on the relative error, $\|\varepsilon\|/\|x\|$, in the input, x, to an operator, T, and we wish to know a bound on the relative error in the output, $T(x)$. This problem, like that with the absolute error, is approached by measuring the factor by which the relative error is magnified, but the relative error magnification factor is more informative than the absolute error magnification factor, since unlike the latter it is independent of magnification of the input by the operator T. That is, if T magnifies the input by a factor of 100, the fact that the input error is also magnified by a factor of 100 does not show bad error propagation since the relative error is the same in the output as in the input. But a relative error magnification by a factor of 100 could be serious indeed.

The relative error magnification factor, $\rho_{x,\varepsilon}$, is given by

$$\rho_{x,\varepsilon} = \frac{\text{output relative error}}{\text{input relative error}} = \frac{\|T(x + \varepsilon) - T(x)\|/\|T(x)\|}{\|\varepsilon\|/\|x\|} \qquad [88]$$

If T is linear,

$$T(x + \varepsilon) - T(x) = T(\varepsilon) \qquad [89]$$

so $\rho_{x,\varepsilon}$ can be rewritten

$$\rho_{x,\varepsilon} = (\|T(\varepsilon)\|/\|\varepsilon\|)(\|x\|/\|T(x)\|) = \beta_\varepsilon \|x\|/\|T(x)\| \qquad [90]$$

For a given x, the maximum of this relative error magnification factor, $\rho_{x,\varepsilon}$, over all input errors $\varepsilon \neq 0$, is called the *sensitivity*† of T at x, which we write $S_x(T)$. So for a linear T

$$S_x(T) = \max_{\varepsilon \neq 0} \rho_{x,\varepsilon} = \max_{\varepsilon \neq 0} \beta_\varepsilon \|x\|/\|T(x)\| \qquad [91]$$

$$= (\|x\|/\|T(x)\|)(\max_{\varepsilon \neq 0} \beta_\varepsilon)$$

But by definition

$$\max_{\varepsilon \neq 0} \beta_\varepsilon \triangleq \|T\| \qquad [92]$$

so

$$S_x(T) = \|T\| \ \|x\|/\|T(x)\| \qquad [93]$$

Note that $\|T\|$ reflects two norms, one in the input space and one in the output space, and these same two norms are used to produce $\|x\|$ and $\|T(x)\|$, respectively. Thus, if we are given that, using the input norm, $\|\varepsilon\|/\|x\| \leq r_x$, then the relative error measured using the output norm will be bounded by $S_x(T)r_x$. A sensitivity less than 10 is normally considered small and one greater than 100 quite large.

As an example consider the addition operator

$$T((u, v)^T) = u + v \qquad [94]$$

at $(u, v) = (2, 3)$. Let us use the ℓ_1 norm on the input space, L_2, and the absolute value norm on the output space, L_1. Then the sensitivity at this input is by equations 50 and 93

$$S_{\begin{bmatrix} 2 \\ 3 \end{bmatrix}}(T) = 1 \cdot (|2| + |3|)/|2 + 3| = 1; \qquad [95]$$

relative error is propagated by a factor no greater than 1.

Replacing either norm by a norm which is a constant multiple of the previously used norm does not affect the sensitivity. But replacing either norm by a norm which differs by other than a constant factor will change the sensitivity. For example, if we use the ℓ_∞ norm on the input space and twice the absolute value norm on the output space, by equations 51 and 93

$$S_{\begin{bmatrix} 2 \\ 3 \end{bmatrix}}(T) = 4 \cdot \max(|2|, |3|)/(2|2 + 3|) = 6/5 \qquad [96]$$

---

† The name ''sensitivity'' was coined by us for this magnification factor, unlike ''operator norm'' and ''condition number,'' which are standard in the literature.

Note that the input relative error bound will also reflect the use of the ℓ_∞ norm.

To be a bit more general, for the addition operator and the first pair of norms above,

$$S_x(T) = 1 \cdot (|u| + |v|)/|u + v| \qquad [97]$$

so if u and v have the same sign, $S_x(T) = 1$. But if u and v have opposite signs, $|u + v| < |u| + |v|$, so $S_x(T) > 1$. For example, if $u = 2$ and $v = -3$, $S_x(T) = 5$, that is, the relative error in the input can be magnified by as much as a factor of 5. For u and v of opposite sign $S_x(T)$ can be made arbitrarily large by letting the magnitudes of u and v be arbitrarily close. Thus, we see that addition is not sensitive for operands of the same sign but is very sensitive (may magnify relative error greatly) if u and v are of opposite sign and are close in magnitude.

The poor behavior of addition for operands of approximately equal magnitude but opposite sign is caused not by the absolute error getting large in magnitude but by the answer getting small in magnitude. Thus if the summands are 0.500000 and −0.499999, each with an absolute error bound of 0.000001, the absolute error in the input is bounded by 0.000002 (using the ℓ_1 norm), so the bound on the input relative error is 0.000002/0.999999 ≈ 2 × 10^{-6}. But though the output absolute error is still only 0.000002, the output relative error is bounded by 0.000002 ÷ 0.000001 = 2, that is, 10^6 times as much as the input relative error bound. The addition of the two inputs causes a loss of significant digits, that is, makes the output much smaller than the inputs. Thus a representation error in a floating point number that was a small relative error becomes (is propagated into) a large relative error when significance is lost due to the addition of a number of about the same magnitude but of opposite sign (or, equivalently, the subtraction of an approximately equal number). Note that the problem is one of error propagation, not of error generation. It was this problem of the sensitivity of the subtraction of approximately equal numbers that caused most of our difficulty in the example in Section 1.4.2.

Probably more time is spent in developing numerical methods that avoid subtraction of approximately equal numbers than on any other problem. This avoidance can often be accomplished by analytical manipulation of the problem. Three examples follow:

1. $e^x - 1$ for $|x| \ll 1$. Instead of computing e^x and then subtracting 1, note

$$e^x = 1 + x + \frac{x^2}{2!} + \cdots \qquad [98]$$

so

$$e^x - 1 = x + \frac{x^2}{2!} + \frac{x^3}{3!} + \cdots \tag{99}$$

The term with largest magnitude is x, but we know that x is much smaller than 1. Not keeping the 1 allows us to keep many more digits of the other terms in a given-size floating-point fraction.

2. $\sqrt{x + 1} - \sqrt{x}$ for $x \gg 0$. Note

$$\sqrt{x + 1} - \sqrt{x} = \frac{1}{\sqrt{x + 1} + \sqrt{x}} \tag{100}$$

a form which does not require the subtraction of two approximately equal values.

3. $x = (-b + \sqrt{b^2 - 4ac})/2a$ [101]

when $b^2 \gg |4ac|$ and $b > 0$. This is the formula for the root of smaller magnitude of the quadratic equation $ax^2 + bx + c = 0$. Using our knowledge that $ax^2 + bx + c = a(x - \text{root}_1)(x - \text{root}_2)$, which implies that $c = a \, \text{root}_1 \, \text{root}_2$, and thus the product of the two roots is c/a, we find

$$x = \frac{-2c}{b + \sqrt{b^2 - 4ac}} \tag{102}$$

a form that produces less error propagation in this case.

Another example of the gains that can be made by attention to the sensitivity of the subtraction of approximately equal numbers is the Gaussian elimination algorithm for the solution of linear equations. Consider the basic computational step of the triangularization:

$$A_{jk} := A_{jk} - A_{ik} \left(\frac{A_{ji}}{A_{ii}} \right) \tag{103}$$

To achieve accuracy, we would prefer that the terms in the subtraction not be approximately equal. However, that situation is not unlikely. Many of the coefficients in a given row, or many of the coefficients in a given column, may be of approximately the same size; and either of these circumstances may lead to subtracting approximately equal terms. How can we avoid the problem of large error propagation? That is, what simple transformations of the system of equations can we make so that this problem will arise as little as possible? Three simple transformations come to mind:

1. Multiply each equation (row) by an appropriate constant. This certainly does not change the solution.
2. Change the order of the equations (rows). As with 1 above, this does not change the solution.
3. Change the order of the variables (columns). This procedure changes only the order of the elements of the solution. If we remember how we change the columns, we can reorder the elements when we are finished with the computation.

Multiplying each row by a different constant (the first approach above) does not improve the error situation. Let us multiply the ith row by k_i for $i = 1, 2, ..., n$. Let A'_{ij} be the matrix values after multiplication. Then the basic operation is

$$A'_{jk} := A'_{jk} - A'_{ik}\left(\frac{A'_{ji}}{A'_{ii}}\right) \qquad [104]$$

that is,

$$k_j A_{jk} := k_j A_{jk} - k_i A_{ik}\left(\frac{k_j A_{ji}}{k_i A_{ii}}\right) \qquad [105]$$

which is equivalent to

$$k_j(A_{jk}) := k_j\left(A_{jk} - A_{ik}\left(\frac{A_{ji}}{A_{ii}}\right)\right) \qquad [106]$$

Thus multiplying the jth row by k_j simply results in the jth row at any step in the triangularization procedure being multiplied by k_j: any subtraction that caused error propagation difficulties in the unmultiplied matrix is unchanged in terms of the relative values of the terms being subtracted. And in floating-point subtraction, it is precisely this relative value that concerns us.

Switching either rows or columns can change the error propagation properties of the computation. Since multiplication of each row by a constant changes neither the roots of the equation nor the error properties of the computation, we can without loss of generality consider the set of equations where each equation has been scaled (multiplied by the reciprocal of the sum of the magnitudes of the coefficients of the equation) so that the average magnitude of each row of the resulting matrix, A'', is the same. Consider the general triangularization operation (relation 103) on the double-primed values. Assume that we are at the ith stage of the triangularization and that the matrix elements referred to are the transformed matrix values at that stage. Since the equations have been scaled, it is not unlikely that the kth element of the jth row (A''_{jk}) is approximately equal to the kth element of the ith row (A''_{ik}). To prevent the terms we subtract from being approximately equal, we would like the factor $|A''_{ji}/A''_{ii}|$ to be as relatively far from 1 as possible. We can arrange this by properly choosing the new ith row from among the present ith through nth rows, all of which have zeros in the first $i - 1$ columns. We would choose as the new row i that row m, such that A''_{mi} is as large in magnitude as possible. Thus, from among the ith through nth rows we find the row for which A''_{mi} is largest in magnitude, and we exchange the ith and mth rows. This process is called *pivoting by rows,* and the element A''_{ii} after the row exchange is called the *pivot element.* Experience shows that pivoting by rows can make a significant improvement in the error properties of the computation in Gaussian elimination.

Note that we also could have considered as the new row i the row such that A''_{mi} is as small as possible in magnitude, but that choice would have allowed us to choose a zero divisor. Furthermore, small matrix elements should be avoided as divisors because they may have a large relative error — due to the fact that they may be the result of the subtraction of approximately equal numbers.

Note also that the argument for pivoting depends strongly on the assumption that the equations have been scaled. Pivoting without scaling makes relatively little sense. In fact, the argument for choosing the largest pivot after scaling is even stronger than that presented above. If A''_{ii} is larger in magnitude than A''_{ji}, then because the equations have been scaled, A''_{ik} tends to be smaller in magnitude than A''_{jk}. This fact contributes further to the objective that the rightmost term in relation 103 should be smaller in magnitude than A''_{jk}.

Suppose we apply Gaussian elimination with scaling and pivoting to the system of simultaneous equations (written in both element and matrix-and-vector notations).

$$
\begin{aligned}
x_1 - x_2 + x_3 &= 2 \\
2x_1 + x_2 + x_3 &= 7 \\
4x_1 + 2x_2 + x_3 &= 11
\end{aligned}
\qquad
\begin{bmatrix} 1 & -1 & 1 \\ 2 & 1 & 1 \\ 4 & 2 & 1 \end{bmatrix} x =
\begin{bmatrix} 2 \\ 7 \\ 11 \end{bmatrix}
\qquad
\begin{aligned} &[107] \\ &[108] \\ &[109] \end{aligned}
$$

which has the solution $x_1 = 1$, $x_2 = 2$, $x_3 = 3$. First we scale each equation by multiplying it by the reciprocal of the sum of the magnitudes of its coefficients (3, 4, and 7, in that order). This produces three equations with the sum of the coefficient magnitudes of each equal to 1:

$$
\begin{aligned}
\tfrac{1}{3}x_1 - \tfrac{1}{3}x_2 + \tfrac{1}{3}x_3 &= \tfrac{2}{3} \\
\tfrac{1}{2}x_1 + \tfrac{1}{4}x_2 + \tfrac{1}{4}x_3 &= \tfrac{7}{4} \\
\tfrac{4}{7}x_1 + \tfrac{2}{7}x_2 + \tfrac{1}{7}x_3 &= \tfrac{11}{7}
\end{aligned}
\qquad
\begin{bmatrix} \tfrac{1}{3} & -\tfrac{1}{3} & \tfrac{1}{3} \\ \tfrac{1}{2} & \tfrac{1}{4} & \tfrac{1}{4} \\ \tfrac{4}{7} & \tfrac{2}{7} & \tfrac{1}{7} \end{bmatrix} x =
\begin{bmatrix} \tfrac{2}{3} \\ \tfrac{7}{4} \\ \tfrac{11}{7} \end{bmatrix}
\qquad
\begin{aligned} &[110] \\ &[111] \\ &[112] \end{aligned}
$$

Then we find the equation with the largest coefficient in magnitude in the first column (112), and we switch that equation and the first (110). After switching equations 112 and 110 (row 1 and 2 of the matrix and vector), we eliminate the first column of the second and third equations (111 and 110), producing

$$\frac{4}{7}x_1 + \frac{2}{7}x_2 + \frac{1}{7}x_3 = \frac{11}{7} \qquad \begin{bmatrix} \frac{4}{7} & \frac{2}{7} & \frac{1}{7} \\ 0 & 0 & \frac{1}{8} \\ 0 & -\frac{1}{2} & \frac{1}{4} \end{bmatrix} x = \begin{bmatrix} \frac{11}{7} \\ \frac{3}{8} \\ \frac{1}{4} \end{bmatrix} \qquad [113]$$

$$\frac{1}{8}x_3 = \frac{3}{8} \qquad\qquad\qquad\qquad\qquad\qquad\qquad [114]$$

$$-\frac{1}{2}x_2 + \frac{1}{4}x_3 = -\frac{1}{4} \qquad\qquad\qquad\qquad\qquad\qquad [115]$$

We now choose the equation from among the second through the last for which the coefficient in the second column has maximum magnitude. We switch this equation with the second. In our example, this results in switching equations 115 and 114. Eliminating the coefficient of the second variable from all equations beyond the second, we find that in our example the third equation (114) is unchanged because its x_2 coefficient is already zero. Had we not switched equations, that zero would have been a divisor and thus would have caused difficulty.

Our triangularized set of equations comprises 113, 115, and 114, in that order. Solving the last equation, we find $x_3 = 3$. Substituting this into equation 115, we obtain

$$-\frac{1}{2}x_2 + \frac{3}{4} = -\frac{1}{4}$$

so $x_2 = 2$. Substituting $x_3 = 3$ and $x_2 = 2$ into equation 113, we obtain

$$\frac{4}{7}x_1 + \frac{4}{7} + \frac{3}{7} = \frac{11}{7}$$

so $x_1 = 1$.

We see that scaling the equations initially does not produce scaled equations after a stage of triangularization. We could rescale the last $n - i + 1$ equations after the ith stage of the triangularization for each i, but the unrescaled coefficients tend to be of the same order of magnitude because of the original scaling. Since this is all we require for the pivoting to work as desired to minimize error propagation, we do not normally rescale at every stage.

We need not explicitly carry out the scaling of the rows to accomplish the pivoting operation. Given a pivot row, the scaling does not affect the result. It only affects which is the pivot row. We can simply compute the scale factors by which each row should be multiplied to make the sum of its coefficient magnitudes equal to 1. Then when we are choosing a pivot, instead of comparing the magnitudes of the candidate values directly, we compare the magnitudes of the product of each candidate value and the scale factor associated with its row. This way we can choose the correct pivot but not incur the arithmetic error of multiplying each element of the original matrix by some number.

If instead of scaling each row we had scaled each column (requiring rescaling of the solution elements at the end of the computation), and if we consider the basic operation on the scaled values (relation 103) in the form $A''_{jk} := A''_{jk} - A''_{ji}(A''_{ik}/A''_{ii})$, a situation parallel to that discussed above with respect to rows pertains. Here we wish to choose as A''_{ii} that element of the ith row which is greatest by switching its column with the ith column. We see that such column pivoting and scaling causes some difficulty in rescaling and reordering the solution elements at the end of a computation. So, generally, row pivoting is preferred. In cases of a matrix producing extremely sensitive subtractions in Gaussian elimination or in a problem with extreme accuracy requirements we may want to do row scaling followed by full pivoting (that is, switch rows and columns so that the largest element in the $n - i + 1$ by $n - i + 1$ right lower matrix of the partially triangularized A is placed at the ith position). But in most situations the gain in accuracy achieved by full pivoting over row pivoting is not great.

It should be noted that the switching of rows does not require physically switching the elements of the matrix in the computer. Rather one can create a permutation vector: a vector of length n where the ith element contains the original number of the row which is now considered to be the ith. The vector is initialized to $(1, 2, 3, ..., n)^T$, and at any point in the elimination the ith row is referenced via the ith element of the permutation vector. To switch two rows, j and k, one simply exchanges elements j and k of the permutation vector. The elimination algorithm with implicit scaling and pivoting by rows and no physical switching of elements is presented in Program 6–1. The gains in accuracy of using this algorithm over Gaussian elimination without pivoting are significant, as is illustrated in Problems 6.10 and 6.11.

Before moving on to further discussion of relative error propagation in general, one final improvement in Gaussian elimination is worth discussing. Often we have to solve $Ax = b$ for many different values of b, for example, when b represents measured data from an experiment and A represents relations between the unknowns and the measurable values, or when we use a technique called iterative improvement (see Pizer, 1975, Section 2.3.4) to decrease the generated arithmetic error in the solution x to $Ax = b$ by solving $Ax^{(i)} = b^{(i)}$ for $i = 1, 2, ...$ When we have more than one b vector, we have to back-substitute for every b, but we would like to avoid retriangularizing A for every b since the

PROGRAM 6-1
Gaussian Elimination with Scaling and Pivoting

```
{*********************************************************************}
{* Solve Ax = b, a set of n linear equations in n unknowns  by    *}
{*     Gaussian Elimination with Scaling and Pivoting             *}
{* Set in an enclosing block is                                   *}
{*     n, number of given equations and unknowns.                 *}
{* Given are                                                      *}
{*     A, the n x n matrix of coefficients; and                   *}
{*     b, an n-vector (right-hand-side of the equation to be      *}
{*         solved).                                               *}
{* Output parameter is                                            *}
{*     x, the solution vector.                                    *}
{*********************************************************************}

PROCEDURE Gauss_pivot(
  A: real_matrix;                {coefficients}
  b: real_n_vector;             {right-hand-side of equation}
  VAR x: real_n_vector          {solution vector}
             );

     {real_matrix is defined as ARRAY[1..n,1..n] OF real}
     {real_n_vector is defined as ARRAY[1..n] OF real}

VAR
  scale: real_n_vector;  {scaling vector}
  p: ARRAY[1..n] OF integer;
                         {permutation vector}
  max: real;             {largest scaled pivot magnitude}
  maxrow: integer;       {row containing max}
  scaleval: real;        {magnitude of scaled pivot candidate}
  amult: real;           {row multiplier}
  temp: integer;         {permutation vector element being switched}
  i,j,k: integer;        {row and column indicators}

BEGIN
{Initialize}
  FOR i:=1 TO n DO
    BEGIN
      {Initialize permutation vector}
        p[i] := i;
      {Compute sum of row magnitudes}
        scale[i] := 0;
        FOR j:=1 TO n DO
          scale[i] := scale[i] + abs(A[i,j])
    END;

  {Triangularize with pivoting}
    FOR i:= 1 TO n-1 DO
      BEGIN
        {Find pivot}
          max := 0;
          FOR j:=i TO n DO
            BEGIN
              scaleval := abs(A[p[j],i]/scale[p[j]]);
              IF (scaleval > max) THEN
                BEGIN
                    max := scaleval;
                    maxrow :=j
                END
            END;
        {Switch rows}
```

```
                temp := p[maxrow];
                p[maxrow] := p[i];
                p[i] := temp;
           {Eliminate subdiagonal elements in i-th column}
              FOR j:=i+1 TO n DO
                 BEGIN
                    {Compute nonzero elements of j-th row}
                       amult := A[p[j],i]/A[p[i],i];
                       FOR k:=i+1 TO n DO
                         A[p[j],k] := A[p[j],k] - amult*A[p[i],k];
                       b[p[j]] := b[p[j]] - amult*b[p[i]]
                 END
        END;

     {Back substitute}
       FOR i:=n DOWNTO 1 DO
         BEGIN
            x[i] := b[p[i]];
            {Subtract terms in already computed x[j]}
              FOR j:=i+1 TO n DO
                x[i] := x[i] - A[p[i],j]*x[j];
            {Solve for x[i]}
              x[i] := x[i]/A[p[i],i]
         END

END; {Gauss_pivot}
```

$n^3/3 + O(n^2)$ M/D's required by the triangularization is much more than the $n^2/2 + O(n)$ M/D's required by back-substitution. The triangularized A matrix does not change for different b, but heretofore we have modified the b vector during the triangularization and we can no longer do this if b is not available at the time of triangularization. To modify b after triangularization, we need only save the row multiplication factors $A_{ji}^{(i)}/A_{ii}^{(i)}$, with $j = i+1, i+2, ..., n$; $i = 1, 2, ..., n-1$ (where the superscript "(i)" indicates that this is the matrix value at the ith stage of triangularization). With these factors the elements of a new b vector can be recurrently multiplied and subtracted from other b elements to produce the modified b vector to which back-substitution is applied. This operation can be indicated concisely as follows.

If we let U be the upper triangular matrix produced by triangularization of A, and L be the lower triangular matrix of multiplying factors with ones added on the diagonal (as shown in equation 116), then it can be shown that

$$
L = \begin{bmatrix}
1 & & & & & \\
\dfrac{A_{21}^{(1)}}{A_{11}^{(1)}} & 1 & & & & \\
\dfrac{A_{31}^{(1)}}{A_{11}^{(1)}} & \dfrac{A_{32}^{(2)}}{A_{22}^{(2)}} & 1 & & & \\
\cdot & \cdot & \dfrac{A_{43}^{(3)}}{A_{33}^{(3)}} & \cdot & & \\
\vdots & \vdots & \vdots & \vdots \ \cdot \ \cdot & & \\
\dfrac{A_{n-1,1}^{(1)}}{A_{11}^{(1)}} & \dfrac{A_{n-1,2}^{(2)}}{A_{22}^{(2)}} & \dfrac{A_{n-1,3}^{(3)}}{A_{33}^{(3)}} & \cdot \ \cdot \ \cdot \ \cdot & 1 & \\
\dfrac{A_{n1}^{(1)}}{A_{11}^{(1)}} & \dfrac{A_{n2}^{(2)}}{A_{22}^{(2)}} & \dfrac{A_{n3}^{(3)}}{A_{33}^{(3)}} & \cdot \ \cdot \ \cdot \ \cdot & \dfrac{A_{n,n-1}^{(n-1)}}{A_{n-1,n-1}^{(n-1)}} & 1
\end{bmatrix}
\qquad [116]
$$

$$LU = A \qquad [117]$$

so this method is called *triangular factorization*. Given L and U, we need to solve

$$LUx = b \qquad [118]$$

If we let

$$Ux = y \qquad \mathbf{[119]}$$

equation 118 can be rewritten

$$Ly = b \qquad [120]$$

which can be solved by front substitution for y. The vector y is simply the modified version of b that would have been produced if b had participated in the triangularization. All that remains is to solve equation 119 for x by back-substitution, as before.

When we add scaling and pivoting to the triangularization, the only thing which changes is that the rows of A and b are permuted. That is, during the triangularization a permutation vector is produced, and in the substitutions the row indices of L, U, and b are referred to via the permutation vector.

Note that each substitution requires $n^2/2 + O(n)$ M/D's, so after triangularization finding x for each b requires $n^2 + O(n)$ M/D's.

Another way to solve $Ax^j = b^j$ for various vectors b^j without requiring retriangularization for each j is to compute A^{-1} once and then for each j to compute $x = A^{-1}b^j$. Multiplying A^{-1} by b^j requires n^2 multiplications, the same number as the two substitutions of the triangular factorization method, if it were used. However, computing A^{-1} involves simultaneously solving the n equations $Ax = e^i$ for $1 \leq i \leq n$, a process that itself requires the triangularization of A. Since finding this solution for the inverse requires n^3 multiplications, and triangular factorization requires only $n^3/3$, the algorithm involving triangular factorization and two substitutions is always more efficient and accurate than computing A^{-1} and multiplying b by it. Therefore, the only reason to compute A^{-1} is if the values of that matrix itself are of interest to us!

As a final example of relative error propagation for a fixed input, let us consider the sensitivity of an operator on a vector space of functions rather than n-vectors. In particular, consider the previously mentioned problem of

$$T(f) = \int_c^d f(u)\, du \qquad [121]$$

Assume we are interested in the behavior of T for the input vector (function) $f(u) = e^u$. Then from equation 76 if we use the $\mathscr{L}_\infty$ norm on the input space and the absolute value norm on the output space,

$$\|T\| = d - c \qquad [122]$$

so

$$S_{e^u}(T) = \frac{(d-c)\|e^u\|_\infty}{\left| \int_c^d e^u\, du \right|} = \frac{(d-c)e^d}{e^d - e^c} = \frac{(d-c)}{1 - e^{c-d}} \qquad [123]$$

assuming $d > c$. Thus a relative error in the function $f(u)$ that is bounded by r_f will result in a bound on the relative error in the integral of $((d - c)/(1 - e^{c-d}))r_f$. Note that the relative error in f is $\|\varepsilon(u)\|/\|f(u)\|$, that is, in our example,

$$\max_{u \in [c,d]} |\varepsilon(u)| / \max_{u \in [c,d]} |f(u)|$$

and not $\|\varepsilon(u)/f(u)\|$.

Condition

The sensitivity is the maximum relative error magnification factor for a given value of the input. Often, when we are analyzing an operator, we do not know the input — it may be any vector in the input vector space. In this case we are interested in the maximum relative error magnification factor not only over all nonzero errors but over all nonzero inputs (nonzero since zero inputs will have infinite input relative error). That is, we are interested in

$$\max_{\substack{\varepsilon \neq 0 \\ x \neq 0}} \rho_{x,\varepsilon} = \max_{x \neq 0} \max_{\varepsilon \neq 0} \rho_{x,\varepsilon} = \max_{x \neq 0} S_x(T) \qquad [124]$$

the maximum sensitivity over all inputs. This value is called the *condition number* of T and is written cond(T).

Like the sensitivity, cond(T) depends on two norms — one in the input space and one in the output space. Given that the relative error in the input (using the input space norm) is bounded by r_x, the norm of the output relative error will be bounded by cond(T)r_x for any input x.

For a linear operator, T, from equation 124 we can obtain a relation between cond(T) and norms of vectors and of the operator T.

$$\text{cond}(T) = \max_{x \neq 0} S_x(T) = \max_{x \neq 0} (\|T\| \, \|x\|/\|T(x)\|)$$

$$= \|T\| \max_{x \neq 0} (\|x\|/\|T(x)\|) \qquad [125]$$

$$= \|T\|/\min_{x \neq 0} (\|T(x)\|/\|x\|)$$

If T has full rank (the output vector space has the same dimension as the input vector space) and thus T^{-1} exists, the minimum ratio between $\|T(x)\|$ and $\|x\|$ is the same as the minimum ratio between $\|y\|$ and $\|T^{-1}(y)\|$ for nonzero y. That is, if y is the vector in the output vector space which minimizes $\|y\|/\|T^{-1}(y)\|$, then if $x = T^{-1}(y)$, x is the vector in the input vector space which minimizes

$$\|y\|/\|T^{-1}(y)\| = \|T(x)\|/\|x\|$$

Thus we have

$$\text{cond}(T) = \|T\|/\min_{y \neq 0}(\|y\|/\|T^{-1}(y)\|)$$

$$= \|T\|/(1/\max_{y \neq 0}(\|T^{-1}(y)\|/\|y\|)) \qquad \textbf{[126]}$$

$$= \|T\| \, \|T^{-1}\|$$

If T is not of full rank, T^{-1} does not exist. But remember that the null space of such a T includes vectors other than 0, that is, there are nonzero vectors which T maps to zero. Thus

$$\min_{x \neq 0} \|T(x)\|/\|x\| = 0 \qquad [127]$$

so from equation 125

$$\text{cond}(T) = \infty; \qquad [128]$$

relative errors may grow arbitrarily large when T is applied.

Let us review what is being said by equation 126. Cond(T) measures the maximum factor by which relative error can be magnified by application of T, over all inputs and all input errors. Relative error will be magnified most when the absolute error is magnified the most and input is magnified the least. The maximum factor by which absolute error can be multiplied is $\|T\|$, and the minimum factor by which the input can be magnified is $1/\|T^{-1}\|$. If the absolute error and input are chosen to achieve these respective magnification factors, the relative error will be multiplied by $\|T\|/(1/\|T^{-1}\|)$, since its numerator, the norm of the input error, is magnified by $\|T\|$, and its denominator, the norm of the input, is magnified by $1/\|T^{-1}\|$. Thus the relative error is magnified at maximum by

$$\frac{\text{the maximum magnification factor of } T}{\text{the minimum magnification factor of } T} = \|T\| \, \|T^{-1}\| = \text{cond}(T).$$

The symmetry under inversion of $\|T\| \, \|T^{-1}\|$ implies that cond(T) = cond(T^{-1}) if cond(T) $\neq \infty$ (T^{-1} exists). Also, since when cond(T) $\neq \infty$, the dimension of the input space is equal to dimension of the output space, usually the same norm is used in both these spaces. In this case "cond" is often subscripted to indicate the norm used, for example, $\text{cond}_\infty(T)$ when the ℓ_∞ norm is used.

Let us evaluate the condition number in the examples we have considered when discussing sensitivity.

The condition number of the addition operator is its maximum sensitivity over all inputs. We saw earlier that this sensitivity could be made arbitrarily large by choosing inputs which were of opposite sign and close in magnitude. Thus cond($+$) $= \infty$. This result could have been obtained by noting that addition is an operator from L_2 (two-dimensional) to L_1 (one-dimensional) and so is not of full rank, and thus $\text{cond}(+) = \infty$.

Sometimes it is useful to take the maximum sensitivity not over all inputs in the input vector space but over some subset to which we know the inputs will be restricted. For example, if we were only interested in the addition of positive operands, we would want to determine the "condition number restricted to positive operands,"

$$\text{cond}_{(\text{inputs}>0)}(+) \stackrel{\Delta}{=} \max_{x>0} S_x(T) = 1 \qquad [129]$$

Consider the condition number of the linear operator T which maps

$$\begin{bmatrix} x \\ y \end{bmatrix} \text{ to } \begin{bmatrix} x + y \\ x + 2y \end{bmatrix}$$

The operator T maps L_2 onto L_2 (the operator has full rank). Thus, $\text{cond}(T) = \|T\| \, \|T^{-1}\|$, where T is the operator obtained by multiplication by the matrix

$$\begin{bmatrix} 1 & 1 \\ 1 & 2 \end{bmatrix}$$

whose inverse is

$$\begin{bmatrix} 2 & -1 \\ -1 & 1 \end{bmatrix}$$

Using the ℓ_∞ norm on L_2, we have seen that the operator norm for a matrix is the maximum row sum of magnitudes, so

$$\left\| \begin{bmatrix} 1 & 1 \\ 1 & 2 \end{bmatrix} \right\| = 3 \text{ and } \left\| \begin{bmatrix} 2 & -1 \\ -1 & 1 \end{bmatrix} \right\| = 3$$

so $\text{cond}_\infty(T) = 9$. Thus, if the input relative error, using the ℓ_∞ norm, is bounded by $r_{x,\infty}$, the output relative error using the ℓ_∞ norm will be bounded by $9r_{x,\infty}$.

If T is a linear operator represented by a matrix A, it is common practice to write $\text{cond}(T)$ as $\text{cond}(A)$ and to write $\|T\|$ as $\|A\|$. If A is a nonsingular, square matrix, that is, one of full rank, $\text{cond}(A)$ will be finite and will be computed as $\|A\| \, \|A^{-1}\|$, using whatever operator norm corresponds to the norm used in the input and output spaces. The condition number of a matrix is of special importance because it arises in the analysis of the error propagated from b to x in the solution of a set of linear equations $Ax = b$. Though strictly we require $\text{cond}(A^{-1})$ for this purpose, the symmetry of the condition number under inversion implies that $\text{cond}(A)$ gives the desired relative error magnification factor as well.

While strictly cond(A) concerns only the propagation of error from b to x and does not relate to the effect of error in the elements of the matrix A or of error generated within the solution process, it turns out (see Problems 6.13 and 6.14) that the size of cond(A) is also an indication of the degree of difficulty experienced due to these errors.

For the operator T mapping $f(u)$ to $\int_c^d f(u)\,du$, we see that the mapping is from an infinite-dimensional vector space of functions on $[c, d]$ to L_1; so the operator is not of full rank, and cond(T) $= \infty$.

Let us consider one final operator, the one mapping functions $f(u)$ on $(-\infty, \infty)$ to functions $g(u)$ on $(-\infty, \infty)$ by $g(u) = T(f(u)) = w(u)\,f(u)$, where $w(u)$ is a function on $(-\infty, \infty)$ such that $w(u) > 0$ for all u. Then $T^{-1}(g(u)) = g(u)/w(u)$, that is, T^{-1} exists. Using the $\mathscr{L}_\infty$ norm,

$$\|T\| = \max_{f(u) \neq 0} \frac{\|T(f)\|_\infty}{\|f\|_\infty} = \max_{f(u) \neq 0} \frac{\max\limits_{u \in (-\infty, \infty)} |w(u)\,f(u)|}{\max\limits_{u \in (-\infty, \infty)} |f(u)|} \quad [130]$$

But

$$\frac{\max\limits_{u \in (-\infty, \infty)} |w(u)\,f(u)|}{\max\limits_{u \in (-\infty, \infty)} |f(u)|} \leq \max_{u \in (-\infty, \infty)} |w(u)| \quad [131]$$

and equality is obtained if $f(u)$ has its maximum where $w(u)$ does. Then

$$\|T\|_\infty = \max_{u \in (-\infty, \infty)} |w(u)| \quad [132]$$

Similarly

$$\|T^{-1}\|_\infty = \max_{u \in (-\infty, \infty)} \frac{1}{|w(u)|} = \frac{1}{\min\limits_{u \in (-\infty, \infty)} |w(u)|} \quad [133]$$

Therefore

$$\text{cond}_\infty(T) = \frac{\max\limits_{u \in (-\infty, \infty)} |w(u)|}{\min\limits_{u \in (-\infty, \infty)} |w(u)|} \quad [134]$$

Note the consistency of this result with the interpretation of the condition number as the ratio of the largest magnification factor under application of T and the smallest magnification factor under application of T.

6.3.4 RELATIVE ERROR: NONLINEAR OPERATORS

For nonlinear operators equation 88 still holds, and equation 90 becomes

$$\begin{aligned} \rho_{x,\varepsilon} &= (\|T(x + \varepsilon) - T(x)\|/\|\varepsilon\|)(\|x\|/\|T(x)\|) \\ &= \beta_{x,\varepsilon}\,\|x\|/\|T(x)\| \end{aligned} \quad [135]$$

Then redefining sensitivity to deal with infinitesimal errors, we have

$$S_x(T) = \max_{\text{tiny } \varepsilon \neq 0} \rho_{x,\varepsilon} = (\max_{\text{tiny } \varepsilon \neq 0} \beta_{x,\varepsilon}) \, \|x\|/\|T(x)\|$$

$$= \|T|_x\| \, \|x\|/\|T(x)\| \qquad\qquad \text{[136]}$$

For example, consider $T((u, v, w)^T) = u \sin(v^2) + w$ at $u = 2$, $v = \sqrt{\pi}$, and $w = 10$, the example used in Section 6.3.2, where we saw $\|T|_{(2,\sqrt{\pi},10)}\| = 4\sqrt{\pi}$ using the ℓ_1 norm in the input space and the absolute value norm in the output space. In this case the sensitivity $S_x(T) = 4\sqrt{\pi}|2 + \sqrt{\pi} + 10|/|2 \sin(\pi) + 10| = 9.8$. Thus a small input relative error bounded, for example, by 1% would produce an output relative error bound 9.8 times as large, that is, 9.8% in this example.

For nonlinear operators the condition number formally is defined as

$$\text{cond}(T) \triangleq \max_{x \neq 0} S_x(T) = \max_{x \neq 0} \|T|_x\| \, (\|x\|/\|T(x)\|) \qquad\qquad \text{[137]}$$

However, the idea of the condition number in this situation makes little sense, because the magnification of input to output by T varies so strongly with $\|x\|$ as well as $x/\|x\|$. In fact $\|x\|/\|T(x)\|$ is usually unbounded over all nonzero x because $T(x) = 0$ for some nonzero x. In contrast to the case with ε_x, we can not solve our problem by restricting x to only small values because there is no reason to believe x will be small. However, the idea of condition number may be useful if we restrict x to some range of interest and find the maximum sensitivity over that range, in which case the maximum in equation 137 will be taken only over this range. In this case the condition number may be used as before, given an input relative error bound, to bound the output relative error over all inputs in the specified range.

6.4 Summary and Complements

We have seen two approaches to the analysis of error propagation: the partial derivative approach and the error magnification factor approach. Both approaches give a bound on propagated error alone, that is, under the assumption that no error is generated. Thus the result is independent of the algorithm used for implementing the operator.

For example, the condition number for solution of the linear equation $Ax = b$ bounds the propagation of relative error in b into relative error in x independent of whether the Gaussian elimination method or the Gauss-Seidel method is used to obtain the solution. Errors due to the subtractions in Gaussian elimination or to incomplete convergence of the Gauss-Seidel method are generated errors with respect to the equation solution operator, though they are propagated errors with respect to the subtraction operator used as part of the Gaussian elimination method and the iterative step operator used as part of the Gauss-Seidel method, respectively. We will discuss this matter at greater length in Chapter 7, but here it should be stressed that error propagation can be understood only if the operator doing the propagating is clearly specified.

We summarize here the principal methods used in analyzing error propagation. If the input is an n-vector and bounds on the error in each component are given, the partial derivative approach is applicable. If absolute error bounds on the magnitudes of the inputs are given, equation 13 should be used; if relative error bounds are given, equation 14 should be used.

If a bound on the norm of the error is given, the error magnification factor approach is indicated. If a bound on the norm of the absolute error in the input vector is given, multiplication of that bound by the operator norm gives a bound on the norm of the output error. If the operator being analyzed is linear, the operator norm is independent of the input value and is given by equations 48 or 58. If the operator is nonlinear, the operator norm depends on the input value. If further the output is real, the operator norm is given by equation 82. In particular, if the input error is bounded according to the ℓ_1 or ℓ_∞ norms, the operator norm is given by equations 83 or 84.

If a bound on the relative error in the input vector is given, multiplying that bound by either the sensitivity or the condition number of the operator gives a bound on the relative error in the output. If the bound desired is for a particular input value, the multiplying factor should be the sensitivity at that input value. The sensitivity for linear operators is given by equation 93 and for nonlinear operators is given by equation 136.

If the bound desired is over all possible inputs, the multiplying factor should be the condition number, which is normally applicable only for linear operators. With these its value is given by equation 126.

REFERENCES

Pizer, S. M., *Numerical Computing and Mathematical Analysis*. Science Research Associates, Chicago, 1975.

PROBLEMS

6.1 Assume that $f(x)$ is twice differentiable on $[x, x^*]$. By expansion of $f(x)$ in a Taylor series about x^*, show that

(a) $b_f = b_x |f'(x^*)| + O(b_x^2)$
(b) $r_f = r_x |x^*f'(x^*)/f(x^*)| + O(r_x^2)$

6.2 Show that the bound on the propagated relative error in the quotient of two variables is approximately the sum of the bounds on the relative errors of the inputs, if these input relative errors are much less than 1 in magnitude.

6.3 Give the bounds for the propagated absolute error and for the propagated relative error in each of the following computations:

(a) $(3 \pm 10^{-2}) + (3.1 \pm 10^{-3})$
(b) $(3 \pm 10^{-2}) - (3.1 \pm 10^{-3})$
(c) $(3 \pm 10^{-2})(3.1 \pm 10^{-3})$
(d) $(3.1 \pm 10^{-3})/(3 \pm 10^{-2})$

6.4 Bound the propagated error in the following functions, given that the error bounds in x and y are $b_x = 10^{-6}$ and $b_y = 10^{-5}$:

(a) $f(x) = \log(x^2 + 5)$ near $x = 2$
(b) $f(x, y) = e^{x^3y} \cos(y)$ near $x = 2, y = \pi/4$

6.5 A step of the modified Euler method is given by $\hat{y}_{i+1} = y_{i-1} + 2h\,f(x_i, y_i)$. Given bounds b_{i-1}, b_i, b_x, and b_h on the magnitudes of the errors in y_{i-1}, y_i, x_i, and h, respectively, give an approximate bound, b_{i+1}, on the magnitude of the propagated error in $\hat{y}_{i+1}$, that is, the error propagated by one step of the modified Euler method.

6.6

(a) Consider multiplying the two matrices A and B where each element of A is positive and has a maximum relative error magnitude r_a and each element of B is positive and has a maximum relative error magnitude r_b. What is the maximum propagated relative error magnitude in each of the elements of C, where $C = AB$? (Assume all errors can be positive or negative.)
(b) Consider computing $\|Ax\|^2 = (Ax)^T Ax$, where A is a matrix with positive elements and x is a vector with positive elements. What is the maximum relative error magnitude in $\|Ax\|^2$ if the elements of A have maximum relative error magnitude r_a and the elements of x have maximum relative error magnitude r_x?

206

6.7 Consider the data (x, y) in the accompanying table. Assume an error of 1 is made in $y(1.5)$, that is, $y^*(1.5) = 91$. The relative error in this value is $1/90 = 1.1\%$. Assume there is no error in any y_i other than $y(1.5)$. What is the relative error in each difference in the divided-difference table? Notice the tendency of the relative error as the order of the difference increases.

x	y
-1.0	107
1.0	96
1.5	90
2.0	83
3.0	67
5.0	29

6.8 Prove that the matrix norm related to $\|x\| = \Sigma_k |x_k|$ (the ℓ_1 norm) is $\|A\| = \max_j \Sigma_i |A_{ij}|$. Compare your method of proof to that used in Section 6.3.1 to prove that the matrix norm related to $\|x\| = \max_k |x_k|$ (the ℓ_∞ norm) is $\|A\| = \max_i \Sigma_j |A_{ij}|$.

6.9

(a) Find the smaller absolute root of the equation $x^2 + 80x + 1$ using equation 101 and symmetrically rounding all results (including intermediate results) to three significant decimal digits.
(b) Repeat part a but use equation 102.
(c) Compare your results in parts a and b with the correct answer to three significant digits: -0.0125.

6.10 Solve the following set of linear equations using two decimal digit arithmetic (with a guard digit for addition), first with no pivoting and then with pivoting by rows (with implicit ℓ_1 scaling).

$$11x + 12y + 16z = 39$$
$$10x + 11y + 12z = 33$$
$$1.2x + 0.1y + 0.1z = 1.4$$

Compare your answers, using the fact that the correct solution is $x = y = z = 1$.

6.11

(a) Apply Program 6–1 to solve $Ax = b$, where

$$A = \begin{bmatrix} 3 & 2.88 & 2.52 & 1.92 \\ 0.72 & 0.691 & 0.3304 & 0.1666 \\ 1.26 & 0.6608 & 1.5 & 0.516 \\ 0.24 & 0.0833 & 0.129 & 0.375 \end{bmatrix} \quad \text{and} \quad b = \begin{bmatrix} 10.32 \\ 1.908 \\ 3.9368 \\ 0.8273 \end{bmatrix}$$

(b) Repeat part a, but use Program 3–14 of this text (Gaussian elimination with no scaling or pivoting).

(c) Compare the results of parts a and b, using the fact that the correct answer is $x = (1, 1, 1, 1)^T$.

6.12 Let $f(x, y) = (x + 2y)^2$.

(a) Bound the propagated error in f at $x = 2$, $y = 3$, given that the error bounds on x and y are $b_x = 10^{-6}$, $b_y = 10^{-5}$.

(b) Evaluate the sensitivity of f at $x = 2$, $y = 3$ using the ℓ_1 norm.

(c) Evaluate the ℓ_1 condition number of f assuming $x > 0$ and $y > 0$.

6.13 Consider the linear equation $Ax = b$, where

$$A = \begin{bmatrix} 3 & 2 \\ 4 & -3 \end{bmatrix} \quad \text{and} \quad b = \begin{bmatrix} 5 \\ 1 \end{bmatrix}$$

This system has the solution $x_1 = x_2 = 1$.

(a) Calculate the solution x^* if b is replaced by $b + \delta b$, where

$$\delta b = \begin{bmatrix} 1 \\ 0 \end{bmatrix}$$

Then, using the ℓ_1 norm, calculate the relative error in b, the relative error in x^*, and the ratio between the latter and the former (the error magnification factor).

(b) Compute $\text{cond}_1 (A)$.

(c) Execute the triangularization step of Gaussian elimination with ℓ_1 scaling and pivoting. Note the magnitude of the scaled diagonal element in the last (second) equation after triangularization. This value is sometimes used as a measure of the error propagation behavior of the equation solution (the smaller the value, the poorer the behavior).

(d) Scale each row of A by dividing it by its ℓ_2 norm, and compute the magnitude of the determinant of the result. This normalized determinant, which can be shown to be no less than 1, is sometimes used as a measure of the error behavior of the equation solution (the larger the value, the poorer the behavior).

(e) Repeat steps a–d if

$$A = \begin{bmatrix} 3 & 2 \\ 4 & 3 \end{bmatrix} \quad \text{and} \quad b = \begin{bmatrix} 5 \\ 7 \end{bmatrix}$$

so the correct solution remains $x_1 = x_2 = 1$.

(f) For the two equations (parts a–d vs. part e), note the actual error magnification factor for the given error and the behavior of the predictors of that behavior: condition number, final scaled pivot, and normalized determinant.

6.14 Show that if b is exact and A is in error by δA in the equation $Ax = b$, then $\|\delta x\|/\|x + \delta x\| \leq \text{cond}(A)(\|\delta A\|/\|A\|)$ for any norm.

6.15 Let P be the vector space of all first degree polynomials, $ax + b$, for $x \in [0, 1]$. Let T be the linear operator mapping P onto P by $T(ax + b) = (b - a)x + (a + b)$.

(a) Using the $\mathcal{L}_\infty$ norm in both the input space and the output space, find $\text{cond}_\infty(T)$.

(b) Assume we wish to apply T to polynomials in the input space P that have relative error, using the $\mathcal{L}_\infty$ norm, bounded by 10^{-10}. Give a bound on the relative error in the output polynomial for any such input polynomial.

6.16 A quadratic approximation function $\hat{f}$ which exact-matches at $x = 0$, 1, and -1 can be written either in the difference form

$$\hat{f}(x) = y_0 + xf[0, 1] + x(x - 1)f[0, 1, -1] \qquad [1]$$

or in the polynomial form

$$\hat{f}(x) = a_0 + a_1x + a_2x^2 \qquad [2]$$

(a) A set of linear equations whose solution gives a_0, a_1, a_2 is obviously

$$a_0 = y_0$$
$$a_0 + a_1 + a_2 = y_1$$
$$a_0 - a_1 + a_2 = y_{-1}$$

Compute the ℓ_∞ condition number of the transformation that takes any $\mathbf{y} = (y_0, y_1, y_{-1})$ to a corresponding $\mathbf{a} = (a_0, a_1, a_2)$.

(b) Combining expressions [1] and [2], one *can* (but you don't need to!) readily show that

$$a_0 = y_0$$
$$a_1 = f[0, 1] - f[0, 1, -1]$$
$$a_2 = f[0, 1, -1],$$

where the $f[\quad]$'s are, of course, really functions of the data ($\mathbf{y}$). If you computed a_0, a_1, a_2 from these relationships, would the condition number of this transformation from $\mathbf{y}$ to $\mathbf{a}$ be the same as for part a? Explain why or why not, in 15 words or less.

6.17 Consider the operator $T(c, d) = \int_c^d f(x)\, dx$ for some function of f which is a parameter of this problem.

(a) Is T linear?

(b) Given that $|\varepsilon_c| \leqslant 10^{-5}$ and $|\varepsilon_d| \leqslant 10^{-6}$, give a bound on the magnitude of the propagated error in $T(c, d)$.

(c) If instead of the information in part b you are given that the relative error in $(c,d)^T$ is bounded by 10^{-7} according to the ℓ_1 norm, give a bound on the relative error in $T(c, d)$ at some particular point $(c, d)^T$ according to the absolute value norm if $f(x) = x^2$ and $d > c > 0$.

CHAPTER SEVEN

Compound Methods: Accumulation of Error

Most operators are not simply applications of a basic operator of arithmetic or approximation. Rather they involve the application of a number of component operators. For example, see Figure 7–1, in which it is illustrated how the calculation of the arithmetic expression

$$y = \frac{u^2 + a}{u^2 + c} \qquad [1]$$

consists of the application of 1 multiplication operator, 2 addition operators, and 1 division operator, applied in a somewhat complex way in that the output from a given operator is in some cases input to more than one following operator. In general, we can consider any compound operator as subdivided into other operators, and these in turn may be subdivided further.

7.1 Error Generation by Successive Operators

A simple but common form of a compound operator involves the successive application of component operators (see Figure 7–2). In this type of operator the output from T_i is simply the input to T_{i+1}.

Successive operators appear in the composition of functions and evaluation of some arithmetic expressions, for example, $y = (a \times b + c) \times d + e$. They also appear when $T_1 = T_2 = \cdots = T_n$, that is, in the repetitive application of a single operator. We have seen that this repetition can be either recurrent or iterative, and these special cases of successive operators will be treated in Chapters 8 and 9. Here

211

FIGURE 7–1

A Compound Operator for an Arithmetic Expression

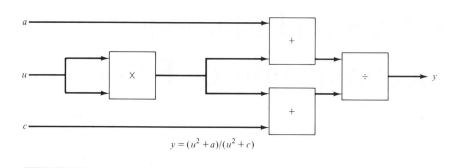

$$y = (u^2 + a)/(u^2 + c)$$

we will assume that the operators T_i are not necessarily the same, and we will have in mind the situation where n is not too large, perhaps a few tens or less. The error analysis for successive operators will be useful not only in its own right but also to provide the basis for extension to the case of more complicated compound operators.

In this section we will be interested in generated error only. That is, we will assume that the inputs to the compound operator are correct. Propagated error due to these inputs will need to be added to the generated error to produce the overall error. This matter will be covered in Section 7.3.

In the situation with successive operators illustrated in Figure 7–2 generated error arises as follows (see Figure 7–3). T_1 generates an error $\varepsilon_{T_1}^{\text{gen}}$, which is added to $T_1(x)$. This sum is then input to T_2, which propagates the error input to it and then adds the error it generates, $\varepsilon_{T_2}^{\text{gen}}$.

FIGURE 7–2

A Compound Operator T Produced by the Successive Operators T_1, T_2, ..., T_n

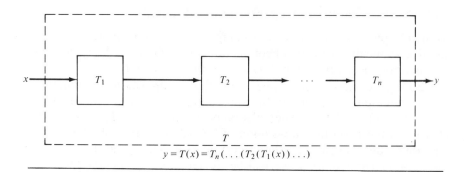

$$y = T(x) = T_n(\ldots(T_2(T_1(x)))\ldots)$$

This process continues, concluding with T_n propagating the error input to it followed by the addition of $\varepsilon_{T_n}^{\text{gen}}$. Algebraically Figure 7–3 is expressed

$$y^* = T_n(\cdots (T_2(T_1(x) + \varepsilon_{T_1}^{\text{gen}}) + \varepsilon_{T_2}^{\text{gen}}) \cdots) + \varepsilon_{T_n}^{\text{gen}} \qquad [2]$$

We wish to express ε_y in terms of the $\varepsilon_{T_i}^{\text{gen}}$ and the operators T_i. Furthermore, we wish to bound both the absolute and the relative error in y^* given corresponding bounds on the errors generated by the individual component operators.

If all of the operators T_i are linear, equation 2 can be rewritten as

$$
\begin{aligned}
y^* = {} & T_n(\cdots (T_2(T_1(x))) \cdots) \\
& + \sum_{i=2}^{n} T_n(\cdots (T_i(\varepsilon_{T_{i-1}}^{\text{gen}})) \cdots) + \varepsilon_{T_n}^{\text{gen}}
\end{aligned} \qquad [3]
$$

If we adopt the convention that $T_n T_{n-1} \cdots T_k$ for $k > n$ is the identity operator, equation 3 can be further rewritten

$$y^* = y + \sum_{i=2}^{n+1} T_n(\cdots (T_i(\varepsilon_{T_{i-1}}^{\text{gen}})) \cdots) \qquad [4]$$

so we have

$$\varepsilon_y^{\text{gen}} = \sum_{i=2}^{n+1} T_n(\cdots (T_i(\varepsilon_{T_{i-1}}^{\text{gen}})) \cdots) \qquad [5]$$

This equation states that the error in the output of the successive application of operators is the sum of the effects of the individual generated errors, where each generated error is propagated by all the operators following the one which generated it.

If not all the operators T_i are linear and the output of each operator is a scalar, the partial derivative approach gives a reasonable approximation. As we have seen it for input errors, this approach states that the output error is the sum of the effects of all the errors contributing to it, where the effect of an error is given approximately by its product with the partial derivative of the result with respect to the value with

FIGURE 7–3
Error Generation and Propagation by Successive Operators

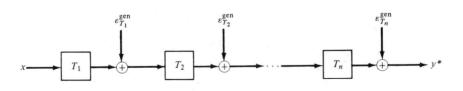

that error. The same relation can be shown to hold (see Problem 7.2) when the errors are generated by component operators.

Let

$$z_i = T_i(T_{i-1}(\cdots (T_1(x)) \cdots)) \tag{6}$$

the result of the application of the first i of the successive operators. Let b_i bound $|\varepsilon_{T_i}^{\text{gen}}|$. Then we have said that

$$\varepsilon_y^{\text{gen}} \approx \sum_{i=1}^{n} \varepsilon_{T_i}^{\text{gen}} \frac{\partial T(x)}{\partial z_i} \tag{7}$$

From this equation we can find a bound on $|\varepsilon_y^{\text{gen}}|$:

$$b_y^{\text{gen}} \approx \sum_{i=1}^{n} b_i \left| \frac{\partial T(x)}{\partial z_i} \right| \tag{8}$$

Note that for $i = n$, $z_i = T(x)$ so $\partial T(x)/\partial z_n = 1$. By its definition

$$z_{i+1} = T_{i+1}(z_i) \tag{9}$$

Thus

$$\frac{\partial T}{\partial z_i} = \frac{\partial T}{\partial z_{i+1}} \frac{\partial z_{i+1}}{\partial z_i} = \frac{\partial T}{\partial z_{i+1}} \frac{\partial T_{i+1}(z_i)}{\partial z_i} \tag{10}$$

so

$$\frac{\partial T}{\partial z_i} = \prod_{j=i+1}^{n} \frac{\partial T_j(z_{j-1})}{\partial z_{j-1}} \tag{11}$$

Using this in equation 8 produces

$$b_{T(x)}^{\text{gen}} = b_n + \sum_{i=1}^{n-1} b_i \prod_{j=i+1}^{n} \left| \frac{\partial T_j(z_{j-1})}{\partial z_{j-1}} \right| \tag{12}$$

Equation 12 gives the bound on the magnitude of the absolute error in the result in terms of bounds on the absolute generated errors at each step and the partial derivatives of the component operators with regard to their arguments. Note that the partial derivative of T_{i+1} is evaluated at its immediate argument, $z_i(x)$. A bound for relative error is obtained by using the relation

$$b_u = |u| r_u \tag{13}$$

for all absolute error bounds in equation 12, producing

$$r_{T(x)}^{\text{gen}} = \frac{1}{|T(x)|} \left(|z_n| r_n + \sum_{i=1}^{n-1} |z_i| r_i \prod_{j=i+1}^{n} \left| \frac{\partial T_j(z_{j-1})}{\partial z_{j-1}} \right| \right) \tag{14}$$

As an example consider the computation of $y = \log(\sin(x) + 2)$ at $x = \pi/6$. Then $z_1 = \sin(x)$, $z_2 = z_1 + 2$, and $z_3 = \log(z_2)$. Thus $dz_2/dz_1 = 1$, and $dz_3/dz_2 = 1/z_2$. These derivatives must be evaluated at $z_1 = \sin(\pi/6) = 0.5$, $z_2 = 0.5 + 2 = 2.5$. Let b_1 be the error generated in evaluating $\sin(\pi/6)$, b_2 be the error generated by adding 2 to this result, and b_3 be the error generated by evaluating $\log(2.5)$ (b_1 and b_3 will involve both arithmetic and approximation error). We see from equation 12 that

$$b_y^{\text{gen}} = b_3 + (1/2.5)b_2 + 1(1/2.5)b_1 = b_3 + 0.4b_2 + 0.4b_1 \qquad [15]$$

A less tight bound than that produced by the partial derivative approach can be produced using an error magnification approach. This approach must be used if the z_i are not scalars and only bounds on the norms of the errors generated in applying the T_i are given. We will develop our results for relative error using sensitivity as the magnification factor. The development for absolute error is parallel except that the magnification factor will be the operator norm.

The approach parallels the development above with scalar z_i using partial derivatives. If r_i is a bound on the magnitude of the relative error generated by the application of T_i to z_{i-1}, the effect of this error on the relative error in the result is bounded by $r_i S_{z_i}(T_n T_{n-1} \cdots T_{i+1})$, and thus, paralleling equation 8, we have

$$r_{T(x)}^{\text{gen}} = \sum_{i=1}^{n} r_i S_{z_i}(T_n T_{n-1} \cdots T_{i+1}) \qquad [16]$$

But for tiny errors

$$
\begin{aligned}
S_{z_j}(T_n T_{n-1} \cdots T_{j+2} T_{j+1}) &\leq S_{z_j}(T_{j+1}) S_{T_{j+1(z_j)}}(T_n T_{n-1} \cdots T_{j+2}) \\
&= S_{z_j}(T_{j+1}) S_{z_{j+1}}(T_n T_{n-1} \cdots T_{j+2})
\end{aligned}
\qquad [17]
$$

(see Problem 7.3), which parallels equation 10. Therefore

$$S_{z_i}(T_n \cdots T_{i+1}) \leq \prod_{j=i+1}^{n} S_{z_{j-1}}(T_j) \qquad [18]$$

which parallels equation 11. Using this in equation 16, we arrive at the parallel to equation 12:

$$
\begin{aligned}
r_{T(x)}^{\text{gen}} &\leq r_n + \sum_{i=1}^{n-1} r_i \prod_{j=i+1}^{n} S_{z_{j-1}}(T_j) \\
&= (\cdots (((r_1 S_{z_1}(T_2) + r_2) S_{z_2}(T_3) + r_3) S_{z_3}(T_4) + r_4) \\
&\quad \cdots + r_{n-1}) S_{z_{n-1}}(T_n) + r_n
\end{aligned}
\qquad [19]
$$

This result involving sensitivities or operator norms is even closer to the result involving partial derivatives, developed above, than might be apparent. It was shown in Chapter 6 that for n-vectors the operator norm and sensitivity involve partial derivatives. Note also that a less tight bound than that given by equation 19 can be obtained by using condition numbers in place of the corresponding sensitivities in this equation.

7.2   Error Generation by General Compound Operators

7.2.1   ARBITRARY OPERATORS

In the previous section, for every operator the result is directly input to just one succeeding operator and is indirectly input to all succeeding operators. However, in a general computation neither of these properties may hold for all operators. For example, consider evaluating

$$y = a(\log(uv + b))^2 + c \log(uv + b) + duv \qquad [20]$$

The computation, using the nested polynomial approach, would proceed as in Figure 7–4. Note that (a) the intermediate result duv is not used at all by any of the operators but the final addition, despite the fact that its computation may precede many of them, and (b) both the intermediate result uv and the intermediate result $\log(uv + b)$ are used as input to two later operators.

A graph such as that shown in Figure 7–4 can be constructed for any computation. We will call this the *calculation graph* for a computation. Let us number the n operators in our calculation graph from T_1 to T_n and give the name z_i to the result of the ith operator T_i, where the indices i are assigned in any way such that the inputs to an operator have lower indices than the output (see Figure 7–5). Thus with n operators $z_n = y$, the final output. For $1 \leq i \leq n$ the operator T_i generates an error ε_i in z_i. The effect of ε_i on y is given by multiplying ε_i by $\partial y / \partial z_i$, that is,

$$\varepsilon_y^{\text{gen}} \approx \sum_{i=1}^{n} \frac{\partial y}{\partial z_i} \varepsilon_i \qquad [21]$$

If $|\varepsilon_i| \leq b_i$, for $1 \leq i \leq n$, then we see that

$$b_y^{\text{gen}} = \sum_{i=1}^{n} \left| \frac{\partial y}{\partial z_i} \right| b_i \qquad [22]$$

Note that in calculating $\partial y / \partial z_i$, y is thought of as a function of z_i and

FIGURE 7–4
Calculation Graph for the Computation of $a(\log(uv + b))^2 + c \log(uv + b) + duv$

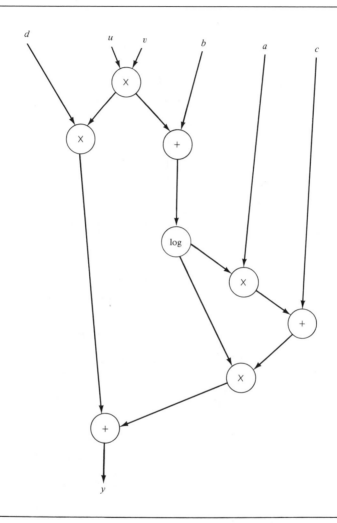

the inputs which contribute to operators other than just those leading to z_i. For example, with the calculation graphed in Figure 7–5,

$$z_4 = \log(uv + b) \qquad [23]$$

and $\partial y / \partial z_4$ is computed from

$$y = az_4^2 + cz_4 + duv \qquad [24]$$

FIGURE 7–5
Labeled Calculation Graph for $a(\log(uv + b))^2 + c \log(uv + b) + duv$

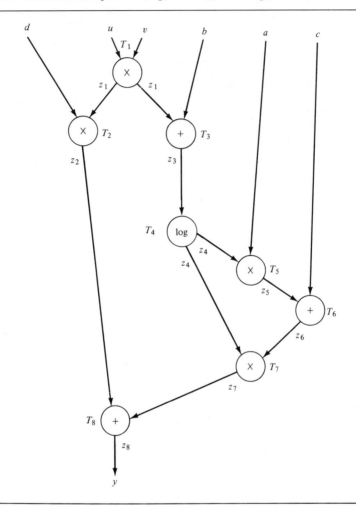

so

$$\frac{\partial y}{\partial z_4} = 2az_4 + c \qquad [25]$$

In particular,

$$\frac{\partial y}{\partial z_n} = \frac{\partial y}{\partial y} = 1 \qquad [26]$$

Thus, equation 22 can be rewritten

$$b_y^{\text{gen}} = b_n + \sum_{i=1}^{n-1} \left| \frac{\partial y}{\partial z_i} \right| b_i \qquad [27]$$

Applying equation 27 to the function graphed in Figure 7–5 produces the following result:

1. $y = a(\log(z_1 + b))^2 + c \log(z_1 + b) + dz_1$

$$\frac{\partial y}{\partial z_1} = \frac{2a \log(z_1 + b) + c}{z_1 + b} + d$$

2. $y = a(\log(uv + b))^2 + c \log(uv + b) + z_2$

$$\frac{\partial y}{\partial z_2} = 1$$

3. $y = a(\log(z_3))^2 + c \log(z_3) + duv$

$$\frac{\partial y}{\partial z_3} = \frac{2a \log(z_3) + c}{z_3}$$

4. $y = az_4^2 + cz_4 + duv$

$$\frac{\partial y}{\partial z_4} = 2az_4 + c$$

5. $y = (z_5 + c) \log(uv + b) + duv$

$$\frac{\partial y}{\partial z_5} = \log(uv + b)$$

6. $y = z_6 \log(uv + b) + duv$

$$\frac{\partial y}{\partial z_6} = \log(uv + b)$$

7. $y = z_7 + duv$

$$\frac{\partial y}{\partial z_7} = 1 \qquad\qquad [28]$$

Thus,

$$
\begin{aligned}
b_y^{\text{gen}} = {} & |(2a \log(uv + b) + c)/(uv + b) + d|b_1 \\
& + b_2 + |(2a \log(uv + b) + c)/(uv + b)|b_3 \qquad [29]\\
& + |2a \log(uv + b) + c|b_4 + |\log(uv + b)|(b_5 + b_6) \\
& + b_7 + b_8
\end{aligned}
$$

In general z_i is an input to some subset of $\{T_j | i < j \leq n\}$. Let S_i be the set of indices j for which z_i is an input to T_j:

$$\frac{\partial y}{\partial z_i} = \sum_{j \in S_i} \frac{\partial y}{\partial z_j} \frac{\partial z_j}{\partial z_i} = \sum_{j \in S_i} \frac{\partial y}{\partial z_j} \frac{\partial T_j}{\partial z_i} \qquad [30]$$

Here z_j is considered a function T_j of its inputs only. Thus in our example $S_4 = \{5, 7\}$,

$$z_5 = a z_4 \qquad [31]$$

and

$$z_7 = z_6 z_4 \qquad [32]$$

so equation 30, with $i = 4$, produces

$$\frac{\partial y}{\partial z_4} = \frac{\partial y}{\partial z_5} \frac{\partial z_5}{\partial z_4} + \frac{\partial y}{\partial z_7} \frac{\partial z_7}{\partial z_4} = \frac{\partial y}{\partial z_5} a + \frac{\partial y}{\partial z_7} z_6 \qquad [33]$$

Equation 27 can thus be rewritten

$$b_y^{\text{gen}} = b_n + \sum_{i=1}^{n-1} \left| \sum_{j \in S_i} \frac{\partial y}{\partial z_j} \frac{\partial z_j}{\partial z_i} \right| b_i \qquad [34]$$

In equation 34 each $\partial y / \partial z_j$ for $j \neq n$ can in turn be rewritten using equation 30 so that ultimately we arrive at a formula involving only derivatives of the form $\partial z_k / \partial z_j$, where z_j is a direct input to T_k. That is, if there are m_i paths from the ith to the nth node and the lth of these paths consists of s_l steps from $i = j_{1l}$ to j_{2l}, from j_{2l} to j_{3l}, ..., and from $j_{s_l-1,l}$ to $j_{s_l l} = n$, then

$$b_y^{\text{gen}} = b_n + \sum_{i=1}^{n-1} b_i \left| \sum_{\substack{\text{all paths } l \\ \text{from } i\text{th to} \\ n\text{th node}}} \prod_{\substack{\text{all steps,} \\ j \text{ to } k, \text{ on} \\ \text{the } l\text{th path}}} \frac{\partial z_k}{\partial z_j} \right|$$

$$= b_n + \sum_{i=1}^{n-1} b_i \left| \sum_{l=1}^{m_i} \prod_{k=1}^{s_l} \frac{\partial z_{j_{k+1,l}}}{\partial z_{j_{kl}}} \right| \qquad [35]$$

An example will clarify this complex result. Consider the function graphed in Figure 7–5. For each z_i for $1 \leq i \leq n - 1$, we compute the partial derivative $\partial z_j / \partial z_i$ for each T_j to which z_i is a direct input.

$$z_1: \quad \frac{\partial z_2}{\partial z_1} = d, \quad \frac{\partial z_3}{\partial z_1} = 1 \qquad\qquad z_5: \quad \frac{\partial z_6}{\partial z_5} = 1$$

$$z_2: \quad \frac{\partial z_8}{\partial z_2} = 1 \qquad\qquad\qquad\qquad z_6: \quad \frac{\partial z_7}{\partial z_6} = z_4$$

$$z_3: \quad \frac{\partial z_4}{\partial z_3} = \frac{1}{z_3} \qquad\qquad\qquad\quad z_7: \quad \frac{\partial z_8}{\partial z_7} = 1 \qquad [36]$$

$$z_4: \quad \frac{\partial z_5}{\partial z_4} = a, \quad \frac{\partial z_7}{\partial z_4} = z_6$$

For each z_i, for $1 \le i \le n - 1$, we list all the paths from z_i to z_n. There is one path from z_7 to z_8; one path from z_6 to z_8, via z_7; one path from z_5 to z_8, via z_6 and z_7; two paths from z_4 to z_8, via z_7 and via z_5, z_6, and z_7; two paths from z_3 to z_8, from z_3 to z_4 and then via either path from z_4 to z_8; one path from z_2 to z_8, directly; and three paths from z_1 to z_8, via z_2 and via z_3 and z_4 and thence via either path from z_4 to z_8.

We use these results to compute

$$
\begin{aligned}
b_y^{\text{gen}} = b_8 &+ b_7 \left|\frac{\partial z_8}{\partial z_7}\right| + b_6 \left|\frac{\partial z_8}{\partial z_7}\frac{\partial z_7}{\partial z_6}\right| + b_5 \left|\frac{\partial z_8}{\partial z_7}\frac{\partial z_7}{\partial z_6}\frac{\partial z_6}{\partial z_5}\right| \\
&+ b_4 \left|\frac{\partial z_8}{\partial z_7}\left(\frac{\partial z_7}{\partial z_4} + \frac{\partial z_7}{\partial z_6}\frac{\partial z_6}{\partial z_5}\frac{\partial z_5}{\partial z_4}\right)\right| + b_3 \left|\frac{\partial z_8}{\partial z_7}\left(\frac{\partial z_7}{\partial z_4} + \frac{\partial z_7}{\partial z_6}\frac{\partial z_6}{\partial z_5}\frac{\partial z_5}{\partial z_4}\right)\frac{\partial z_4}{\partial z_3}\right| \\
&+ b_2 \left|\frac{\partial z_8}{\partial z_2}\right| + b_1 \left|\frac{\partial z_8}{\partial z_2}\frac{\partial z_2}{\partial z_1} + \frac{\partial z_8}{\partial z_7}\left(\frac{\partial z_7}{\partial z_4} + \frac{\partial z_7}{\partial z_6}\frac{\partial z_6}{\partial z_5}\frac{\partial z_5}{\partial z_4}\right)\frac{\partial z_4}{\partial z_3}\frac{\partial z_3}{\partial z_1}\right| \\
= b_8 &+ b_7 + b_6|z_4| + b_5|z_4| + b_4|z_6 + z_4 a| + b_3|(z_6 + z_4 a)/z_3| \qquad [37] \\
&+ b_2 + b_1|d + (z_6 + z_4 a)/z_3| \\
= b_8 &+ b_7 + (b_6 + b_5)|\log(uv + b)| \\
&+ (b_4 + b_3/|uv + b|)|2a \log(uv + b) + c| \\
&+ b_2 + b_1|d + (2a \log(uv + b) + c)/(uv + b)|
\end{aligned}
$$

Notice that the terms are computed in reverse order to take advantage of the fact that a path from z_i to z_n will often include as subpaths the paths from z_j to z_n for $j > i$, and thus the multiplier of b_i will involve terms including factors which have multiplied b_j for $j > i$.

The approach just discussed, which produces an error bound using equation 35, gives the same bound as that given by equation 27. The approach given by equation 27 has the advantage of being somewhat more direct, but that given by equation 35 has the advantage of involving derivatives of an operator's output only with respect to its direct inputs.

7.2.2 ARITHMETIC ERROR

The evaluation of arithmetic expressions is just a special case of the compound operators discussed in section 7.2.1. The operators are restricted to $+$, $-$, $*$, and $\div$, so the error generated by an operator T_i with result z_i is bounded by $R|z_i|$. Here R is the bound, given in section 5.3, on the relative error generated by an arithmetic operator:

$$
R = \frac{1}{B^{n'-1}} \qquad [38]
$$

where B is the base of the floating-point number system used and n' is the number of digits in the fraction. That is,

$$b_i = R|z_i| \quad \text{for} \quad 1 \le i \le n \tag{39}$$

so from equation 22

$$b_y^{\text{arith}} = R \sum_{i=1}^{n} |z_i| \left| \frac{\partial y}{\partial z_i} \right| \tag{40}$$

Note that unary minus has not been included as an operation because it generates no error and can therefore be combined with a preceding or following operator.

As an example, consider error generation by the expression given in equation 1, $y = (u^2 + a)/(u^2 + c)$. As shown in the calculation graph in Figure 7–6, errors are generated in the multiplication to calculate u^2, in the addition to calculate $u^2 + a$, in the addition to calculate $u^2 + c$, and in the division to calculate $(u^2 + a)/(u^2 + c)$. As shown in Figure 7–6, we let

$$z_1 = u^2, \quad z_2 = u^2 + a, \quad z_3 = u^2 + c, \quad \text{and} \quad z_4 = \frac{u^2 + a}{u^2 + c} \tag{41}$$

FIGURE 7–6
Calculation Graph for $y = (u^2 + a)/(u^2 + c)$

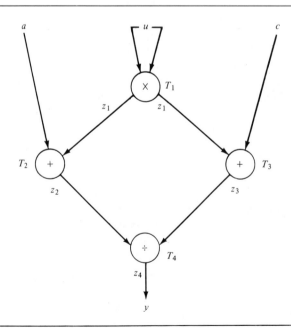

Then

$$\frac{\partial y}{\partial z_1} = \frac{\partial}{\partial z_1}\left(\frac{z_1 + a}{z_1 + c}\right) = \frac{c - a}{(z_1 + c)^2} = \frac{c - a}{(u^2 + c)^2}$$

$$\frac{\partial y}{\partial z_2} = \frac{\partial}{\partial z_2}\left(\frac{z_2}{u^2 + c}\right) = \frac{1}{u^2 + c}$$

$$\frac{\partial y}{\partial z_3} = \frac{\partial}{\partial z_3}\left(\frac{u^2 + a}{z_3}\right) = -\frac{u^2 + a}{z_3^2} = -\frac{u^2 + a}{u^2 + c^2}$$

[42]

$$\frac{\partial y}{\partial z_4} = \frac{\partial y}{\partial y} = 1$$

Thus

$$b_y^{\text{arith}} = R\left|(u^2)\left(\frac{c - a}{(u^2 + c)^2}\right)\right| + R\left|(u^2 + a)\frac{1}{u^2 + c}\right|$$

$$+ R\left|(u^2 + c)\frac{u^2 + a}{(u^2 + c)^2}\right| + R\left|\frac{u^2 + a}{u^2 + c}\right|$$

[43]

$$= \frac{R}{|u^2 + c|}\left(\left|\frac{u^2(c - a)}{u^2 + c}\right| + 3|u^2 + a|\right)$$

Alternatively the step-by-step analysis given by equation 35 can be used. Since only four operators are possible, we can list the partial derivatives required.

$$z_j = z_k + z_\ell: \qquad \frac{\partial z_j}{\partial z_k} = 1, \qquad \frac{\partial z_j}{\partial z_\ell} = 1 \qquad\qquad [44]$$

$$z_j = z_k - z_\ell: \qquad \frac{\partial z_j}{\partial z_k} = 1, \qquad \frac{\partial z_j}{\partial z_\ell} = -1 \qquad\qquad [45]$$

$$z_j = z_k z_\ell: \qquad \frac{\partial z_j}{\partial z_k} = z_\ell, \qquad \frac{\partial z_j}{\partial z_\ell} = z_k \qquad\qquad [46]$$

$$z_j = \frac{z_k}{z_\ell}: \qquad \frac{\partial z_j}{\partial z_k} = \frac{1}{z_\ell}, \qquad \frac{\partial z_j}{\partial z_\ell} = -\frac{z_k}{z_\ell^2} \qquad\qquad [47]$$

From the example graphed in Figure 7–6, equation 35 gives

$$b_y^{\text{arith}} = R|z_4| + R|z_3|\left|\frac{\partial z_4}{\partial z_3}\right| + R|z_2|\left|\frac{\partial z_4}{\partial z_2}\right|$$

$$+ R|z_1|\left|\frac{\partial z_4 \partial z_3}{\partial z_3 \partial z_1} + \frac{\partial z_4 \partial z_2}{\partial z_2 \partial z_1}\right|$$

[48]

The z_i, for $2 \leq i \leq 4$, in terms of their inputs are given by

$$z_4 = \frac{z_2}{z_3}$$

$$z_3 = z_1 + c \tag{49}$$

$$z_2 = z_1 + a$$

so the partial derivatives required, reverse ordered by operation, are

$$z_3: \quad \frac{\partial z_4}{\partial z_3} = -\frac{z_2}{z_3^2}$$

$$z_2: \quad \frac{\partial z_4}{\partial z_2} = \frac{1}{z_3} \tag{50}$$

$$z_1: \quad \frac{\partial z_3}{\partial z_1} = 1, \quad \frac{\partial z_2}{\partial z_1} = 1$$

Thus,

$$\begin{aligned}
b_y^{\text{arith}} &= R|z_4| + R|z_3|\,|-z_2/z_3^2| + R|z_2|\,|1/z_3| \\
&\quad + R|z_1|\,|-z_2/z_3^2 + 1/z_3| \\
&= R|(u^2 + a)/(u^2 + c)| + R|(u^2 + a)/(u^2 + c)| \\
&\quad + R|(u^2 + a)/(u^2 + c)| + R|u^2(c - a)/(u^2 + c)^2| \\
&= \frac{R}{|u^2 + c|}(|u^2(c - a)/(u^2 + c)| + 3|u^2 + a|)
\end{aligned} \tag{51}$$

the same result as given by equation 43.

To summarize, arithmetic error is analyzed using a calculation graph, the fact that T_i generates an error bounded by $R|z_i|$, and either partial derivatives of the result with respect to each z_i (see equation 40) or the step-by-step approach (see equation 35) with the partial derivatives limited to those of the forms given by equations 44–47.

7.2.3 REPRESENTATION ERROR

Until now, we have been implicitly assuming not only that the inputs are correct, since only error generation is being analyzed, but also that there is no error generated in representing the inputs. If this is not the case, a bound on the representation error propagated into the result must be added to the generated error bound due to arithmetic or other operators to produce the overall generated error:

$$b_y^{\text{gen}} = b_y^{\text{gen by operators}} + b_y^{\text{rep}} \tag{52}$$

Since for each input x_i the error generated by representation is bounded by $R|x_i|$, a bound on the effect of representation error on the final result y is given by

$$b_y^{rep} = \sum_{\substack{i \ni: \\ x_i \text{ has} \\ \text{representation error}}} R|x_i| \left| \frac{\partial y}{\partial x_i} \right| \qquad [53]$$

For example, in the function $y = (u^2 + a)/(u^2 + c)$, if all of u, a, and c are given as real numbers and thus error is generated in their computer representation,

$$\begin{aligned} b_y^{rep} &= R|u| \, |(u^2 + c)(2u) - (u^2 + a)(2u)|/(u^2 + c)^2 \\ &\quad + R|a|/|u^2 + c| + R|c| \, |u^2 + a|/(u^2 + c)^2 \qquad [54] \\ &= R(2u^2 \, |c - a| + |c(u^2 + a)|)/(u^2 + c)^2 \\ &\quad + R|a/(u^2 + c)| \end{aligned}$$

This must be added to the arithmetic error bound given in equation 43 or 51 to produce the overall computation error — that generated by computing y given the real inputs u, a, and c.

7.3 Overall Error

In Chapter 5 we showed that

$$\varepsilon_y = \varepsilon_y^{comp} + \varepsilon_y^{approx} + \varepsilon_y^{prop} \qquad [55]$$

$$b_y = b_y^{comp} + b_y^{approx} + b_y^{prop} \qquad [56]$$

and

$$r_y = r_y^{comp} + r_y^{approx} + r_y^{prop} \qquad [57]$$

We defined ε^{comp} as the error due to arithmetic and representation of inputs, that is, the error generated by the operator that takes the inputs before representation, assumed correct, and carries out the desired computation, except that it uses computer representation and arithmetic rather than real representation and arithmetic. As discussed in Section 7.2.3, if the computation is considered to be a two-step process, representation of inputs followed by arithmetic, the computation error is the sum of the error generated by the arithmetic and that generated by the input representation and propagated by the arithmetic:

$$\varepsilon_y^{comp} = \varepsilon_y^{arith} + \varepsilon_y^{rep} \qquad [58]$$

Therefore,

$$\varepsilon_y = \varepsilon_y^{\text{rep}} + \varepsilon_y^{\text{arith}} + \varepsilon_y^{\text{approx}} + \varepsilon_y^{\text{prop}} \qquad [59]$$

$$b_y = b_y^{\text{rep}} + b_y^{\text{arith}} + b_y^{\text{approx}} + b_y^{\text{prop}} \qquad [60]$$

and

$$r_y = r_y^{\text{rep}} + r_y^{\text{arith}} + r_y^{\text{approx}} + r_y^{\text{prop}} \qquad [61]$$

The difficulty is that with compound operators it is not clear what to call certain errors. For example, the error due to approximation in one component operator is then propagated by operators which follow. In the result is this error approximation error or propagated error? A similar confusion occurs for arithmetic error and representation error. Such confusion is unimportant as long as each source of error is thought of in only one category. To illustrate this point and at the same time show a full-blown error analysis of a class of real numerical algorithms, let us treat the issue of bounding the error associated with various numerical integration methods and of choosing among these methods.

Consider a class of methods for evaluating

$$I = \int_a^c f(x) \, dx$$

where each method is of the form

$$\hat{I} = h \sum_{i=0}^{n} w_i f(x_i) \qquad [62]$$

with

$$x_0 = a, \qquad x_n = c, \qquad h = \frac{c - a}{n} \qquad [63]$$

and

$$x_i = a + ih \qquad (1 \leqslant i \leqslant n) \qquad [64]$$

Assume that n, c, and a, and thus h are fixed, that f is given analytically, and that the same subroutine for computing f at any argument is given to all of the methods, so the only difference among the methods are the values of the weights w_i. Assume further that all of the methods to be compared have no approximation error for polynomials up to some degree m, at least 1. Finally, assume that $c > a$ and that $f(x)$ and $f'(x)$ are quite smooth and $f(x)$ does not change sign on $[a, c]$.

The representation error is that due to computer representation of the inputs and constants, which here are a, c, n, and the w_i. Being a small integer, n can be assumed to have no representation error. The w_i are normally the quotient of two small integers, so we will write

$$w_i = \frac{\beta_i}{\gamma_i} \qquad [65]$$

We will assume that β_i and γ_i have no representation error and will decide that the error due to dividing γ_i into β_i will be included under arithmetic error and thus not here. Representation of a and c generates errors bounded by $R|a|$ and $R|c|$, respectively, so

$$
\begin{aligned}
b_I^{\text{rep}} = {} & R|a|\left|\frac{\partial \hat{I}}{\partial a}\right| + R|c|\left|\frac{\partial \hat{I}}{\partial c}\right| \\
= {} & R|a|\left|\frac{\partial h}{\partial a}\sum_{i=0}^{n} w_i f(x_i) + h\sum_{i=0}^{n} w_i \frac{\partial f(x_i)}{\partial x_i}\frac{\partial x_i}{\partial a}\right| \\
& + R|c|\left|\frac{\partial h}{\partial c}\sum_{i=0}^{n} w_i f(x_i) + h\sum_{i=0}^{n} w_i \frac{\partial f(x_i)}{\partial x_i}\frac{\partial x_i}{\partial c}\right|
\end{aligned}
\tag{66}
$$

Since

$$
-\frac{\partial h}{\partial a} = \frac{\partial h}{\partial c} = \frac{1}{n}
\tag{67}
$$

$$
\frac{\partial x_i}{\partial a} = 1 + i\frac{\partial h}{\partial a} = 1 - \frac{i}{n}
\tag{68}
$$

and

$$
\frac{\partial x_i}{\partial c} = i\frac{\partial h}{\partial c} = \frac{i}{n}
\tag{69}
$$

we have

$$
\begin{aligned}
b_I^{\text{rep}} = {} & R|a|\left|\frac{1}{n}\sum_{i=0}^{n} w_i f(x_i) + h\sum_{i=0}^{n} w_i f'(x_i)\left(-1 + \frac{i}{n}\right)\right| \\
& + R|c|\left|\frac{1}{n}\sum_{i=0}^{n} w_i f(x_i) + h\sum_{i=0}^{n} w_i f'(x_i)\frac{i}{n}\right|
\end{aligned}
\tag{70}
$$

Let us move on to the propagated error. The inputs a, c, and n are assumed to have no error, so they contribute nothing to b_y^{prop}. Error in the w_i is being handled under arithmetic error. As for the $f(x_i)$ values, each such value will have error from four sources: (1) representation error in x_i; (2) approximation error associated with the algorithm for evaluating f; (3) error generated by the calculations of the subroutine for evaluating f; and (4) representation error for any constants used in the subroutine for evaluating f. We have included error of type 1 in b_I^{rep}, so it should not be included here. Errors of types 2, 3, and 4 could be included in b_I^{approx}, b_I^{arith}, and b_I^{rep}, respectively, but for those errors it is convenient to consider the $f(x_i)$ as inputs to the numerical integration algorithm, each with a single error bound b_i associated with these three sources combined. Thus we will include errors from these three sources not in other categories but as propagated error. The partial derivative approach applied to equation 62 produces

$$b_I^{\text{prop}} = \sum_{i=0}^{n} |hw_i| b_i \qquad [71]$$

Since approximation error in evaluating the $f(x_i)$ has been included in b_I^{prop}, the only source of approximation error remaining is that due to the numerical integration formula. For any given method

$$b_I^{\text{approx}} = \alpha(c - a)h^{m+1} \max_{\phi \in [a,c]} |f^{(m+1)}(\phi)| \qquad [72]$$

where α and m depend on the method.

Assuming that errors associated with evaluating the $f(x_i)$ have been handled elsewhere, b_I^{arith} results from (1) the calculation of h (see equation 63); (2) the calculation of the w_i (see equation 65); (3) the multiplication of the w_i and $f(x_i)$ values; (4) the summation of the $w_i f(x_i)$ values; and (5) the multiplication of the sum by h. Thus if

$$v_i \overset{\Delta}{=} w_i f(x_i) \qquad [73]$$

and

$$s_k \overset{\Delta}{=} \sum_{i=0}^{k} w_i f(x_i) \qquad (1 \leq k \leq n), \qquad [74]$$

the partial sums produced if the terms of the sum in equation 62 are added in order of index, then

$$
\begin{aligned}
b_I^{\text{arith}} = {} & \left| \frac{\partial \hat{I}}{\partial h} \right| b_h^{\text{arith}} + \sum_{i=0}^{n} \left| \frac{\partial \hat{I}}{\partial w_i} \right| b_{w_i}^{\text{arith}} \\
& + \sum_{i=0}^{n} \left| \frac{\partial \hat{I}}{\partial v_i} \right| b_{w_i f(x_i)}^{\text{mult}} + \sum_{i=1}^{n} \left| \frac{\partial \hat{I}}{\partial s_i} \right| b_{s_i}^{\text{plus}} + b_I^{\text{mult by } h}
\end{aligned}
\qquad [75]
$$

The two arithmetic operations in computing h may each generate an error, with the result that

$$b_h^{\text{arith}} = \frac{R(c - a)}{n} + Rh = 2Rh \qquad [76]$$

The remaining bounds required in equation 75 are for computations consisting of a single arithmetic operation, so each bound is given by R times the result of the operation. Thus

$$
\begin{aligned}
b_I^{\text{arith}} = {} & 2Rh \left| \sum_{i=0}^{n} w_i f(x_i) \right| + Rh \sum_{i=0}^{n} |w_i f(x_i)| \\
& + Rh \sum_{i=0}^{n} |w_i f(x_i)| + Rh \sum_{i=1}^{n} \left| \sum_{k=0}^{i} w_k f(x_k) \right| + R|\hat{I}|
\end{aligned}
$$

$$= 3Rh \left| \sum_{i=0}^{n} w_i f(x_i) \right| + 2Rh \sum_{i=0}^{n} |w_i f(x_i)|$$

$$+ Rh \sum_{i=1}^{n} \left| \sum_{k=0}^{i} w_k f(x_k) \right| \tag{77}$$

Summing the bounds given by equations 70, 71, 72, and 77 gives b_I^{overall}. Comparison of different numerical integration methods involves comparing their b_I^{overall} values. Certain definitions and approximations will ease this comparison.

We define $\bar{f}$, $\bar{f}_{(i)}$, $\bar{f'}$, and $\bar{\bar{f'}}$ by

$$\sum_{i=0}^{n} w_i f(x_i) = \bar{f} \sum_{i=0}^{n} w_i \tag{78}$$

$$\sum_{k=0}^{i} w_k f(x_k) = \bar{f}_{(i)} \sum_{k=0}^{i} w_k \tag{79}$$

$$\sum_{i=0}^{n} w_i f'(x_i) = \bar{f'} \sum_{i=0}^{n} w_i \tag{80}$$

and

$$\sum_{i=0}^{n} i w_i f'(x_i) = \bar{\bar{f'}} \sum_{i=0}^{n} i w_i \tag{81}$$

Since $f(x)$ and $f'(x)$ are assumed to be smooth, $\bar{f}$, the $\bar{f}_{(i)}$, $\bar{f'}$, and $\bar{\bar{f'}}$ do not change much from method to method, so we will approximate them as constant across methods. Furthermore, since $f(x)$ is assumed not to change sign in $[a, c]$, we can define $\bar{\bar{f}}$ by

$$\sum_{i=0}^{n} |w_i f(x_i)| = |\bar{\bar{f}}| \sum_{i=0}^{n} |w_i| \tag{82}$$

where $\bar{\bar{f''}}$ does not change much from method to method. Finally assuming that the error bound in $f(x_i)$ changes little with i, we can define $\bar{b}$ by

$$\sum_{i=0}^{n} |h w_i| b_i = \bar{b} h \sum_{i=0}^{n} |w_i| \tag{83}$$

where $\bar{b}$ does not change much from method to method. Substituting these forms and the relation

$$\sum_{i=0}^{n} |w_i| = \sum_{i=0}^{n} (|w_i| - w_i) + \sum_{i=0}^{n} w_i = 2 \sum_{\substack{i \ni: \\ w_i < 0}} |w_i| + \sum_{i=0}^{n} w_i \tag{84}$$

into b_I^{rep}, b_I^{prop}, and b_I^{arith} gives

$$b_I^{\text{rep}} \approx R|a| \left| (\bar{f}/n) \sum_{i=0}^{n} w_i - h\bar{f}' \sum_{i=0}^{n} w_i + (h/n)\bar{\bar{f}}' \sum_{i=0}^{n} iw_i \right|$$

$$+ R|c| \left| (\bar{f}/n) \sum_{i=0}^{n} w_i + (h/n)\bar{\bar{f}}' \sum_{i=0}^{n} iw_i \right| \qquad [85]$$

$$b_I^{\text{prop}} \approx \bar{b}h \sum_{i=0}^{n} |w_i| = 2\bar{b}h \sum_{\substack{i\ni: \\ w_i<0}} |w_i| + \bar{b}h \sum_{i=0}^{n} w_i \qquad [86]$$

and

$$b_I^{\text{arith}} \approx 3Rh|\bar{f}| \left| \sum_{i=0}^{n} w_i \right| + 2Rh|\bar{\bar{f}}| \sum_{i=0}^{n} |w_i| + Rh \sum_{i=1}^{n} |\bar{f}_{(i)}| \left| \sum_{k=0}^{i} w_k \right|$$

$$= 3Rh|\bar{f}| \left| \sum_{i=0}^{n} w_i \right| + 2Rh|\bar{\bar{f}}| \sum_{i=0}^{n} w_i + 4Rh|\bar{f}| \sum_{\substack{i\ni: \\ w_i<0}} |w_i| \qquad [87]$$

$$+ \sum_{i=1}^{n} |\bar{f}_{(i)}| \left| \sum_{k=0}^{i} w_k \right|$$

We note that $\sum_{i=0}^{n} w_i$ and $\sum_{i=0}^{n} iw_i$ are the same for all methods which have no approximation error for polynomials of degree 0 or 1, since they are the formulas produced by integrating $f(x) = 1/h$ and $f(x) = (x - a)/h$, respectively. Furthermore we may reasonably assume that $\sum_{k=0}^{i} w_k \approx i/n$ for all methods. Therefore, the only terms in b_I^{overall} which differ significantly among methods are $2\bar{b}h\sum_{i\ni:w_i<0}|w_i|$ from b_I^{prop}, $4Rh|\bar{f}| \sum_{i\ni:w_i<0}|w_i|$ from b_I^{arith}, and $\alpha(c - a)h^{m+1} \max_{\phi\in[a,c]}|f^{(m+1)}(\phi)|$ from b_I^{approx}.

We see that two properties are to be avoided:

1. $\sum_{i\ni:w_i<0}|w_i|$ is large, making the arithmetic error and the errors propagated from the $f(x_i)$ evaluations larger than necessary, and
2. $\alpha h^{m+1}\max_{\phi\in[a,c]}|f^{(m+1)}(\phi)|$ is large, making the approximation error large.

The first property can be avoided by using only formulas with positive weights. It can be shown that the Newton-Cotes rules based on polynomials of degree greater than 7 all have some negative w_i, so these rules should be avoided on these grounds alone. Furthermore, considerations of approximation error would lead us to avoid integration methods, like the high-order Newton-Cotes rules, for which αh^{m+1} is not small enough to counteract the fact that $|f^{(m+1)}(\phi)|$ is usually large for large m. With our previous result that even-order Newton-Cotes rules are preferable to odd-order rules (see Section 5.4.2), we see that the only

Newton-Cotes rules which remain candidates for use are those of order 0 (in the midpoint form), 2, 4, and 6, with the last somewhat suspect from the point of view of approximation error.

It should be noted that there exist families of integration methods with only positive w_i such that the approximation error decreases monotonically with m. These families are quite attractive. We will see such a family, produced by the method called Romberg integration, in Section 7.4.2.

We return now to the general question of error analysis of complicated methods and in particular to an approach that avoids confusion of errors among the four categories. We saw in the above example that it was useful to put off the analysis of the error in the evaluation of the $f(x_i)$ to a separate step and to consider the $f(x_i)$ as inputs. It often avoids confusion to carry this approach further by subdividing any given operator into only a few component operators and temporarily considering the inputs to these operators as the inputs to the problem, with error bounds to be calculated later but now considered as input error. These operators can then be further subdivided and analyzed, and the operators into which they have been subdivided can be still further subdivided and analyzed, and so on. This approach can be best clarified by an example.

Assume we wish to compute

$$y = (u/v + w)^2 f(u, v, w) \tag{88}$$

and we choose to approximate $f(u, v, w)$ by $\hat{f}(u, v, w)$ such that the approximation relative error is bounded in magnitude by 10^{-3} and the relative error made in evaluating $\hat{f}$ (including the representation of u, v, and w) is bounded in magnitude by 10^{-5}. Assume further that the relative error in $[u \; v \; w]^T$ defined using the ℓ_∞ norm is bounded by 10^{-6}, that

$$S_{\begin{bmatrix} 10^{-5} \\ 10^{-4} \\ 10^{-1} \end{bmatrix}}(f) = 10$$

according to the ℓ_∞ norm, and that all operations are done on a hexadecimal floating-point computer with a 6-digit fraction. Assume we wish to bound the magnitude of the overall relative error in y at $u = 10^{-5}$, $v = 10^{-4}$, $w = 10^{-1}$.

We first note that y is the product of one term with arithmetic and representation error but no approximation error and another term with arithmetic, representation, and approximation error. It is therefore useful to write

$$y = g(u, v, w) f(u, v, w) = gf \tag{89}$$

where

$$g(u, v, w) = (u/v + w)^2 \tag{90}$$

We analyze equation 89 by thinking of it as a simple multiplication with two inputs g and f. There is no functional approximation involved in this multiplication and thus no approximation error. Nor is there representation error, as g and f are already computer represented. Thus we can write

$$r_y = r_y^{\text{arith}} + r_y^{\text{prop}} \qquad [91]$$

r_y^{arith} is the bound on the relative error generated by the multiplication, which we know to be $R = 16^{-5}$. By the rule for propagation of multiplication

$$r_y^{\text{prop}} = r_g + r_f \qquad [92]$$

We now must analyze r_g and r_f separately, first treating r_g.

Since g is an arithmetic expression involving no functional approximation,

$$r_g = r_g^{\text{arith}} + r_g^{\text{rep}} + r_g^{\text{prop}} \qquad [93]$$

The arithmetic error bound can be produced using the methods of Section 7.2.2, giving

$$r_g^{\text{arith}} = 3R + 2R|u/(u + vw)| \qquad [94]$$

evaluated at the given values of u, v, and w. The approach of Section 7.2.3 gives the effect of error in the representation of u, v, and w:

$$r_g^{\text{rep}} = R(|4u| + |2vw|)/|u + vw| \qquad [95]$$

The propagated error in g must be produced using the error magnification approach since bounds are given on the relative error in the vector rather than in the individual inputs:

$$r_g^{\text{prop}} = S_{\begin{bmatrix} 10^{-5} \\ 10^{-4} \\ 10^{-1} \end{bmatrix}} \frac{((u/v + w)^2)r}{\begin{bmatrix} u \\ v \\ w \end{bmatrix}} \qquad [96]$$

$r_{[u\ v\ w]^T}$, is given to be 10^{-6}, and the sensitivity required can be calculated by equations 84 and 136 of Chapter 6 as

$$\frac{\max(10^{-5}, 10^{-4}, 10^{-1})}{(u/v + w)^2} 2|u/v + w|(|1/v| + |u/v^2| + 1)$$

$$= 0.2(|1/v| + |u/v^2| + 1)/|u/v + w|$$

evaluated at the given values of u, v, and w.

Having completed the analysis of r_g, we move on to the analysis of r_f. We know that

$$r_f = r_f^{\text{arith}} + r_f^{\text{rep}} + r_f^{\text{approx}} + r_f^{\text{prop}} \qquad [97]$$

where $r_f^{\text{arith}} + r_f^{\text{rep}}$ and r_f^{approx} are given. By the error magnification approach, r_f^{prop} is simply the product of the given sensitivity of f at the inputs in question and the given relative error bound on the inputs.

By appropriate subdivision of the problem, the error analysis has become quite simple. At each step each overall error bound required was written as the sum of the corresponding bound for arithmetic, representation, approximation, and propagated errors. A representation error is calculated only if the particular operator being analyzed at that step involves initial inputs or constants in which representation error is made. An approximation error is calculated only if there is approximation in the particular operator being analyzed at a given step, not simply in a component operator. Propagated error is analyzed for inputs that may not be the ultimate inputs but only the input to the step being analyzed.

At any subdivision an operator may consist of recurrent or iterative repetition of some operator. The error generated by such an operator can be bounded by methods to be discussed in Chapters 8 and 9. One point about iterative methods should be mentioned here. In Chapter 9 we will analyze iterative methods as if an infinite number of iterative steps are carried out. More precisely, we will assume that enough iterations are carried out so that the accuracy is limited only by error generated in the iterative steps and the error propagated from the inputs. In real life we may stop the iteration sooner out of concern for efficiency. If this is the case, the nonconvergence error can be thought of as a sort of approximation error, which must be added to the arithmetic error to produce the overall generated error.

As we have seen, one use of the above technique is to compare two methods for solving the same problem, $y = T(x)$, in terms of their overall accuracy. For both methods $\varepsilon_y = \varepsilon_y^{\text{prop}} + \varepsilon_y^{\text{arith}} + \varepsilon_y^{\text{rep}} + \varepsilon_y^{\text{approx}}$. Since $\varepsilon_y^{\text{prop}}$ depends only on T and not on the approximation to T used by the method, it is necessary to compare the methods only on the basis of generated error, $\varepsilon_y^{\text{gen}} = \varepsilon_y^{\text{approx}} + \varepsilon_y^{\text{arith}} + \varepsilon_y^{\text{rep}}$. For example, in comparing the Gauss-Seidel and Gaussian elimination methods for solving the linear equations $Ax = b$, we can analyze the methods as if the inputs b and A were accurate. Only when we choose a particular method need we analyze the propagated error to obtain the overall error bound.

Of course, accuracy is not the only criterion by which a method is judged. Efficiency is also important. It is worth noting that, everything else being equal, increased efficiency is not only desirable for itself but because the arithmetic error tends to increase with the number of arithmetic operations carried out.

7.4 Strategies for Reduction of Error

It is often the case that we can produce an estimate of a particular value by two or more different methods. With these estimates we can usually produce an estimate of better accuracy than any of the original estimates. Intuitively, this idea has much in common with the idea of measuring a given value independently n times and then averaging these n estimates to get a more accurate estimate. Here, however, we will find an approximate relation between the errors in our estimates and use this to produce an estimate of higher accuracy. In all cases we will assume arithmetic error is negligible compared to approximation error.

7.4.1 AVERAGING IN SOLUTION OF DIFFERENTIAL EQUATIONS

For example, consider the modified-Euler–Heun predictor-corrector method for the solution of the differential equations $y' = f(x, y)$, $y(x_0) = y_0$. The predictor and the corrector are both estimates of y_{i+1}. Discussing only the approximation error in a step (assuming y_i and y_{i-1} are known exactly and error generated by representation, arithmetic operations, and evaluations of $f(x, y)$ is negligible compared to approximation error), we know from Section 5.4.1 that the error in the predictor is given by

$$\varepsilon^{(p)}_{y_{i+1}} = -(h^3/3)y_i''' + O(h^4) \tag{98}$$

and that the error in the corrector, assuming even the predictor is exact, is

$$\varepsilon^{(c)}_{y_{i+1}} = (h^3/12)y_i''' + O(h^4) \tag{99}$$

To this approximation error in the corrector we must add the error propagated from an error in the predictor to get the overall error, assuming still that y_i and y_{i-1}, the inputs to the full step, are correct. This error propagated from the predictor is obtained using the partial derivative approach, giving

$$\varepsilon^{(c)\text{prop from predictor}}_{y_{i+1}} = \varepsilon^{(p)}_{y_{i+1}} \frac{\partial y^{(c)}_{i+1}}{\partial y^{(p)}_{i+1}}$$

$$= \varepsilon^{(p)}_{y_{i+1}} \frac{\partial}{\partial y^{(p)}_{i+1}}\left(y_i + \frac{h}{2}(f(x_i, y_i) + f(x_{i+1}, y^{(p)}_{i+1}))\right) \tag{100}$$

$$= \varepsilon^{(p)}_{y_{i+1}} \frac{h}{2} \frac{\partial f(x_{i+1}, y^{(p)}_{i+1})}{\partial y^{(p)}_{i+1}}$$

From equation 98 we see $\varepsilon^{(p)}_{y_{i+1}}$ is $O(h^3)$, so the error in the corrector propagated from the predictor is $O(h^4)$. Adding this propagated error to the approximation error in equation 99 gives the overall error in the corrector, assuming the inputs to the step are correct:

$$\varepsilon^{(c)}_{y_{i+1}} = (h^3/12)\, y_i''' + O(h^4) \tag{101}$$

Note the importance of using a predictor with error of the same order as the corrector. Had we used one of lower order, say the Euler predictor, the effect of the error in the predictor on the corrector would be, from equation 100, of the same order as the approximation error, but the term added would involve a partial derivative. Not only would we rather not have to evaluate this derivative to determine the most significant term of the error in the corrector, but we will see that the appearance of this derivative would complicate the error reduction.

Rewriting equations 98 and 101, we have

$$y_{i+1}^{(p)} - y_{i+1} = -(h^3/3)\, y_i''' + O(h^4) \qquad [102]$$

and

$$y_{i+1}^{(c)} - y_{i+1} = (h^3/12)\, y_i''' + O(h^4) \qquad [103]$$

Subtracting equation 102 from equation 103 gives us

$$y_{i+1}^{(c)} - y_{i+1}^{(p)} = (5h^3/12)\, y_i''' + O(h^4) \qquad [104]$$

from which we can see that

$$(1/5)(y_{i+1}^{(c)} - y_{i+1}^{(p)}) = \varepsilon_{i+1}^{(c)} + O(h^4) \qquad [105]$$

Therefore we can produce y_{i+1} by subtracting the error in the corrector from the value of the corrector, producing

$$
\begin{aligned}
y_{i+1} = y_{i+1}^{(c)} - \varepsilon_{i+1}^{(c)} &= y_{i+1}^{(c)} - (1/5)(y_{i+1}^{(c)} - y_{i+1}^{(p)}) + O(h^4) \\
&= (4/5)y_{i+1}^{(c)} + (1/5)y_{i+1}^{(p)} + O(h^4)
\end{aligned}
\qquad [106]
$$

Equation 106 tells us that the weighted average of the corrector and the predictor, with weights 4/5 and 1/5, has error of order 4 in h, that is, of one higher order than the error in the predictor and the corrector. For small h this weighted average of the separate estimates of y_{i+1} given by $y_{i+1}^{(c)}$ and $y_{i+1}^{(p)}$ is more accurate than either estimate. It is therefore recommended that the calculation of a predictor and a corrector in a step of the modified-Euler–Heun method be followed by a *mop-up*, a very short operation computing

$$y_{i+1}^{(m)} = (4/5)y_{i+1}^{(c)} + (1/5)y_{i+1}^{(p)} \qquad [107]$$

A step consisting of prediction, correction, and mop-up will have greater accuracy by one order of h than one consisting of prediction and correction only.

Notice the source of the weights in the mop-up equation. They involve the constants multiplying the low-order term in the expressions for the error in the predictor and corrector, namely $-1/3$ and $1/12$, respectively. The weight of the corrector in the mop-up equation is a fraction whose numerator is the negative of the ratio of these two constants and whose denominator is 1 plus the numerator. The weight of the predictor is the complement of the weight of the corrector. Note that the factor that was $1/12$ would not have been a constant had the Euler predictor been used, since this would have caused the low-order term in the corrector to include a partial derivative.

Mop-up can also be done on the predictor at the $(i + 1)$th step, using $(y_i^{(c)} - y_i^{(p)})$, but it is not very useful as it would cause the corrector to improve only in the $O(h^4)$ term, that is, it will not affect the low-order term in the corrector error. Moreover, the mop-up of the corrector would have to be done with the unmopped-up predictor, because the term used for mop-up must have an error of the same order in h as the corrector. Therefore, mop-up on the predictor is normally not done.

The fact that we can calculate an estimate of the error in the corrector from the values of the predictor and corrector leads to the important additional ability of being able to adjust the step size as we proceed in the solution of a differential equation, without having to do the significant extra computation of the value of a high derivative of the solution across the relevant interval. The appropriate value of h for succeeding steps is determined by comparing the estimate to the error tolerance for one step. If it is too large, h can be halved, or if it is too small, h can be doubled to decrease the number of steps, thus increasing efficiency and decreasing arithmetic error. Of course, the error in the mopped-up value of y_{i+1}, which we actually use, is less than the error in the corrector, but we have no easy way to estimate the error in the mopped-up value, and keeping the error in the corrector appropriately small will usually keep the error in the mopped-up value even smaller.

Doubling h is no problem if we have saved a few previous y_j values. If the formula calls for y_{i-k}, with the new double value of h we use y_{i-2k} instead. On the other hand, halving h requires new values of y at arguments midway between those of values already computed. To obtain these, we can move forward with a step of size half that of the previous h using a method that requires no previous values of y_j until we have enough to get started again with the predictor–corrector method. A class of methods, called Runge-Kutta methods, requiring no previous values of y_j will be discussed shortly.

What is the effect of halving h? For the modified-Euler–Heun method with mop-up, we have seen that $\varepsilon_{i+1}^{(m)} = O(h^4)$, so halving h multiplies the error by $(1/2)^4 = 1/16$. Similarly, doubling h multiplies the error in a step by 16. But halving the interval doubles the number of steps to get to x_{goal}. Assuming that the error propagation into $y(x_{\text{goal}})$ of error generated at each step is such that the error in $y(x_{\text{goal}})$ is approximately equal to the sum of the errors generated at the steps, we see that halving h causes the effect on $\varepsilon_{y(x_{\text{goal}})}$ of the errors in the remaining steps to be multiplied by $1/8$ and doubling h multiplies this effect by 8. Furthermore, halving h requires the tolerance of the error in each step to be halved and doubling h requires this tolerance to be doubled. Therefore, if we have a tolerance on the error per step for the particular value of h used, when our error estimate gets greater than this tolerance, we should halve h and halve our tolerance. When the error estimate becomes smaller than 1/8 of the tolerance, we should double h and double the tolerance.

The modified-Euler–Heun method with mop-up has an error of order h^4. Experiment shows this not to be small enough: such an approximation error requires too many steps and thus too low efficiency and too large an arithmetic error generation. We can get a lower approximation error per step without severely increasing the work per step or suffering propagation misbehavior (to be discussed in Chapter 8) if we use methods such that the predictor and corrector have error of order h^5 and thus the mopped-up value has error of order h^6. The predictor-corrector method of choice, the Adams method, consisting of the Adams-Bashforth predictor and the Adams-Moulton corrector, followed by its mop-up (see Problem 7.11c for its mop-up formula), has this property. We will not discuss this method beyond noting that $y_{i+1}^{(p)}$ involves y_i, y_{i-1}, y_{i-2}, and y_{i-3} and $y_{i+1}^{(c)}$ involves $y_{i+1}^{(p)}$, y_i, y_{i-1}, and y_{i-2}. That is, the Adams method requires more previous y_j values than the modified-Euler–Heun method and therefore storage of more previous values to allow interval doubling and more time using a Runge-Kutta method for start-up or when interval halving is desired. Furthermore, interval doubling is indicated when the error estimate becomes $2^{-5} = 1/32$ times the tolerance, since the mop-up error is $O(h^6)$.

It remains for us to discuss how to compute start-up and half-interval values by a method requiring no previous values of y_j. These Runge-Kutta methods are also based on the notion of averaging different estimates to get an improved estimate for a desired value. In particular, we found in Section 3.3 that

$$y_{i+1} = y_i + \int_{x_i}^{x_{i+1}} y'(x)\, dx \qquad [108]$$

The integral in this equation can be rewritten as

$$\int_{x_i}^{x_{i+1}} y'(x)\, dx = (x_{i+1} - x_i)\overline{y_i'} = h\overline{y_i'} \qquad [109]$$

where $\overline{y_i'}$ is the *average value* of $y'(x)$ over $[x_i, x_{i+1}]$, producing

$$y_{i+1} = y_i + h\overline{y_i'} \qquad [110]$$

Thus we can take our objective in calculating y_{i+1} from y_i as determining the value of $\overline{y_i'}$. We will compute a number of estimates of y' in the interval $[x_i, x_{i+1}]$ and determine the appropriate average of these to agree with $\overline{y'_i}$ to as high a degree in h as possible. For example, we might compute one value of y' at x_i, two estimates of y' at $x_{i+1/2}$, and one estimate of y' at x_{i+1}. We begin with $y'_i = f(x_i, y_i)$. The first y' estimate at $x_{i+1/2}$ is given by $f(x_{i+1/2}, y^{(p)}_{i+1/2})$, where $y^{(p)}_{i+1/2}$ is obtained by applying Euler's method with step $h/2$ beginning at y_i. We can compute another estimate of $y'_{i+1/2}$ by reevaluating $f(x_{i+1/2}, y^{(p)}_{i+1/2})$ but here using a different predictor, again produced by Euler's method beginning at y_i with step $h/2$ but using the previously computed slope at $x_{i+1/2}$ (at the right end of the half-interval) in place of $f(x_i, y_i)$, the slope at the left end of the interval. We can compute the estimate of y' at x_{i+1} by computing $f(x_{i+1}, y^{(p)}_{i+1})$, with $y^{(p)}_{i+1}$ computed by Euler's method beginning at y_i and using step h but using the slope at the midinterval point computed at the previous step. We now have four values of y' symmetrically spaced in the interval in question. By appropriate expansion of everything in Taylor series (see Problem 7.15), we can choose weights of these slopes such that their weighted sum agrees with $\overline{y'_i}$ to as high a power in h as possible. Using this weighted average in place of $\overline{y'_i}$ in equation 110, we produce an estimate of y_{i+1} which has error $O(h^5)$:

$$\hat{y}_{i+1} = y_i + (h/6)(f(x_i, y_i) + 2f(x_{i+1/2}, y^{(p1)}_{i+1/2}) \qquad [111]$$
$$+ 2f(x_{i+1/2}, y^{(p2)}_{i+1/2}) + f(x_{i+1}, y^{(p)}_{i+1}))$$

where

$$y^{(p1)}_{i+1/2} = y_i + \frac{h}{2}f(x_i, y_i) \qquad [112]$$

$$y^{(p2)}_{i+1/2} = y_i + \frac{h}{2}f(x_{i+1/2}, y^{(p1)}_{i+1/2}) \qquad [113]$$

$$y^{(p)}_{i+1} = y_i + hf(x_{i+1/2}, y^{(p2)}_{i+1/2}) \qquad [114]$$

The above describes one Runge-Kutta method. By computing more slope estimates, Runge-Kutta methods with errors of order higher than 5 in h can be produced. Normally the method with error of order 5 or 6 is used for start-up or interval halving with the Adams predictor-corrector method. Of course, a Runge-Kutta method could be used by itself. In fact, doing so has an advantage in the propagation of approximation errors by later steps (stability, discussed in Chapter 8). However, the Runge-Kutta methods with error of order 5 or more in h require at least four evaluations of f per step as contrasted with two for predictor-corrector methods, so the predictor-corrector methods have an advantage in efficiency over Runge-Kutta methods. Therefore, except where the propagation over many steps of approximation errors in each step is an especially acute problem, or in higher dimensional equations where storage of the previous values of the y_j becomes prohibitive, a predictor-corrector method is used basically, with the Runge-Kutta method used only for start-up and interval halving.

7.4.2 RICHARDSON EXTRAPOLATION: ROMBERG INTEGRATION

Let us continue with examples of combining a number of estimates of a value to get an improved estimate of the value. Consider the problem of numerical integration, for which

$$I = \int_a^b f(x)\, dx \qquad [115]$$

One estimate of I can be computed by applying the composite trapezoidal rule with interval h, the result of which we will call $I_{0,h}$, where the first subscript indicates the number of estimate combinations that have contributed to the value and the second subscript indicates the smallest interval width used to produce the result. We can compute another estimate of I by applying the composite trapezoidal rule with interval $h/2$, producing $I_{0,h/2}$. But we know from Chapter 5 that

$$I_{0,h} - I = ((b - a)/12)h^2 f''(a) + O(h^3) \qquad [116]$$

and therefore that

$$\begin{aligned} I_{0,h/2} - I &= ((b - a)/12)(h/2)^2 f''(a) + O(h^3) \\ &= ((b - a)/48)h^2 f''(a) + O(h^3) \end{aligned} \qquad [117]$$

Just as with the predictor and corrector in our example in the early part of this section, we subtract equation 117 from equation 116 producing

$$\begin{aligned} I_{0,h} - I_{0,h/2} &= (3/48)(b - a)h^2 f''(a) + O(h^3) \\ &= 3(I_{0,h/2} - I) + O(h^3) \end{aligned} \qquad [118]$$

Solving for I produces

$$I = (4/3)I_{0,h/2} - (1/3)I_{0,h} + O(h^3) \qquad [119]$$

That is, if

$$\begin{aligned} I_{1,h/2} &\triangleq (4/3)I_{0,h/2} - (1/3)I_{0,h} \\ &= I_{0,h/2} + (1/3)(I_{0,h/2} - I_{0,h}) \end{aligned} \qquad [120]$$

$I_{1,h/2}$ has error of order h^3, one higher order in h than $I_{0,h}$ and $I_{0,h/2}$.

But the process of interval halving can be repeated. We can compute $I_{0,h/4}$ and combine it with $I_{0,h/2}$ by the same rule used to compute $I_{0,h/2}$, to produce

$$I_{1,h/4} = I_{0,h/4} + (1/3)(I_{0,h/4} - I_{0,h/2}) \qquad [121]$$

which has error of order h^3, but with a constant one-eighth the size of the constant in the error term of $I_{1,h/2}$, since $I_{1,h/4}$ was computed with half the value of h. That is, for some value k_3,

$$I_{1,h/2} - I = k_3 h^3 + O(h^4) \qquad [122]$$

and

$$I_{1,h/4} - I = k_3 (h/2)^3 + O(h^4) \qquad [123]$$

from which we can produce

$$\begin{aligned} I_{2,h/4} &\triangleq (8/7)I_{1,h/4} - (1/7)I_{1,h/2} \\ &= I_{1,h/4} + (1/7)(I_{1,h/4} - I_{1,h/2}) \end{aligned} \qquad [124]$$

which has error $O(h^4)$. The interval halving can be repeated once more, producing $I_{0,h/8}$, $I_{1,h/8}$, $I_{2,h/8}$, and

$$I_{3,h/8} \triangleq I_{2,h/8} + (1/15)(I_{2,h/8} - I_{2,h/4}) \qquad [125]$$

which has error $O(h^5)$.

For small h we should expect that of two estimates with the same second subscript (smallest interval involved) the one with the higher first subscript gives greater accuracy, and this is in fact the case. This notion of successive interval halving and taking weighted averages of the form

$$I_{i+1,h/2} \triangleq I_{i,h/2} + \frac{1}{2^{i+1} - 1}(I_{i,h/2} - I_{i,h}) \qquad [126]$$

is called *Richardson extrapolation* and is applicable quite generally to equal-interval methods (see Problem 7.19).

In the case of its application to the result of the trapezoidal rule the effect, after a slight modification, is even more powerful than is indicated above. The Taylor series for the trapezoidal rule error can be shown to be of the form

$$\varepsilon_{I_{0,h}} \triangleq I_{0,h} - I = c_2 h^2 + c_4 h^4 + c_6 h^6 + \cdots \qquad [127]$$

that is, the coefficients of the odd powers of h are all zero. This is due to the same effect as that discussed in Section 5.4.2 which causes the even order Newton-Cotes rules to have an error of one higher order than expected. The result is that when $I_{1,h/2}$ is calculated to eliminate the term in h^2 in the error the result has error of order 4 in h instead of order 3, as would happen in the general case. Thus a further interval halving will multiply the error by $1/16$, rather than $1/8$, so for the application of Richardson extrapolation to trapezoidal rule integration, we have

$$I_{i+1,h/2} = I_{i,h/2} + \frac{1}{4^{i+1} - 1}(I_{i,h/2} - I_{i,h}) \qquad [128]$$

Applying this relation with interval halving will cause the error to improve at a rate much faster than in normal Richardson extrapolation. This application of Richardson extrapolation to composite trapezoidal rule integration is called *Romberg integration*.

Let us clarify how Romberg integration is carried out (see also Program 7–1). We start the algorithm by evaluating

$$I_{0,b-a} = \frac{b-a}{2}(f(b) + f(a)) \qquad [129]$$

We then evaluate

$$I_{0,(b-a)/2} = (1/2)(I_{0,b-a} + (b-a)f((b+a)/2)) \qquad [130]$$

PROGRAM 7–1
Romberg Integration

```
{****************************************************************************}
{* Romberg integration to integrate f(x) from xmin to xmax                 *}
{* Given are                                                               *}
{*    f, the function to be integrated;                                    *}
{*    xmin and xmax, limits of integration;                                *}
{*    maxit, the maximum number of iterations (assumed <= 100); and        *}
{*    tol, error tolerance.                                                *}
{****************************************************************************}


FUNCTION Romberg(
    FUNCTION f(z:real): real;        {function to be integrated}
    xmin,xmax: real;                 {integration limits}
    maxit: integer;                  {maximum number of iterations}
    tol: real                        {error tolerance}
           ): real;

VAR
    h: real;                         {interval width}
    arg: real;                       {evaluation argument}
    rombrow: ARRAY[0..100] OF real;  {last row of table}
    newtab,old1,old2: real;          {temporary table entries}
    denom: real;                     {4, 3, 2 or 1}
    n: integer;                      { (xmax-xmin)/h }
    k: integer;                      {current table row}
    i,j: integer;


BEGIN

    {Initialize}
      k := 0;
      h := xmax-xmin;
      n := 1;
      {Simple trapezoidal rule}
        rombrow[0] := (h/2)*(f(xmax)+f(xmin));

    {Compute as many table rows as necessary}
```

PROGRAM 7–1 (*continued*)

```
REPEAT
  k := k + 1;
  old1 := rombrow[0];
  {Compute new trapezoidal rule for interval h/2}
    newtab := 0;
    arg := xmin + h/2;
    FOR j:=1 TO n DO
      BEGIN
        newtab := newtab + f(arg);
        arg := arg + h
      END;
    rombrow[0] := (old1 + h*newtab)/2;
  {Compute remainder of k-th row}
    denom := 0;
    FOR i:=1 TO k-1 DO
      BEGIN
        {Save value in previous row}
          old2 := rombrow[i];
        rombrow[i] := rombrow[i-1]+(rombrow[i-1]-old1)/(denom-1);
        old1 := old2;
        denom := denom + 4
      END;
    rombrow[k] := rombrow[k-1]+(rombrow[k-1]-old1)/(denom-1);
    {Set up next row computation}
      h := h/2;
      n := 2*n;
  UNTIL ((abs(rombrow[k] - rombrow[k-1]) < tol) OR (k = maxit));

  {Return final result}
    Romberg := rombrow[k]

END; {Romberg}
```

Note that the evaluation of the trapezoidal rule at half the interval does not require reusing any of the previously used function values (see equation 92 of Chapter 3). It can be computed as an average of the value of the trapezoidal rule for the previous interval and h times the sum of the data values with arguments halfway between all of the arguments of the previously used values:

$$I_{0,h/2} = \frac{1}{2}\left[I_{0,h} + h \sum_{i=1}^{(b-a)/h} f\left(a + \left(i - \frac{1}{2}\right)h\right)\right] \qquad [131]$$

Having computed $I_{0,b-a}$ and $I_{0,(b-a)/2}$, we compute

$$I_{1,(b-a)/2} = I_{0,(b-a)/2} + \frac{I_{0,(b-a)/2} - I_{0,b-a}}{3} \qquad [132]$$

Note the form used to compute $I_{i,h/2}$:

$$I_{i,h/2} = I_{i-1,h/2} + \frac{I_{i-1,h/2} - I_{i-1,h}}{4^i - 1} \qquad [133]$$

This form generates less arithmetic error than the form $(4^i/(4^i - 1))I_{i-1,h/2} + (1/(4^i - 1))I_{i-1,h}$ (see Problem 7.16). If the value $I_{1,(b-a)/2}$ is accurate enough, we stop the process. If not, we halve the interval once more, computing $I_{0,(b-a)/4}$ from equation 131 and applying equation 133 for $h = (b - a)/2$ and $i = 1$ to produce $I_{1,(b-a)/4}$. Reapplying equation 133 with $h = (b - a)/2$ and $i = 2$, we produce $I_{2,(b-a)/4}$. The overall process can be visualized as developing elements of the table in Figure 7–7 row by row. Saving only the previous row, we compute a new row and decide whether the diagonal element is accurate enough. If so, we stop; if not, we add another row.

How do we know whether a given value is accurate enough? Since we know that the process converges in both the horizontal and vertical directions, we can simply continue the process until two adjacent vertical or horizontal values agree to the number of significant digits required; then we stop.

As an illustration of the above algorithm, let us apply it to evaluate $\int_0^\pi \sin(x)\, dx$ (which has the value 2) with an error tolerance of 0.01. We start by computing $I_{0,\pi} = (\pi/2)(0 + 0) = 0$. We then compute $I_{0,\pi/2} = (1/2)(I_{0,\pi} + \pi \sin(\pi/2)) = 1.571$, and $I_{1,\pi/2} = 1.571 + (1.571 - 0)/3 = 2.094$. Since $|2.094 - 1.571| > 0.01$, we continue our computation process:

$$I_{0,\pi/4} = (1/2)(I_{0,\pi/2} + (\pi/2)(\sin(\pi/4) + \sin(3\pi/4))) = 1.895$$
$$I_{1,\pi/4} = 1.895 + (1.895 - 1.571)/3 = 2.004$$

and

$$I_{2,\pi/4} = 2.004 + (2.004 - 2.094)/15 = 1.999$$

FIGURE 7–7
Romberg Integration

h	$m:$ 0	1	2	3
$b - a$	•			
$\dfrac{b - a}{2}$	•	•		
$\dfrac{b - a}{4}$	•	•	•	$I_{m,h}$
$\dfrac{b - a}{8}$	•	•	•	•
$\vdots$				

Since $|1.999 - 2.004| < 0.01$, we stop with the approximation for the integral, 1.999. Note how much better this approximation is than the result (1.895) of the trapezoidal rule using the same number of evaluation arguments.

It might be surprising that we applied Richardson extrapolation to trapezoidal rule integration rather than Simpson's rule integration when the latter is known to be more accurate. The reason for this is that it turns out (see equation 93 of Chapter 3) that the values $I_{1,h}$ are the Simpson's rule results, computed by a slightly more efficient algorithm than was given in Chapter 3. It is not, however, the case that any of the Romberg integration results $I_{i,h}$ for $i > 1$ are equivalent to a Newton-Cotes rule.

Finally, it turns out that any result of Romberg integration is equivalent to a positively weighted sum of the function values at the tabular arguments. In Section 7.3 it was shown that integration rules with all positive weights are well behaved in regard to propagation of arithmetic errors and errors made in evaluating f. For all these reasons Romberg integration is a method of choice and is always used in preference to Newton-Cotes methods.

Before we leave Richardson extrapolation, of which Romberg integration is a special case, we should note that improvement can be expected from Richardson extrapolation only for values of h for which approximation error remains dominant. For trapezoidal integration this is true for the values of h required by common error tolerances, but it is not true, for example, below some relatively large value of h for numerical differentiation (see Problem 7-4). Thus we would expect Richardson extrapolation to be of limited use with numerical differentiation.

7.4.3 HORIZONTAL PATHS IN DIFFERENCE TABLES

Another example of combining more than one estimate of a value to produce an improved estimate is in equal interval polynomial exact-matching approximation. We have seen that for a value of x indicated by the arrow in Figure 7–8, the path drawn with a solid line is the preferred path. But the zigzag paths given by the dashed line and the dotted line are almost as good. We would therefore expect the appropriate average of the results given by two different paths would do better than that for a single path. This is true for a pair of paths that end at a different point in a column in question, but it is of course not the case for paths that end at the same point in the column in which we want to stop, because two paths that end at the same point give the same result. Analysis bears out our intuition that if we wish to end our path on the column of an even-order difference we should average the results of the paths given by the path starting at the tabular argument immediately

FIGURE 7–8
Zigzag Paths in a Difference Table

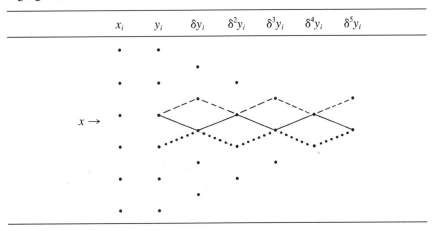

	x_i	y_i	δy_i	$\delta^2 y_i$	$\delta^3 y_i$	$\delta^4 y_i$	$\delta^5 y_i$

above (less than) the evaluation argument and going downward and the path starting at the tabular argument immediately below the evaluation argument and going upward; if we wish to end on an odd-order difference we should average the results of the paths beginning at the nearest tabular argument with one going upward and the other going downward. It turns out that when we do not know at which column we will end we should use the two paths starting at the closest tabular argument if the evaluation argument is closer to a tabular argument than a midinterval point and we should use two paths starting at adjacent tabular arguments and going toward each other if the opposite is the case.

What are the appropriate weights in the average? When the two path endings in question are one difference position apart in the same column of the difference table, with corresponding approximations $\hat{f}_1(x)$ through tabular points n through m and $\hat{f}_2(x)$ through tabular points $n + 1$ through $m + 1$,

$$\varepsilon_{\hat{f}_1(x)} = \frac{1}{(m - n + 1)!} \prod_{i=n}^{m} (x - x_i) f^{(m-n+1)}(\xi_1) \tag{134}$$

$$\text{for some} \quad \xi_1 \in [x_n, x_m]$$

and

$$\varepsilon_{\hat{f}_2(x)} = \frac{1}{(m - n + 1)!} \prod_{i=n+1}^{m+1} (x - x_i) f^{(m-n+1)}(\xi_2) \tag{135}$$

$$\text{for some} \quad \xi_2 \in [x_{n+1}, x_{m+1}]$$

for some values of m and n. We thus have

$$(x - x_{m+1})\varepsilon_{\hat{f}_1(x)} \approx (x - x_n)\varepsilon_{\hat{f}_2(x)} \tag{136}$$

so the weighted average given by equation 137 will have an error of one higher order than $\hat{f}_1$ and $\hat{f}_2$:

$$\hat{f}_{\text{avg}} = \frac{(x - x_n)\hat{f}_2 - (x - x_{m+1})\hat{f}_1}{(x - x_n) - (x - x_{m+1})} = \frac{(x - x_n)\hat{f}_2 - (x - x_{m+1})\hat{f}_1}{x_{m+1} - x_n} \tag{137}$$

But Problem 3.3a shows that the result given by equation 137 of linearly interpolating between $\hat{f}_1$ and $\hat{f}_2$ is precisely the same as carrying the exact-matching approximation to one further difference, the one which both paths would have in common.

This method of iterated linear interpolation is an inefficient way of doing polynomial exact-matching (see Problem 3.3b). Note, however, that

$$x - x_n \approx -(x - x_{m+1}) \approx \tfrac{1}{2}(x_{m+1} - x_n) \tag{138}$$

since we have assumed equal interval approximation and that points are added in order of their closeness to x, so x is about halfway between x_n and x_{m+1}, the extreme arguments used. Thus, $\tfrac{1}{2}(\hat{f}_2 + \hat{f}_1)$ can be expected to have a smaller error than either $\hat{f}_2$ or $\hat{f}_1$ [namely, $\tfrac{1}{2}(\varepsilon_{\hat{f}_2(x)} + \varepsilon_{\hat{f}_1(x)})$] when the two paths end at adjacent differences.

This averaging of two polynomial formulas of the same degree is said to correspond to a straight-line path where the path is the average between the two zigzag paths on which it is based. The formulas which result from these paths are the Stirling formula, in the case when the two paths being averaged begin at the same point, y_0:

$$\hat{f}(x) = y_0 + \binom{s}{1}\mu\delta y_0 + \frac{\binom{s}{2} + \binom{s+1}{2}}{2}\delta^2 y_0 + \binom{s+1}{3}\mu\delta^3 y_0$$

$$+ \frac{\binom{s+1}{4} + \binom{s+2}{4}}{2}\delta^4 y_0 + \cdots \tag{139}$$

$$= y_0 + s\mu\delta y_0 + \frac{s^2}{2!}\delta^2 y_0 + \frac{s(s^2 - 1)}{3!}\mu\delta^3 y_0 + \frac{s^2(s^2 - 1)}{4!}\delta^4 y_0$$

$$+ \frac{s(s^2 - 1)(s^2 - 4)}{5!}\mu\delta^5 y_0 + \frac{s^2(s^2 - 1)(s^2 - 4)}{6!}\delta^6 y_0 + \cdots$$

and the Bessel formula if the two paths being averaged start at y_0 and y_1:

$$\hat{f}(x) = \mu y_{1/2} + \frac{\binom{s}{1} + \binom{s-1}{1}}{2} \delta y_{1/2} + \binom{s}{2} \mu\delta^2 y_{1/2}$$

$$+ \frac{\binom{s}{3} + \binom{s+1}{3}}{2} \delta^3 y_{1/2} + \cdots$$

$$= \mu y_{1/2} + (s - \tfrac{1}{2})\delta y_{1/2} + \frac{s(s-1)}{2!} \mu\delta^2 y_{1/2} \qquad [140]$$

$$+ \frac{(s - \tfrac{1}{2})(s)(s-1)}{3!} \delta^3 y_{1/2}$$

$$+ \frac{(s+1)(s)(s-1)(s-2)}{4!} \mu\delta^4 y_{1/2}$$

$$+ \frac{(s - \tfrac{1}{2})(s+1)(s)(s-1)(s-2)}{5!} \delta^5 y_{1/2} + \cdots$$

where

$$s = \frac{x - x_0}{h} \qquad [141]$$

In both of these formulas we have used the notation

$$\mu\delta^i y_j \triangleq \frac{\delta^i y_{j+1/2} + \delta^i y_{j-1/2}}{2} \qquad [142]$$

The Stirling formula should be used if we wish to end on an odd-order difference or if we do not know which difference we will end on and $|x - x_0| \leq h/4$, where x_0 is the nearest tabular argument to the evaluation argument x. Otherwise the Bessel formula should be used, with x_0 chosen to be the largest tabular argument less than x.

7.4.4 AITKEN'S δ^2-ACCELERATION

Our final example of combining more than one estimate of a value to produce an improved estimate is in the case of iteration with so-called linear convergence to a fixed point ξ. Consider the Picard iteration

$$x_{i+1} = g(x_i) \qquad [143]$$

where g is differentiable at all points of interest. The objective of this iteration is to reach the fixed point ξ, so the error in x_i is

$$\varepsilon_i = x_i - \xi \qquad [144]$$

Considering step i given by equation 143, we have the normal situation that

$$\varepsilon_{i+1} = \varepsilon_{i+1}^{\text{gen}} + \varepsilon_{i+1}^{\text{prop}}. \qquad [145]$$

$\varepsilon_{i+1}^{\text{prop}}$ is the error due to the fact that $x_i \neq \xi$. Assuming that $\varepsilon_{i+1}^{\text{gen}}$, the error generated by evaluating g at x_i, is negligible compared to $\varepsilon_{i+1}^{\text{prop}}$, we have, using the partial derivative approach, the following:

$$\begin{aligned}
\varepsilon_{i+1} &= \frac{\partial g}{\partial x_i} \varepsilon_i + O(\varepsilon_i^2) = g'(x_i) \varepsilon_i + O(\varepsilon_i^2) \\
&= (g'(\xi) + O(\varepsilon_i)) \varepsilon_i + O(\varepsilon_i^2) = g'(\xi) \varepsilon_i + O(\varepsilon_i^2) \qquad [146] \\
&= k_1 \varepsilon_i + O(\varepsilon_i^2)
\end{aligned}$$

Since $g'(\xi)$ is a constant with respect to i, we have written it as k_1. This constant gives the factor by which error propagates from step to step. Clearly k_1 must be no greater than 1 in magnitude if convergence is to occur. If $k_1 \neq 0$, the convergence is called linear. Let us assume this is the case.

We note that equation 146 is the relation between the error in one estimate of ξ, x_{i+1}, and another estimate of ξ, x_i. It is just such a relation that we need to appropriately combine estimates to produce an improved estimate. As with previous developments of error reduction, we need another equation, here for ε_{i+2}. But equation 144 holds for $i + 1$ in place of i:

$$\varepsilon_{i+2} = k_1 \varepsilon_{i+1} + O(\varepsilon_{i+1}^2) \qquad [147]$$

Rewriting equations 146 and 147 produces

$$(x_{i+1} - \xi) = k_1(x_i - \xi) + O(\varepsilon_i^2) \qquad [148]$$

and

$$(x_{i+2} - \xi) = k_1(x_{i+1} - \xi) + O(\varepsilon_{i+1}^2) \qquad [149]$$

Since $O(\varepsilon_{i+1}^2) = O(\varepsilon_i^2)$, subtracting these equations produces

$$x_{i+2} - x_{i+1} = k_1(x_{i+1} - x_i) + O(\varepsilon_i^2) \qquad [150]$$

Solving equation 150 for k_1 and solving equation 149 for ξ using this value of k_1 produces

$$\xi = x_{i+2} - \frac{(x_{i+2} - x_{i+1})^2}{(x_{i+2} - x_{i+1}) - (x_{i+1} - x_i)} + O(\varepsilon_i^2) \qquad [151]$$

The improved estimate,

$$x_{i+2}^{\text{acc}} \triangleq x_{i+2} - \frac{(x_{i+2} - x_{i+1})^2}{(x_{i+2} - x_{i+1}) - (x_{i+1} - x_i)} \qquad [152]$$

can be rewritten in shorthand as

$$x_{i+2}^{\text{acc}} = x_{i+2} - \frac{(\delta x_{i+3/2})^2}{\delta^2 x_{i+1}} \qquad [153]$$

so the formula, due to Aitken, is often called δ^2-*acceleration*.

The method for obtaining the improved estimate of ξ given by x_{i+2}^{acc} was obtained in much the same way as the previous methods of obtaining improved estimates from more than one previous estimate: a relation was found between the various estimates ignoring all but the low-order error term, and then one solved for the desired value, here ξ. The result was an improved estimate with error of order of the ignored error terms in the original estimates.

For Picard iteration if x_i, x_{i+1}, and x_{i+2} are close enough to ξ, the accelerated value x_{i+2}^{acc} will be closer to ξ than x_i, x_{i+1}, and x_{i+2}. Thus, if x_{i+2}^{acc} is used as a new starting point for the Picard iteration, after two more iterations we will have another triple of estimates to which Aitken's δ^2-acceleration can be applied. But this new triple will be closer to ξ and therefore will have the property that the ignored terms in ε^2 will be even more negligible than before. That is, we can expect the ξ estimate from this new acceleration to be even better than the original accelerated value, and in fact repetition of this process will make the successive accelerated values converge very quickly to ξ. Thus, the approach in using Aitken's δ^2-acceleration is to start with initial values relatively close to ξ, apply the Picard iteration twice, accelerate, apply the Picard iteration twice to the accelerated result, accelerate, and so on until convergence to within the desired tolerance occurs.

In developing Aitken's δ^2-acceleration, nowhere did we assume that $g'(\xi)$, that is, the factor k_1, was less than 1 in magnitude. It is possible that the original Picard iteration was diverging rather than converging. The only thing we did assume was that for all three iterates used in the acceleration the $O(\varepsilon^2)$ terms in equations 148 and 149 were negligible compared to the $O(\varepsilon)$ terms. This will be true if x_i is appropriately close to ξ. Thus, Aitken's δ^2-acceleration can be used to cause a sequence diverging linearly from a fixed point to converge to the fixed point as well as to accelerate the convergence of a linearly converging sequence. It is clearly more dangerous to apply it to the former case because it is less clear whether the acceleration has been applied with appropriately small ε_i, but such application is possible.

7.5   Summary and Complements

Sections 7.1–7.3 of this chapter can be summarized as follows. Every error input to an operator or generated in the application of the operator produces a contribution to the error in the output of the operator. A bound on the magnitude of that contribution can be obtained by multiplying the bound on the input or generated error in question by the magnitude of the partial derivative of the final result with regard to the value in which the error in question initially appears. The overall error is bounded in magnitude by the sum of the bounds on the contributions.

It is important to include every contribution and to avoid including a given contribution more than once. If one approaches the problem in an organized way, one can ensure that each input or generated error is included and is included only once in a contribution to the overall error. In particular, it is easy to forget contributions of errors in representing constants, and it is easy to count more than once contributions of error in representing inputs. In the latter case one must decide whether this representation error is to be included in the input error, that is, whether the input is assumed to occur after the input is represented or whether the representation is considered to be part of the operator itself, in which case the representation error is a generated error. Either decision is reasonable. The partial derivative which multiplies the representation error is the same as that multiplying the input error since the error is in the same value.

Arithmetic errors depend on the algorithm used to do the computation. That is, a prescription as given by a program or the corresponding calculation graph is required, not just a formula. From the calculation graph the errors made can be identified and bounded and their propagation through later operations can be determined (see equation 35 or 40).

Approximation errors arise for every function other than $+$, $-$, $*$, and $\div$ in the calculation graph. These include evaluations by both direct and iterative methods. With direct methods the approximation error is associated with the approximating function used, and with iterative methods it is associated with incomplete convergence. In both cases there will be arithmetic as well as approximation error associated with the operator, and the sum of these will be the error generated by evaluating the function. This error will, of course, be propagated by operators following the function evaluation (see equation 27 or 35).

With the computing strategy (discussed in Section 7.4) of reducing error by using more than one estimate of a value, we have completed the list of computing strategies to be covered in this book. Remember that these strategies are as follows: approximate and operate, divide and conquer, recur, iterate, and combine multiple estimates to produce more accurate estimates.

PROBLEMS

7.1 Let A and B each be operators. Show that $\text{cond}(AB) \leq \text{cond}(A)\,\text{cond}(B)$.

7.2 Let x be a scalar and let T_i, for $i = 1, 2, \ldots, n$, be differentiable functions from scalars to scalars. Let

$$y = T(x) = T_n(T_{n-1}(T_{n-2}(\cdots (T_1(x)) \cdots)))$$

$$y^* = T_n^*(T_{n-1}^*(T_{n-2}^*(\cdots (T_1^*(x)) \cdots)))$$

$$z_i = T_i(T_{i-1}(\cdots (T_1(x)) \cdots))$$

Assuming that

$$\frac{\partial T_n^* T_{n-1}^* \cdots T_{i+1}^*(z_i)}{\partial z_i} \approx \frac{\partial T_n T_{n-1} \cdots T_{i+1}(z_i)}{\partial z_i}$$

for $i = 0, 1, \ldots, n-1$, show by adding and subtracting appropriate terms that

$$\varepsilon_y \approx \sum_{i=1}^{n} \varepsilon_{T_i}^{\text{gen}} \frac{\partial y}{\partial z_i}$$

7.3 Let y, y^*, and z_i, for $i = 1, 2, \ldots, n$, be defined as in Problem 7.1 except that $x = z_0$ and z_i is a vector in a vector space V_i, for $i = 0, 1, \ldots, n$. Assume that

$$S_{z_i}(T_n^* T_{n-1}^* \cdots T_{i+1}^*) \approx S_{z_i}(T_n T_{n-1} \cdots T_{i+1}).$$

(a) If r_i is a bound on the relative error in z_i generated by T_i, show that

$$r_y^{\text{gen}} \approx \sum_{i=1}^{n} r_i S_{z_i}(T_n T_{n-1} \cdots T_{i-1})$$

(b) Let P be an operator from vector space U to vector space V and let Q be an operator from vector space V to vector space W. Show that $S_x(QP) \leq S_x(P)S_{P(x)}(Q)$.

(c) Using the results of parts a and b, show that

$$r_{T(x)}^{\text{gen}} \lesssim r_n + \sum_{i=1}^{n-1} r_i \prod_{j=i+1}^{n} S_{z_{j-1}}(T_j)$$

7.4 Consider the formula for numerical differentiation $f'(x_0) = (f_1 - f_{-1})/(2h)$. You have shown the approximation error to be bounded by

$$b_{f_0'}^{\text{approx}} = \frac{h^2}{6} \max_{\xi \in [x_{-1}, x_1]} |f'''(\xi)|.$$

(a) Assume that $|\varepsilon_{f_1}| \leq b$ and $|\varepsilon_{f_{-1}}| \leq b$, that h has only representation error, and that arithmetic is done using floating-point numbers with 6-digit hexadecimal fractions. Show that the overall error in $f'(x_0)$ is bounded in magnitude by $b_{f_0'}^{\text{approx}} + b/h + 4 \times 16^{15}|f_0'|$.

(b) Assume that

$$b = 2.5 \times 16^{-4} \max_{\xi \in [x_{-1}, x_1]} |f'''(\xi)| = 60 \times 16^{-4}$$

and

$$f_0' \triangleq \frac{f_1 - f_{-1}}{2h} = 0.04 \quad \text{for all } h$$

What value of h minimizes the bound on the overall error magnitude?

(c) What is the relative error bound in f_0' corresponding to the "best" value of h found in part b? Note that this is 751 times the bound on the relative error due to arithmetic and representation, despite the fact that the approximation error and the propagated error can each be made as small as desired by choosing h appropriately (unfortunately not by the same h).

7.5 Consider the calculation $y = (e^{uv - u/v} + e^{u/v})/(e^{uv - u/v} - e^{u/v})$ as composed of the basic arithmetic operations and the exponentiation operator. Assume that no simplification is done. Assume that each arithmetic operator generates a relative error bounded by R and that the exponential operator generates a relative error bounded in magnitude by R_{exp}.

(a) Assume that the inputs u and v can be represented exactly. Give a bound on the magnitude of the absolute error generated by the computation.

(b) What must be added to the bound if representing u and v can generate error?

7.6 Assume we wish to calculate $\cos[(\pi/2)u^2]$ at $u = 1/\sqrt{3} \approx 0.577$.

(a) Give a bound on the magnitude of the propagated absolute error if b_u, the bound on the magnitude of the error in u, is equal to the maximum representation error in a 3-digit decimal floating-point computer.

(b) Assume we tabulate $\cos[(\pi/2)u^2]$ at $u = 0$, 1, and 2, producing

x	y
0	1
1	0
2	1

and thus the exact-matching approximation $1 - 2u + u^2$. Give a formula for a bound on the magnitude of the absolute approximational error at $u = 1/\sqrt{3}$. You need not evaluate the formula.
(c) Note that $(1 - 2u) + u^2 = 1 + u(u - 2) = (1 - u)^2$. Bound the magnitude of the generated arithmetic absolute error in each of these forms, assuming the computing is done on a 3-digit decimal floating-point computer with truncation and a guard digit.
(d) Assuming the best of the forms in part c is used, give a formula for a bound on the magnitude of the overall error in the result of evaluating $\cos[(\pi/2)u^2]$ at $1/\sqrt{3}$ by evaluating the approximation given above at $1/\sqrt{3}$ on the computer specified.

7.7 This whole problem has to do with the integration of continuous functions over the interval [2,4], the Simpson's rule operator which approximates this integration using tabular argument interval $h = 1$, and the relationship between these two operators. To be more precise, the following definitions hold: Let F be the set of functions $f(x)$ which are continuous for $x \in [2,4]$; R be the set of real numbers; L_3 be the set of 3-vectors with real elements [these 3-vectors will have elements which are the values of $f(x)$ at the tabular arguments $x = 2$, $x = 3$, and $x = 4$, respectively]; T be an operator from F to R for which $T(f)$ is defined to be $\int_2^4 f(x)\, dx$, that is, T is the integration operator in question; and Q be an operator from L_3 to R for which

$$Q\left(\begin{bmatrix} y_0 \\ y_1 \\ y_2 \end{bmatrix}\right) = \frac{1}{3}y_0 + \frac{4}{3}y_1 + \frac{1}{3}y_2$$

that is, if

$$y \equiv \begin{bmatrix} y_0 \\ y_1 \\ y_2 \end{bmatrix} = \begin{bmatrix} f(2) \\ f(3) \\ f(4) \end{bmatrix}$$

$Q(y)$ is the result of approximating $T(f)$ by applying Simpson's rule with $h = 1$.

In some parts of this problem, we will be dealing with the integration of a specific function $g(x) \in F$, where we define $g(x) = x^4$. We will also be dealing with the approximation to $T(g)$ given by Simpson's rule with $h = 1$, namely, $Q(z)$, where

$$z \equiv \begin{bmatrix} 2^4 \\ 3^4 \\ 4^4 \end{bmatrix} = \begin{bmatrix} 16 \\ 81 \\ 256 \end{bmatrix}$$

Note that

$$T(g) = \left. \frac{x^5}{5} \right|_2^4 = \frac{992}{5} = 198.4$$

and

$$Q(z) = \frac{1}{3}(16 + 4 \times 81 + 256) = \frac{596}{3} = 198.666...$$

In all but part a, when a norm is used on F it will be the $\mathcal{L}_1$ norm, and when a norm is used on either L_3 or R it will be the ℓ_1 norm. That is,

$$\text{for } f \in F, \quad \|f\| \equiv \int_2^4 |f(x)| dx;$$

$$\text{for } y \in L_3, \quad \|y\| \equiv |y_0| + |y_1| + |y_2|;$$

$$\text{and for } \alpha \in R, \quad \|\alpha\| \equiv |\alpha|.$$

Make sure you understand the above before you go on.

(a) Let $f \in F$, and let $N(f) \equiv |f(2)| + |f(3)| + |f(4)|$. Is N a norm on F? Show why or why not.

(b) Show that T is a linear operator.

(c) Assume we compute $T(f^*)$ instead of $T(f)$, where

$$f^*(x) = f(x) + \varepsilon(x) \quad \text{and} \quad \|\varepsilon(x)\|_1 \le 10^{-3}$$

Give a bound on the magnitude of the absolute error in $T(f^*)$. Note that

$$\|T\|_1 = \max_{f \in F} \frac{\|T(f)\|_1}{\|f\|_1} = \max_{f \in F} \frac{\left\| \int_2^4 f(x) dx \right\|_1}{\|f\|_1} = \max_{f \in F} \frac{\left| \int_2^4 f(x) dx \right|}{\int_2^4 |f(x)| dx}$$

Since

$$\frac{\left| \int_2^4 f(x) dx \right|}{\int_2^4 |f(x)| dx} \le 1$$

and is equal to 1 if $f(x) > 0$ for all $x \in [2,4]$, $\|T\| = 1$.

(d) Show that the sensitivity of T at $f(x) = g(x) = x^4$, using the $\mathscr{L}_1$ norm, is equal to 1.

(e) Using the ℓ_1 norm, give the sensitivity of Q at

$$y = \begin{bmatrix} 16 \\ 81 \\ 256 \end{bmatrix}$$

[Note in passing that T at g and Q at z do not have approximately equal sensitivity even though $Q(x) = T(f)$.]

(f) In previous parts we were interested in error propagated from error in f or y. In this part we are interested in the error generated by computing $Q(z)$ to approximate $T(g)$. Give a bound on the magnitude of this generated error. That is, give a bound on the magnitude of the approximation error in the result of Simpson's rule for the integrand x^4 where $h = 1$ and the integration interval is [2,4]. Show that your error bound is equal to the error. Explain why.

(g) Assume you were given the accompanying table for x^4 and did not know the underlying function.

x	y
1	1
2	16
3	81
4	256
5	625

Estimate the generated approximation error in the result of Simpson's rule applied to this tabular data with $h = 1$ for the interval [2,4]. Your estimate should come out to be the exact error, namely, the result of part f. Explain why.

(h) Consider the Simpson's rule formula in the form $Q(y) = \frac{1}{3}(y_0 + 4y_1 + y_2)$. Assume all computations are done on a decimal floating-point computer with a 6-digit fraction, with truncation and a guard digit, and assume $|\varepsilon_{y_0}| \le b_y$, $|\varepsilon_{y_1}| \le b_y$, $|\varepsilon_{y_2}| \le b_y$, $\varepsilon_{1/3}$ is known, and $\varepsilon_4 = 0$. Give a bound on the magnitude of the absolute error in $Q(y)$ computed using the above formula. (Note that the error generated in multiplication by 4 has no special properties.)

7.8 Let $h(z) = ze^z$. Assume that we are using System 360, computing in double precision (the fraction part of floating-point numbers have 14 hexadecimal digits). Assume our subroutine for computing e^z produces a computed value with a relative error bounded in magnitude by 10^{-6} with error-free z. Assume z has relative error bounded in magnitude by

10^{-8}. Give a bound on the relative error in the computed value of $h(z)$. You need not collect terms.

7.9 Give a mop-up formula for the predictor-corrector pair produced in Problem 5.11.

7.10 In terms of A_1 and C_1 give a mop-up formula for the predictor-corrector pair produced in Problem 5.13.

7.11 **(a)** Show that the Adams-Bashforth predictor,

$$y_{i+1}^{(p)} = y_i + \frac{h}{24}(55y_i' - 59y_{i-1}' + 37y_{i-2}' - 9y_{i-3}')$$

for the solution of a first-order initial-value differential equation has approximation error $-(251/270)h^5y_i^{(5)} + O(h^6)$.
(b) Show that the Adams-Moulton corrector,

$$y_{i+1}^{(c)} = y_i + \frac{h}{24}(9y_{i+1}' + 19y_i' - 56y_{i-1}' + y_{i-2}')$$

for the solution of a first-order initial-value differential equation has approximation error $(19/720)h^5y_i^{(5)} + O(h^6)$.
(c) Show that the mop-up formula for the Adams-Bashforth predictor and Adams-Moulton corrector is

$$y_{i+1}^{(m)} = \frac{251y_{i+1}^{(c)} + 19y_{i+1}^{(p)}}{270}$$

7.12 We know that a constant, C, times the difference between the values, at a given step, of the corrector and the predictor of a predictor-corrector method gives the low-order term (in h) of the truncation error in the corrector at that step in cases where the predictor and corrector have truncation errors of the same order in h.

(a) Show that a constant (say, D) times this difference gives the low-order term of the truncation error in the predictor of the next step.
(b) What is the value of D for the modified Euler predictor and Heun corrector?
(c) We can produce a doubly mopped-up predictor-corrector sequence by the following procedure:

(1) Compute $y_{i+1}^{(p)}$ from the modified Euler predictor.
(2) Compute $y_{i+1}^{(pm)} = y_{i+1}^{(p)} - D(y_i^{(c)} - y_i^{(p)})$.
(3) Compute $y_{i+1}^{(c)}$ from the Heun corrector using $y_{i+1}^{(pm)}$ as the estimate of y_{i+1}.
(4) Compute $\hat{y}_{i+1} = y_{i+1}^{(cm)} = y_{i+1}^{(c)} - C(y_{i+1}^{(c)} - y_{i+1}^{(p)})$.

What is the order of h in the truncation error in $y_{i+1}^{(pm)}$? In $y_{i+1}^{(c)}$? In $\hat{y}_{i+1}$? Why is the above method better than the singly mopped-up modified-Euler–Heun method? Why cannot $y_{i+1}^{(pm)}$ appear in the corrector mop-up (step 4 above) in place of $y_{i+1}^{(p)}$?

7.13 Use the modified-Euler–Heun method to compute the solution of $y' = -2xy^2$, $y(0) = 1$ [which has the solution $y(x) = 1/(1 + x^2)$] on $[0,4]$ with $h = 0.4$ as follows:

(a) Using the predictor only.
(b) Using the predictor and corrector with no mop-up.
(c) As in part b but with single mop-up.
(d) As in part b but with double mop-up (see Problem 7.12).

7.14 Show that the elements of y_{i+1} given by $y_{i+1}^{(p)}$ in equation 114 has error of order 3 in h.

7.15 Show that the weights 1, 2, 2, 1 of the slope terms in equation 111 maximize the order of h in the approximation error in $\hat{y}_{i+1}$ thus produced.

7.16 Compare the arithmetic error bounds for the expressions $y = x_1 + (x_1 - x_2)/(4^i - 1)$ and $y = (4^i/(4^i - 1))x_1 + x_2/(4^i - 1)$, assuming that $|x_1 - x_2| \ll |x_1|$ and $|x_1 - x_2| \ll |x_2|$ and that 4^i and 4^{i-1} can be calculated exactly, as they can on a binary computer if i is not too large.

7.17 Romberg integration for $I = \int_a^b f(x)\,dx$ is specified by the following relations:

(1) $T^0(b - a) = \dfrac{b - a}{2}(f(b) + f(a))$

(2) $T^0(h/2) = \dfrac{1}{2}\left[T^0(h) + h\sum_{i=1}^{N} f(a + (i - \tfrac{1}{2})h)\right]$ where $h = (b - a)/N$

(3) $T^{i+1}(h/2) = T^i(h/2) + \dfrac{T^i(h/2) - T^i(h)}{4^{i+1} - 1}$

Assume that the absolute error generated in evaluating $f(x)$ is less than C and that the error generated elsewhere is negligible compared to the errors propagated from errors in $f(x)$.

(a) Show that the propagated error in $T^0(h/2)$ is bounded by the same constant for all h used in a Romberg integration (all values in the T^0 column of the Romberg table have the same error bound). What is the constant?
(b) Show that the error bound for $T^{i+1}(h/2)$ is bounded by a function of i only (not of h). What is the function? Show that this error bound approaches a limit as $i \to \infty$. Use the hints on the next page.

Hints: $\displaystyle\prod_{i=1}^{\infty} a_i = exp\left(\sum_{i=1}^{\infty} \log(a_i)\right)$, and $\log(1 + x/(y - 1)) < x/y$

      if $(x - 2)y > -2$.

(c) Assume the following:

(1) $T^i(h) = I + \varepsilon_1$ and $T^i(h/2) = I + \varepsilon_2$ where $|\varepsilon_1| < b_1$ and $|\varepsilon_2| < b_2$, where b_1 and b_2 are known.

(2) Generated error is *not* negligible. Rather, the relative generated error for all arithmetic operations is bounded by R.

(3) Products of error bounds are negligible (for example, neglect Rb_1).

(4) $4^{i+1} - 1$ can be evaluated exactly.

Give a bound on the error in $T^{i+1}(h/2)$ as computed in equation 3 above.

(d) How does a generated error E in $T^i(h)$ at step i of the Romberg integration propagate into the result $T^j(h/2^k)$ of step $j > i$? (Obtain an approximate answer only.) Note that the bound on the contribution of E into the jth step is greater than E. Therefore, an error E at each step would eventually cause the overall error bound (the sum of the contributions due to each step) to $\to\infty$; the Romberg integration process might diverge.

(e) What property of digital computer arithmetic will nullify the behavior described in part d, that is, will cause the error propagation due to using equation 3 to stop?

7.18   In Problem 3.18, you showed that to integrate $\int_0^1 e^{-x}\,dx$ with an approximation error of magnitude less than 2×10^{-5}, the trapezoidal rule required 66 data points and Simpson's rule required 7 points. Carry out Romberg integration for this integral with the above tolerance. How many points were required by Romberg integration?

7.19   Show how Richardson extrapolation can be applied to the Runge-Kutta method.

7.20   Given a function $f(x)$ tabulated at the equally spaced points $\{x_0, x_1, ..., x_N\}$, equation 61 of Chapter 5 presents a three point differentiation formula approximating the derivative at a tabular point x_i:

$$D_h(x_i) = \frac{f_{i+1} - f_{i-1}}{2h}$$

It has approximation error

$$E_h(x_i) = \frac{h^2}{6}f'''(x_i) + O(h^4)$$

Devise an algorithm which produces an approximation error of order h^4 or higher — using the approach of error reduction by averaging.

7.21 Assume you are given the accompanying table for the function $y(x)$.

x	y
0	4
1	7
2	12
3	20
4	33
5	54

(a) Using difference techniques, find the polynomial of lowest degree passing through the points of the table. Identify the numerical method used.

(b) Which standard polynomial exact-matching interpolation method should be used to find $y(2.4)$ with an approximation error of magnitude less than 2? Why?

7.22 Assume you are given the accompanying table for the function $y(x)$.

x	y
0	1
1	9
2	31
3	61
4	69
5	1

(a) Using difference techniques, find the polynomial of lowest degree passing through the points of the table. Identify the numerical methods used.

(b) Which standard polynomial exact-matching interpolation method should be used to find $y(2.4)$ with an approximation error of magnitude less than 2? Why?

7.23 Consider the following tabulated function where the x_i are equally spaced: $y_0 = 150$; $y_1 = 122$; $y_2 = 102$; $y_3 = 76$; $y_4 = 58$; $y_5 = 35$; $y_6 = 12$; $y_7 = -5$; $y_8 = -12$; $y_9 = -3$; $y_{10} = 27$.

(a) At which difference column should you end to do polynomial exact-matching approximation with greatest accuracy for $y_{2.5}$?

(b) For $y_{7.5}$?

(c) What would be the path ending for the approximation in part a?

(d) For the approximation in part b?

7.24   Consider the accompanying table with differences indicated.

x	y	δ	δ^2	δ^3	δ^4	δ^5
0.20	0.20134					
		•				
0.21	0.21155		•			
		•		•		
0.22	0.22178		•		•	
		•		•		•
0.23	0.23203		•		•	
		•		•		•
0.24	0.24231		•		•	
		•		•		•
0.25	0.25261		•		•	
		•		•		•
0.26	0.26294		•		•	
		•		•		
0.27	0.27329		•			
		•				
0.28	0.28367					

(a)   For what values of x would you use this specific path, assuming you do not know beforehand at what difference you wish to stop?
(b)   Give an expression [in terms of $s = (x - 0.26)/0.01$ and derivatives of f] for the error when this path is used out to the last difference indicated.
(c)   Which formula would you use to compute $f(0.234)$ to within some error bound to be given?
(d)   Which formula would you use to compute $f(0.234)$ using up to third differences?
(e)   Bound the error in part d, given that $f(x) = \sinh(x)$.

7.25   **(a)**   Show by differentiation of Stirling's formula that

$$y_0' = \frac{1}{h}\left(\mu\delta y_0 - \frac{\mu\delta^3 y_0}{3!} + \frac{1(4)}{5!}\mu\delta^5 y_0 - \frac{1(4)(9)}{7!}\mu\delta^7 y_0 + \cdots\right)$$

(b)   Use this formula with the accompanying table for $f(x) = \sinh(x)$ to compute $f'(0.4)$:

(1) With $h = 0.002$ using up to first differences.
(2) With $h = 0.001$ using up to first differences.
(3) With $h = 0.001$ using up to third differences.

Which is the most accurate? Note $f'(0.4) = \cosh(0.4) = 1.081072$.

x	$\sinh(x)$
0.398	0.408591
0.399	0.409671
0.400	0.410752
0.401	0.411834
0.402	0.412915

(c) Show that twice differentiating Stirling's formula and truncating at the second difference gives $y_0'' = \delta^2 y_0 / h^2$.

7.26 Apply the method of false position to find the root in $[-1, 0]$ of the equation $x + \frac{1}{2} + \sin^{-1}(x) = 0$

(a) Using the method of false position alone.

(b) Applying Aitken's δ^2 acceleration every two iterations after one interval endpoint becomes frozen (and rejecting the accelerated value if it falls outside the present interval).

7.27 Consider the function $g(x) = e^{-x}f(x)$, where $g(x)$ is not polynomial-like but $f(x)$ is, over the interval $[0,50]$.

(a) Assume we wish a subroutine to evaluate $g(x)$ for any x in $[0,50]$, but $f(x)$ is too complicated to evaluate every time we need a value of $g(x)$. A good solution is to tabulate $f(x)$ and for the given input argument, x_{inp}, for our subroutine, interpolate a value y from the table by polynomial exact-matching interpolation or a variant, and estimate $g(x_{inp})$ as $e^{-x_{inp}}y$. Briefly and without mathematical analysis, why is this a better scheme than tabulating $g(x)$ and doing polynomial exact matching in that table?

(b) Assume we tabulate f at $x = 0, 1, 2, \ldots, 50$ as shown here.

x	$f(x)$
0	-5.1
1	-1.0
2	4.9
3	12.9
4	22.8
5	34.7
.	.
.	.
.	.

(1) By examining the difference table, decide which difference to stop on.

(2) Carry out the approximation for $g(2.8)$, explaining what you are doing.

(3) Estimate the approximation error magnitude in your answer.

(c)   Assume we wish to evaluate $\int_0^{50} g(x)\,dx$. One possibility for computing this value is as follows: for each tabular interval in the table of part b, fit f by an exact-matching line and integrate the approximation to g produced by weighting the resulting line by e^{-x}, and then sum the results for all of the intervals. Show that the error in the integral over a single tabular interval is $O(h^3)$, where h is the tabular interval width.

(d)   We can further show that the error in the approximation for the integral over [0,50] is $O(h^2)$. Better still, we can show that it is $kh^2 + O(h^3)$ for some constant k independent of h. A variation on Romberg integration is applicable here.

(1) Set up this algorithm for two steps.

(2) Compare the improvement from step to step to that achieved in normal Romberg integration. (Discuss; do not do numerical calculations.)

CHAPTER EIGHT

Recurrent Application of Operators: Stability

In Chapters 5 and 6 we discussed the error generated and propagated by simple operators, and in Section 7.1 we began to discuss the errors produced when a number of these simple operators are applied in sequence. A special case of the sequential application of simple operators occurs, as we have seen in Chapter 3 and others, when a single operator is repetitively applied:

$$y_{i+1} = T(y_i, y_{i-1}, ..., y_{i-n+1}) \qquad [1]$$

We will sometimes refer to this equation as the recurrence relation to be applied (solved).

We saw that this repetitive application of an operator could be of two forms: recurrent and iterative. With the recurrent application of an operator we may be interested in the answer obtained after every step or we may be interested only in the answer obtained after some fixed number of steps, but in either case the answer at an intermediate step is not an approximation to some final answer but an answer in its own right. That is, we have a notion of the correct answer, y_i, to be obtained after i steps, and if due to error generation or propagation we do not obtain this answer but rather y_i^*, we will thus have an error in y_i^* which is $y_i^* - y_i$.

In contrast is the iterative form of the repetitive application of an operator, in which the desired answer is the limit of a sequence obtained by repetitively applying the operator in question. In this case the result of the ith step is of no interest for its own sake but only as an approximation to the final limit. Therefore, for iterative repetition the error in y_i^* is its distance from the limit ξ, that is, $y_i^* - \xi$.

Among the examples of the recurrent form of operator repetition are the various methods for the solution of ordinary differential equations, in which the result of the ith step of the recurrence, $\hat{y}_i$, is an approximation to $y(x_i)$, and the triangularization step of Gaussian elimination, where the operation of zeroing the first column of a matrix is repetitively applied. Examples of the iterative form of repetition are methods we have seen for solving nonlinear equations, such as Newton's method and the method of false position, and the method of iterative improvement applied with Gaussian elimination to solve linear equations.

In this chapter we will analyze the error with the recurrent form of operator repetition. In the next chapter we will deal with the iterative form. We should not be surprised to find, however, that the error analyses for these two forms have much in common.

Because a prime example in this chapter will be the solution of differential equations, in which not only y_i, y_{i-1}, etc. but also x_i change from step to step, we will find it useful to analyze the somewhat more general recurrence,

$$y_{i+1} = T(y_i, y_{i-1}, \ldots, y_{i-n+1}; x_i) \tag{2}$$

in which a parameter x_i can change with each application of T. Where there is no such parameter, we can take x_i to be a constant.

8.1   Stability and Propagated Error Bounds

At each step of the repetitive application of an operator error will be generated and propagated. Let us consider the sources of error in the result of a number of steps. Assume that we have a problem whose solution is given by recurrently applying equation 2. Here we assume that T is not the operator actually applied, such as the Euler-Heun predictor-corrector mop-up operator in the case of the solution of differential equations. Rather T is the operator which produces the correct answer; in our example the operator

$$T(y_i; x_i) = y_i + \int_{x_i}^{x_i + h} f(x, y(x)) \, dx \tag{3}$$

We will call the operator actually applied recurrently by the name $\hat{T}$, and we will call the results of its recurrent application $\{v_i\}$:

$$v_{i+1} = \hat{T}(v_i, v_{i-1}, \ldots, v_{i-n+1}; x_i) \tag{4}$$

In some problems $\hat{T} = T$, but in many, such as the solution of differential equations, $\hat{T} \neq T$.

In a real situation, we are given initial values $v_0, v_1, v_2, \ldots, v_{n-1}$ and

$\hat{T}$ is applied $i - n + 1$ times to produce v_k for $k = n, n+1, ..., i$. There may be error in the n initial v_i values which are input to the process, and each application of $\hat{T}$ generates not only approximation error due to the fact that $\hat{T} \neq T$ but also arithmetic error due to the fact that even $\hat{T}$ is not being exactly applied. Let us distinguish four results of $i - n + 1$ applications of T or $\hat{T}$:

y_i, the result of recurrently applying T with the correct inputs $y_0, y_1, ..., y_{n-1}$; this is the desired result

u_i, the result of recurrently applying $\hat{T}$ with the correct inputs $y_0, y_1, ..., y_{n-1}$

v_i, the result of recurrently applying $\hat{T}$ with the inaccurate inputs $v_0, v_1, ..., v_{n-1}$

v_i^*, the same as v_i except that arithmetic error is generated with each application of $\hat{T}$; this is the real result

The relative error τ_i in the real result v_i^* is

$$\tau_i = \frac{v_i^* - y_i}{y_i} \qquad [5]$$

By an approach we have seen in Section 5.1 τ_i can be rewritten as

$$\tau_i = \frac{v_i^* - v_i}{y_i} + \frac{v_i - u_i}{y_i} + \frac{u_i - y_i}{y_i} \qquad [6]$$

The first term of equation 6, $(v_i^* - v_i)/y_i$, is the relative error due to the arithmetic errors made in each step and propagated through all later steps. Similarly, the third term of equation 6, $(u_i - y_i)/y_i$, is the relative error due to the approximation errors made in each step and propagated through all later steps. To understand each of these we must understand how errors generated in any step, that is, in the input to the next step, are propagated by some number of applications of the operator $\hat{T}$. But that is just the issue which the second term in equation 6, $(v_i - u_i)/y_i$, forces us to face: what is the error in v_i due to errors in the inputs to some number (here $i - n + 1$) of applications of $\hat{T}$? Therefore, if we can develop an approach for determining a bound on $(v_i - u_i)/y_i$, this approach will be used not only directly to produce the second term in the error decomposition in equation 6 but to produce the first and third terms as well.

Therefore, for now let us analyze the relative error $(v_i - u_i)/y_i$. That is, let us assume that $\hat{T}$ is the operator that we wish to apply and that it is exactly applied, so that the only source of error is that from the inputs to the first application of $\hat{T}$, though the denominator of the relative error is y_i, the correct answer, and not u_i.

To simplify the notation, we will rewrite $\hat{T}$ as g, so that the recurrence relation being analyzed is

$$v_{i+1} = g(v_i, v_{i-1}, \ldots, v_{i-n+1}; x_i) \qquad [7]$$

We will define

$$\varepsilon_i = v_i - u_i \qquad [8]$$

the numerator of our relative error, $(v_i - u_i)/y_i$. For the solution of differential equations y_i is the solution at x_i to the differential equation; u_i is the solution at x_i to the recurrence relation (equation 7) given by the numerical method, with accurate starting values; and v_i is the solution at x_i to the recurrence relation with inaccurate starting values.

We have assumed that the initial values, v_j, for $0 \leqslant j \leqslant n - 1$, have errors, ε_j, from y_j, and that no other error is generated. That is, we are interested in the propagation of $\varepsilon_0, \varepsilon_1, \varepsilon_2, \ldots, \varepsilon_{n-1}$ into v_i for any i.

It will be useful to define $\mathbf{v}^{(i)} \triangleq (v_i, v_{i-1}, \ldots, v_{i-n+1})^T$. Thus,

$$v_{i+1} = g(\mathbf{v}^{(i)}; x_i) \qquad [9]$$

Then any propagation of the error in the initial iterates $\mathbf{v}^{(n-1)}$ into v_i occurs by the propagation through equation 9 from $\mathbf{v}^{(n-1)}$ to v_n, thence from $\mathbf{v}^{(n)}$ into v_{n+1}, and so on — ending with the propagation from $\mathbf{v}^{(i-1)}$ into v_i. To understand this multiple propagation, it is necessary first to characterize the propagation due to a single recurrence step.

By the partial derivative approach of Section 6.2 the error propagated from a vector-valued variable $\mathbf{z}$ (multiple variables $z_1, z_2, \ldots, z_n$) to an output w by the application of an operator S is given by

$$\varepsilon_w = \sum_{j=1}^{n} \varepsilon_j \frac{\partial S(z)}{\partial z_j}\bigg|_{\mathbf{z}=\zeta} \qquad [10]$$

where ε_k is the error in z_k^* and, for $1 \leqslant k \leqslant n$, $\zeta_k \in [z_k, z_k^*]$. Applying this result directly to equation 9, we obtain

$$\varepsilon_{i+1} = \sum_{j=0}^{n-1} \varepsilon_{i-j} \frac{\partial g(\mathbf{v}^{(i)}; x_i)}{\partial v_{i-j}}\bigg|_{\mathbf{v}^{(i)}=\boldsymbol{\eta}^{(i)}} \qquad [11]$$

where ε_k is the error in v_k and $\eta_k^{(i)} \in [u_{i-k+1}, v_{i-k+1}]$.

Equation 11 gives us a recurrence relation for the error in v_{i+1} given the errors in the n previous v_k values. We are interested in a bound on this error for any i. To compute this bound we would like, starting with the recurrence relation given by equation 11, to find an explicit function s of i, $\varepsilon_0, \varepsilon_1, \ldots, \varepsilon_{n-1}$ which gives ε_i. Given such an s, the "solution to the recurrence relation," and bounds b_j on ε_j, for $0 \leqslant j \leqslant n - 1$, we could find the values of these ε_j within their bounds which maximize $|\varepsilon_i|$, and set b_i to this maximum.

Since the operator g, that is, $\hat{T}$, will be applied many times, we are interested in the behavior of ε_i for large i. We will therefore examine†

$$\lim_{i \to \infty} \varepsilon_i = \lim_{i \to \infty} s(i; \varepsilon_0, \varepsilon_1, ..., \varepsilon_{n-1})$$

If this limit is either $+\infty$ or $-\infty$, we say that the recurrence is *absolutely unstable:* the absolute error becomes very large in magnitude when the operator is applied many times. If the limit is 0, we say that the recurrence is *absolutely stable:* the absolute error dies out as the operator is repetitively applied. If the ε_i do not approach zero but are bounded as $i \to \infty$, we say the recurrence is neither stable nor unstable. (Sometimes we will be slightly sloppy and say that the operator is stable, unstable, or neither instead of that its recurrence has these properties.)

It is of course not the case that a method is fully characterized by being described as stable or unstable. Over a not too large number of steps, a slightly unstable method (one for which the errors get large only slowly as the recurrence takes place) may not be unacceptable, and a slightly stable method may be inadequate when the errors generated over very many steps are combined. Still, since an unstable method normally produces unacceptable answers, the categorization is useful.

As noted many times, we are more often interested in relative error than in absolute error. Thus we need to discuss how to bound ε_i/y_i for any i and how to characterize the behavior as $i \to \infty$ of this relative error propagated from the initial errors. To find ε_i/y_i, we must both solve for ε_i, normally using equation 11, and also solve for y_i. Given these two solutions, we examine $\lim_{i \to \infty} \varepsilon_i/y_i$. If this limit of the relative error is infinite in magnitude we say that the recurrence is *relatively unstable;* if the limit is 0, we say that the recurrence is *relatively stable;* if the relative error is bounded but nonzero as $i \to \infty$, we say the recurrence is neither relatively stable nor relatively unstable. Note that an absolutely unstable operator can be relatively stable if the absolute error magnitude goes to ∞ more slowly than the correct solution magnitude as $i \to \infty$, and an absolutely stable recurrence can be relatively unstable if the absolute errors go to 0 more slowly than the correct answers as $i \to \infty$.

8.2 Single-Step Recurrent Methods

The simplest recurrences are those for which $n = 1$ in equation 7, that is, those for which

$$v_{i+1} = g(v_i; x_i) \tag{12}$$

---

† To handle some unusual situations, we should strictly write $\lim \sup s(i; \varepsilon_0, \varepsilon_1, ..., \varepsilon_n)$.

For these equation 11 becomes

$$\varepsilon_{i+1} = \varepsilon_i \frac{\partial g(v_i; x_i)}{\partial v_i}\bigg|_{v_i = \eta_i} \tag{13}$$

where $\eta_i \in [u_i, v_i]$. The solution $s(i; \varepsilon_0)$ to equation 13 is clearly

$$\varepsilon_i = \varepsilon_0 \prod_{j=0}^{i-1} \frac{\partial g(v; x_i)}{\partial v}\bigg|_{v = \eta_j} \tag{14}$$

It should be clear that if

$$\lim_{i \to \infty} \left| \frac{\partial g(v; x_i)}{\partial v}\bigg|_{v = \eta_i} \right| > 1,$$

$|\varepsilon_i| \to \infty$ as $i \to \infty$; the recurrence will be absolutely unstable. Similarly, if

$$\lim_{i \to \infty} \left| \frac{\partial g(v; x_i)}{\partial v}\bigg|_{v = \eta_i} \right| < 1$$

the recurrence will be absolutely stable. If

$$\lim_{i \to \infty} \left| \frac{\partial g(v; x_i)}{\partial v}\bigg|_{v = \eta_i} \right| = 1$$

or if the limit does not exist, further analysis is necessary.

Using equation 13, we can produce a recurrence relation for the relative error

$$\frac{\varepsilon_{i+1}}{y_{i+1}} = \frac{\varepsilon_i \dfrac{\partial g(v; x_i)}{\partial v}\bigg|_{v = \eta_i}}{y_{i+1}} = \frac{\varepsilon_i}{y_i} \left(\frac{y_i}{y_{i+1}} \frac{\partial g(v; x_i)}{\partial v}\bigg|_{v = \eta_i} \right) \tag{15}$$

Thus if

$$\lim_{i \to \infty} \left| \frac{y_i}{y_{i+1}} \frac{\partial g(v; x_i)}{\partial v}\bigg|_{v = \eta_i} \right| < 1$$

the recurrence will be relatively stable; if the same limit is greater than 1, the recurrence will be relatively unstable.

For example, assume we wish to recur with

$$y_{i+1} = \frac{y_i^2}{2} \tag{16}$$

Here $y_i = u_i$ (that is, $T = \hat{T}$) and x_i is constant. Equation 14 gives

$$\varepsilon_i = \varepsilon_0 \prod_{j=0}^{i-1} \frac{\partial y^2/2}{\partial y}\bigg|_{y=\eta_j} = \varepsilon_0 \prod_{j=0}^{i-1} \eta_j \qquad [17]$$

where $\eta_j \approx y_j$. If $\lim_{i\to\infty}|y_i| < 1$, the product in equation 17 will approach zero as $i \to \infty$, so we will have absolute stability. Similarly, if $\lim_{i\to\infty}|y_i| > 1$, we will have absolute instability. From equation 16 we see that

$$y_i = 2(y_0/2)^{2^i} \qquad [18]$$

so $\lim_{i\to\infty}|y_i| < 1$ if $|y_0| < 2$ and $\lim_{i\to\infty}|y_i| > 1$ if $|y_0| > 2$. Thus we have absolute stability for $|y_0| < 2$ and absolute instability for $|y_0| > 2$.

A bound b_i on $|\varepsilon_i|$, given b_0, can be obtained as follows. From equations 16 and 18

$$\varepsilon_i \approx \varepsilon_0 \prod_{j=0}^{i-1} y_j = \varepsilon_0 \prod_{j=0}^{i-1} 2(y_0/2)^{2^j}$$

$$= \varepsilon_0 2^i(y_0/2) \sum_{j=0}^{i-1} 2^j = \varepsilon_0 2^i(y_0/2)^{2^i - 1} \qquad [19]$$

Thus

$$b_i = b_0 2^i|y_0/2|^{2^i - 1} \qquad [20]$$

From equation 15 the recurrence relation for the relative error is given by

$$\frac{\varepsilon_{i+1}}{y_{i+1}} = \frac{\varepsilon_i}{y_i}\left(\frac{y_i\eta_i}{y_i^2/2}\right) \approx 2\frac{\varepsilon_i}{y_i} \qquad [21]$$

It follows that the recurrence is unstable for all y_0 and that the bound on the relative error at the ith step is given by

$$r_i = 2^i r_0 \qquad [22]$$

The same result could have been obtained by using equations 18 and 20.

As another example, consider Euler's method for the solution of differential equations:

$$\hat{y}_{i+1} = \hat{y}_i + hf(x_i, \hat{y}_i) \qquad [23]$$

In this case

$$g(v; x_i) = v + hf(x_i, v) \qquad [24]$$

so

$$\frac{\partial g(v; x_i)}{\partial v} = 1 + h\frac{\partial f(x_i, v)}{\partial v} \qquad [25]$$

and thus

$$\frac{\partial g(v; x_i)}{\partial v}\bigg|_{v=\eta_i} = 1 + h\frac{\partial f(x_i, y)}{\partial y}\bigg|_{y=\eta_i} \qquad [26]$$

where $\eta_i \in [\hat{y}_i, \hat{y}_i^*]$. It is clear that if as $i \to \infty$ $\partial f(x, y)/\partial y|_{x_i, \eta_i}$ is of the same sign as h, Euler's method will be absolutely unstable, and that if $\partial f(x, y)/\partial y|_{x_i, \eta_i}$ is bounded as $i \to \infty$, then for appropriately small h of opposite sign from the partial derivative as $i \to \infty$, Euler's method will be absolutely stable. Thus, for example, if we are using Euler's method to solve the equation $y' = 3x^2y$, $y(0) = 1$ on $x \in [0, 1]$, $\partial f/\partial y = 3x^2 \geqslant 0$ for $x \in [0, 1]$. Thus, since in this problem $h > 0$, Euler's method will be absolutely unstable. If, however, the solution had been wanted for $x \in [-1, 0]$, as long as $|h| < 2/\max_{x \in [-1, 0]} 3x^2 = 2/3$, Euler's method will be absolutely stable. Note that it is not necessary that as i gets large, x gets large in magnitude. We can take many steps because h is small rather than because x is large.

The analysis of the relative stability of Euler's method requires knowledge of y_i. Since we can not give the solution, $y(x)$, to the differential equation in general, let us restrict the problem to the case where $\partial f/\partial y$ does not change much from step to step and thus we can approximate the problem by analyzing the case where $\partial f/\partial y = K$, a constant. Restricting ourselves to analyzing the stability of Euler's method applied to this class of differential equations makes the mathematics straightforward. From equations 14 and 26 we have

$$\varepsilon_i = \varepsilon_0(1 + hK)^i \qquad [27]$$

so

$$b_i = b_0 |1 + hK|^i \qquad [28]$$

Furthermore, since

$$\frac{\partial f(x, y)}{\partial y} = K \Rightarrow f(x, y) = Ky + a(x) \qquad [29]$$

we can analytically solve the differential equation

$$y'(x) = Ky + a(x), \ y(x_0) = y_0 \qquad [30]$$

to produce

$$y(x) = y_0\, e^{K(x - x_0)} + e^{Kx} \int_{x_0}^{x} e^{-Ku}\, a(u)\, du \qquad [31]$$

Thus,

$$y_i \overset{\Delta}{=} y(x_i) = y_0 \, e^{K(x_i - x_0)} + e^{Kx_i} \int_{x_0}^{x_i} e^{-Ku} \, a(u) \, du$$

$$= y_0 \, e^{Kih} + e^{K(x_0 + ih)} \int_{x_0}^{x_i} e^{-Ku} \, a(u) \, du \qquad [32]$$

$$= (y_0 + e^{Kx_0} \int_{x_0}^{x_i} e^{-Ku} \, a(u) \, du)(e^{hK})^i$$

so

$$\frac{\varepsilon_i}{y_i} = \frac{\varepsilon_0 (1 + hK)^i}{(y_0 + e^{Kx_0} \int_{x_0}^{x_i} e^{-Ku} \, a(u) \, du)(e^{hK})^i}$$

$$\qquad [33]$$

$$= \frac{\varepsilon_0}{y_0 + e^{Kx_0} \int_{x_0}^{x_i} e^{-Ku} \, a(u) \, du} \left(\frac{1 + hK}{e^{hK}}\right)^i$$

The denominator of the left term of the last expression for ε_i / y_i in equation 33 is normally bounded away from both 0 and ∞ as $i \to \infty$. Thus, ε_i / y_i will behave as

$$\left(\frac{1 + hK}{e^{hK}}\right)^i$$

as $i \to \infty$. Since for $hK \geqslant -1.3$,

$$e^{hK} = 1 + hK + (hK)^2/2 \, e^\theta > 1 + hK, \text{ where } \theta \in [0, hK] \qquad [34]$$

$$\lim_{i \to \infty} \left(\frac{1 + hK}{e^{hK}}\right)^i = 0 \qquad [35]$$

so Euler's method is relatively stable for any member of the class of equations under consideration for which $hK \geqslant -1.3$. So we have finally shown that at least over intervals for which $\partial f / \partial y$ is approximately constant and small in magnitude, we need not be concerned with the stability of Euler's method.

8.3 Multistep Recurrent Methods

In multistep methods the value of the result at the ith step depends explicitly not only on the result of the previous step but also on the results of even earlier steps ($n > 1$ in equation 1). The result is that the errors in a number of steps combine in a more complicated way than that seen in Section 8.2 (see equation 11), and we intuitively fear, correctly, that stability may be a more significant problem.

As an example, assume we wish to recur with

$$y_{i+1} = y_i - (15/64)y_{i-1} \qquad [36]$$

given starting values y_0 and y_1. Then from equation 11,

$$\varepsilon_{i+1} = 1\varepsilon_i - (15/64)\varepsilon_{i-1} \qquad [37]$$

To go further we must solve this recurrence relation, that is, find a function s such that $\varepsilon_i = s(i; \varepsilon_0, \varepsilon_1)$.

As another example, let us consider the repetitive application of the modified Euler method,

$$\hat{y}_{i+1} = \hat{y}_{i-1} + 2hf(x_i, \hat{y}_i) \qquad [38]$$

applied in recurrence to approximate the solution of the differential equation

$$y'(x) = f(x, y), y(x_0) = y_0 \qquad [39]$$

Note that we are discussing the repetitive application of the modified Euler method alone, not the use of that method as a predictor with a following corrector and mop-up as is commonly done. In this case the function g in equation 7 is given by

$$g(v_i, v_{i-1}; x_i) = v_{i-1} + 2hf(x_i, v_i) \qquad [40]$$

and thus the error propagated from the inputs for the ith step to the output of the ith step is given, following equation 11, by

$$\varepsilon_{i+1} = \varepsilon_i 2h \left. \frac{\partial f(x, y)}{\partial y} \right|_{x_i, \eta_i} + \varepsilon_{i-1} \cdot 1 \qquad [41]$$

If, as in our analysis of Euler's method, we restrict the class of functions f to those for which $\partial f/\partial y$ is a constant K and thus approximate the class of methods for which $\partial f/\partial y$ does not vary strongly over the interval of interest, we obtain

$$\varepsilon_{i+1} = \varepsilon_{i-1} + 2hK\varepsilon_i \qquad [42]$$

Here again we must solve a recurrence relation, that is, express ε_i as a function of i, ε_0, and ε_1 alone so that we can find the limit of the error propagated from ε_0 and ε_1 as $i \to \infty$. Equations 37 and 42 can each be written in the form of a linear homogeneous difference equation with constant coefficients:

$$\varepsilon_{i+1} - \varepsilon_i + (15/64)\varepsilon_{i-1} = 0 \qquad [43]$$

and

$$\varepsilon_{i+1} - 2hK\varepsilon_i - \varepsilon_{i-1} = 0 \qquad [44]$$

Such equations are called *difference equations* because they can be written in terms of undivided differences of ε_k, *homogeneous* because there are no terms in the equation which do not involve ε_k for some k, *linear* because the set of solutions is closed under addition and multiplication by a constant, and *with constant coefficients* because none of the coefficients are functions of i. There exist methods to solve such difference equations. We must take a slight detour to discuss their solution.

8.3.1 SOLUTION OF LINEAR HOMOGENEOUS DIFFERENCE EQUATIONS WITH CONSTANT COEFFICIENTS

As stated above, a set of solutions to the general nth order linear homogeneous difference equation with constant coefficients,

$$z_{i+1} - \sum_{j=0}^{n-1} \alpha_j z_{i-j} = 0 \qquad [45]$$

is closed under multiplication by a constant and addition. $z_i \equiv 0$ is a solution, and all the other rules for vector spaces apply, so the set of solutions to equation 45 is a vector space. It can be shown that the dimension of this vector space is n, so to find all solutions, all that remains is to find the n members of a basis of this vector space of solutions.

Without motivating the choice, consider the possibility of a solution of the form

$$z_i = \rho^i \qquad [46]$$

where ρ is some constant. For equation 46 to give a solution, it must satisfy equation 45, that is

$$\rho^{i+1} - \sum_{j=0}^{n-1} \alpha_j \rho^{i-j} = 0 \qquad [47]$$

We are looking for a solution other than $z_i \equiv 0$, that is, other than that for $\rho = 0$, so we can divide equation 47 by ρ^{i+1-n}, producing

$$\rho^n - \sum_{j=0}^{n-1} \alpha_j \rho^{n-1-j} = 0 \qquad [48]$$

a polynomial equation in ρ called the *characteristic equation* corresponding to the difference equation. That is, equation 48 gives those values of the constant for which $z_i = \rho^i$ is a solution to the difference equation, 45.

Equation 48, being an nth degree polynomial equation, has n solutions, $\rho_1, \rho_2, \ldots, \rho_n$, called the *characteristic roots* corresponding to the difference equation. If all of these roots are distinct, the solutions ρ_1^i, $\rho_2^i, \ldots, \rho_n^i$ are n linearly independent functions of i, and thus they form a basis for the vector space of solutions. If any root, ρ_k, is of multiplicity $m > 1$, it can be shown that not only is ρ_k^i a solution but so is $i\rho_k^i$, $i^2\rho_k^i, \ldots, i^{m-1}\rho_k^i$, and thus we still have n linearly independent solutions when those corresponding to the distinct roots are collected. For the remainder of this section we will assume that the roots are distinct, but the reader can make the appropriate substitutions if this assumption does not hold.

Assume we have a basis, $\{\rho_1^i, \rho_2^i, \ldots, \rho_n^i\}$, for the vector space of solutions for z_i. We know that all members of the vector space of solutions are a linear combination of these basis vectors, that is,

$$z_i = \sum_{k=1}^{n} \beta_k \rho_k^i \qquad [49]$$

If, along with the difference equation we are given initial conditions, $z_0, z_1, \ldots, z_{n-1}$, these can be used to determine the values of the β_k by solving the set of equations produced by equation 49 with $i = 0, 1, 2, \ldots, n-1$.

8.3.2 LINEAR MULTISTEP METHODS

We now return to the difference equations for errors given by equations 43 and 44, treating equation 43 first. By Section 8.3.1 the basis solutions corresponding to it are the ith powers of the roots of

$$\rho^2 - \rho + 15/64 = 0 \qquad [50]$$

These roots are

$$\rho_{1,2} = \frac{1 \pm 1/4}{2} = 5/8, 3/8 \qquad [51]$$

Thus

$$\varepsilon_i = \beta_1(5/8)^i + \beta_2(3/8)^i \qquad [52]$$

with β_1 and β_2 chosen so that equation 52 produces the given value ε_0 for $i = 0$ and the given value ε_1 for $i = 1$. That is, β_1 and β_2 are the solutions of

$$\varepsilon_0 = \beta_1 + \beta_2 \qquad [53]$$

$$\varepsilon_1 = \beta_1(5/8) + \beta_2(3/8)$$

that is,

$$\beta_1 = 4\varepsilon_1 - (3/2)\varepsilon_0 \qquad \beta_2 = (5/2)\varepsilon_0 - 4\varepsilon_1 \qquad [54]$$

Since $|5/8| < 1$ and $|3/8| < 1$, $\lim_{i \to \infty} \varepsilon_i = 0$, so the recurrence given by equation 36 is absolutely stable. In the general case, we have absolute stability if all the characteristic roots are less than 1 in magnitude and absolute instability if any characteristic root is greater than 1 in magnitude.

Equations 52 and 54 can also be used to produce b_i, a bound on $|\varepsilon_i|$, given b_0 and b_1. Substituting equation 54 into equation 52 gives $s(i; \varepsilon_0, \varepsilon_1)$:

$$\varepsilon_i = \varepsilon_0((5/2)(3/8)^i - (3/2)(5/8)^i) + \varepsilon_1(4(5/8)^i - 4(3/8)^i) \qquad [55]$$

For $i > 1$, ε_i is maximum when ε_0 is as negative as possible ($\varepsilon_0 = -b_0$) and ε_1 is as positive as possible ($\varepsilon_1 = b_1$). Thus

$$b_i = b_0((3/2)(5/8)^i - (5/2)(3/8)^i) + 4b_1((5/8)^i - (3/8)^i) \qquad [56]$$

It can be seen that very quickly the terms in $(5/8)^i$ in equation 56 dominate the terms in $(3/8)^i$. That is, for all but the smallest i

$$b_i \approx ((3/2)\, b_0 + 4b_1)(5/8)^i \qquad [57]$$

In fact, for any difference equation with a single characteristic root of maximum magnitude ρ_{max}, ρ_{max}^i dominates for all but the smallest, i. Thus, in this common case

$$b_i \approx \left(\sum_{j=0}^{n-1} |\lambda_j| b_j \right) \rho_{max}^i \qquad [58]$$

where λ_j is the coefficient of ε_j in the β_k corresponding to ρ_{max}. When there is more than one root of maximum magnitude, for example, when there is a complex conjugate pair, matters complicate somewhat, but it remains the case that the error grows in magnitude by a factor of approximately $|\rho_{max}|$ at each step. Because of its simplicity, the approximation based on dominance of the characteristic roots of maximum magnitude should usually be used as the value of b_i in the analysis of recurrent methods.

The relative stability analysis is not very interesting for linear recurrence relations where the desired result is that given by the recurrence relation. In these cases we see that as with equations 36 and 37, the error ε_i and the value y_i satisfy the same recurrence relation and thus their solutions differ only in the coefficients of ρ_j^i:

$$\varepsilon_i = \sum_{j=1}^{n} \beta_j \rho_j^i \qquad y_i = \sum_{j=1}^{n} \phi_j \rho_j^i \qquad [59]$$

The β_j are determined by $\varepsilon_0, \varepsilon_1, \ldots, \varepsilon_{n-1}$, and likewise the ϕ_j are determined by $y_0, y_1, \ldots, y_{n-1}$. In the case with a single characteristic root of maximum magnitude ρ_k, ε_i is dominated by $\beta_k \rho_k^i$ and y_i is dominated by $\phi_k \rho_k^i$, so

$$\lim_{i \to \infty} \frac{\varepsilon_i}{y_i} = \lim_{i \to \infty} \frac{\beta_k \rho_k^i}{\phi_k \rho_k^i} = \frac{\beta_k}{\phi_k} \qquad [60]$$

That is, the recurrence is neither relatively stable nor relatively unstable, a more or less satisfactory situation. Here again matters are a bit complicated if the characteristic root of maximum magnitude is not unique, but the result that there is neither relative stability nor relative instability remains.

Let us now return to the analysis of cases where $v_i \neq y_i$ and in particular to the recurrent application of the modified Euler method. The difference equation for the error in this case was given by equation 44. The corresponding characteristic equation is

$$\rho^2 - 2hK\rho - 1 = 0 \qquad [61]$$

the roots of which are

$$\rho_{1,2} = hK \pm \sqrt{(hK)^2 + 1} \qquad [62]$$

so the general solution is

$$\varepsilon_i = \beta_1 (hK + \sqrt{1 + (hK)^2})^i + \beta_2 (hK - \sqrt{1 + (hK)^2})^i \qquad [63]$$

Note that the characteristic roots and thus the general solution are functions of hK.

Let us forego the analysis of the absolute stability of the modified Euler method and all future methods we will discuss, because we are principally interested in the relative stability of these methods. Thus we are interested in the value of ε_i/y_i. Since we have already restricted ourselves to those differential equations for which $\partial f/\partial y$ is constant, y_i is given by

$$y_i = \gamma_i e^{hKi} \qquad [64]$$

where

$$\gamma_i = y_0 + e^{hK} \int_{x_0}^{x_i} e^{-Ku} a(u) \, du$$

(see equation 32), so

$$\frac{\varepsilon_i}{y_i} = \frac{\beta_1}{\gamma_i} \left(\frac{hK + \sqrt{1 + (hK)^2}}{e^{hK}} \right)^i + \frac{\beta_2}{\gamma_i} \left(\frac{hK - \sqrt{1 + (hK)^2}}{e^{hK}} \right)^i \qquad [65]$$

β_1 and β_2 do not depend on i but only on ε_0 and ε_1, and γ_i is normally bounded away from zero and ∞ as $i \to \infty$. Thus, the behavior of the relative error given by equation 65 as $i \to \infty$ depends upon whether or not both of the parenthesized expressions in that equation are less than 1 in magnitude, that is whether both numerators are less in magnitude than e^{hK}. Since

$$\max(|hK + \sqrt{1 + (hK)^2}|, |hK - \sqrt{1 + (hK)^2}|) \\ = |hK| + \sqrt{1 + (hK)^2}, \qquad [66]$$

we can see that the modified Euler method for this class of differential equations is stable if and only if $|hK| + \sqrt{1 + (hK)^2} < e^{hK}$. This condition can be tested for any value of hK.

We are normally interested in applying methods with small values of $|h|$ and thus small values of $|hK|$. Let us thus analyze the stability condition for the modified Euler method for small values of $|hK|$. At $hK = 0$ both sides of the inequality to be tested are 1. Thus graphically, here and with many methods for the solution of differential equations, there are four possibilities for the case of small nonzero $|hK|$ as illustrated in Figure 8–1. In case 1 there is instability for both negative and positive hK. In cases 2 and 3 there is instability for one sign of hK and stability for the other sign, and in case 4 there is stability for any small $|hK|$. Instabilities in cases 1–3 are not as undesirable as in the case where the quotient even at $hK = 0$ is greater than 1.

In the method under analysis, we know that for hK small in magnitude and negative, $e^{hK} < 1$, and $|hK| + \sqrt{1 + (hK)^2} > 1$, so the method

FIGURE 8–1
Relative Stability Analysis for Small $|hK|$

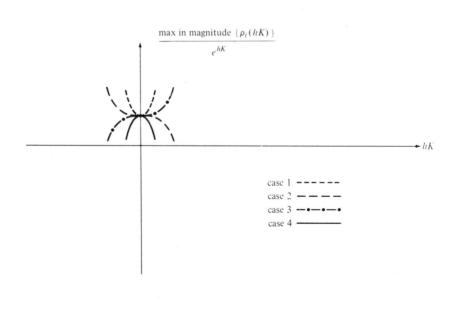

is relatively unstable for such hK. For small positive hK we know that

$$|hK| + \sqrt{1 + (hK)^2} = hK + 1 + \frac{(hK)^2}{2} + O(hK)^4 \qquad [67]$$

whereas

$$e^{hK} = 1 + hK + \frac{(hK)^2}{2} + \frac{(hK)^3}{6} + O(hK)^4 \qquad [68]$$

so e^{hK} is the larger of the two and thus the method is relatively stable. Thus we have shown the modified Euler method to be in case 2 in Figure 8-1.

For any other method we proceed similarly. We first find a recurrence relation for ε_i on the assumption that $\partial f/\partial y = K$, and we find the characteristic roots of the corresponding difference equation. We then know that the general solutions for the error propagated to ε_i from ε_0, ε_1, ..., ε_{n-1} is $\Sigma_{k=1}^{n} \beta_k \rho_k^i (hK)$ and that for this class of equations y_i is given by equation 64. Thus,

$$\frac{\varepsilon_i}{y_i} = \sum_{k=1}^{n} \frac{\beta_k}{\gamma_i} \left(\frac{\rho_k(hK)}{e^{hK}} \right) \qquad [69]$$

So the relative stability of the method depends upon whether all the characteristic roots are less in magnitude than e^{hK} for the value of hK in question.

Let us illustrate the above with the analysis of a more realistic method than that analyzed above, the modified-Euler/Heun predictor-corrector method with mop-up. We must first write as a single equation the recurrence relation corresponding to a single step:

$$
\begin{aligned}
\hat{y}_{i+1} &= (4/5)y_{i+1}^{(c)} + (1/5)y_{i+1}^{(p)} \\
&= (4/5)(\hat{y}_i + (h/2)(f(x_i, \hat{y}_i) + f(x_{i+1}, y_{i+1}^{(p)}))) + (1/5)y_{i+1}^{(p)} \\
&= (4/5)(\hat{y}_i + (h/2)f(x_i, \hat{y}_i) + (h/2)f(x_{i+1}, \hat{y}_{i-1} + 2hf(x_i, \hat{y}_i))) \\
&\quad + (1/5)(\hat{y}_{i-1} + 2hf(x_i, \hat{y}_i)) \\
&= (4/5)\hat{y}_i + (1/5)\hat{y}_{i-1} + (4h/5)f(x_i, \hat{y}_i) \\
&\quad + (2h/5)f(x_{i+1}, \hat{y}_{i-1} + 2hf(x_i, \hat{y}_i))
\end{aligned} \qquad [70]
$$

Thus the function g in equation 9 is

$$
\begin{aligned}
g(v_i, v_{i-1}; x_i) &= (4/5)v_i + (1/5)v_{i-1} + (4h/5)f(x_i, v_i) \\
&\quad + (2h/5)f(x_{i+1}, v_{i-1} + 2hf(x_i, v_i))
\end{aligned} \qquad [71]
$$

so

$$\left.\frac{\partial g}{\partial v_i}\right|_{v^{(i)}=\eta^{(i)}} = \frac{4}{5} + \frac{4}{5}h\left.\frac{\partial f(x, y)}{\partial y}\right|_{x_i, \eta_i}$$

$$+ \left.\frac{2h}{5}\frac{\partial f(x, y)}{\partial y}\right|_{x_{i+1}, \zeta_{i+1}} \cdot 2h\left.\frac{\partial f(x, y)}{\partial y}\right|_{x_i, \eta_i} \qquad [72]$$

where

$$\zeta_{i+1} \in [y_{i+1}^{(p)}, y_{i+1}^{(p)*}] \quad \text{and} \quad \eta_i \in [\hat{y}_i, \hat{y}_i^*]$$

and

$$\left.\frac{\partial g}{\partial v_{i-1}}\right|_{v^{(i)}=\eta^{(i)}} = \frac{1}{5} + \left.\frac{2h}{5}\frac{\partial f(x, y)}{\partial y}\right|_{x_{i+1}, \zeta_{i+1}} \qquad [73]$$

If $\partial f / \partial y = K$, equation 11 becomes

$$\varepsilon_{i+1} = ((4/5) + (4/5)hK + (4/5)(hK)^2)\varepsilon_i + ((1/5) + (2/5)hK)\varepsilon_{i-1} \quad [74]$$

The characteristic roots corresponding to this equation are

$$(2/5)(1 + hK + (hK)^2) \pm \sqrt{(4/25)(1 + hK + (hK)^2)^2 + (1/5) + (2/5)hK}$$

For $hK = 0$ the root with the negative square root has value $-1/5$, which is less than 1 in magnitude, so we need not be concerned with its characteristic root causing instability for small $|hK|$. The root with the positive square root has value 1 at $hK = 0$ and can be shown to be less than e^{hK} for small positive or negative hK. Thus we are in case 4 in Figure 8–1: the method is relatively stable for appropriately small h of either sign.

The characteristic root analysis also shows how small h must be for the recurrence to be stable, a question whose importance was illustrated in section 1.4.3. For example, from the formula just computed for the characteristic roots corresponding to the modified-Euler–Heun method, we see that if $K < 0$ and $x_{\text{goal}} > x_0$, so $h > 0$, the larger characteristic root is greater than e^{hK} for h greater than the positive root of the following function of h:

$$\frac{2}{5}(1 + hK + (hK)^2)$$

$$+ \sqrt{(4/25)(1 + hK + (hK)^2)^2 + (1/5) + (2/5)hK} - e^{hK}$$

For the recurrence to remain stable, h must be chosen to be smaller than this root.

8.4   Accumulation of Error

We have discussed only the propagation of initial errors by the repeated application of an operator. In reality each application of the operator not only propagates error but also generates error to be propagated by later applications of the operator. For example, with methods for the solution of differential equations, each step generates both approximation error and arithmetic error. The overall error in the result of any step is therefore produced by the propagation of the errors generated at all steps, not just the initial steps.

We assume that by solving the recurrence relation for error propagation (equation 11) we have produced a function $s(i; \varepsilon_0, \varepsilon_1, \ldots, \varepsilon_{n-1})$ that gives the effect of errors in n inputs on the result of recurrently applying the operator in question $i - n + 1$ times. We assume further that from this function s we can produce a function $t(i; b_0, b_1, \ldots, b_{n-1})$ such that if $|\varepsilon_i| \leq b_i$ for $0 \leq i \leq n - 1$, then the error in v_i due to errors in $v_0, v_1, \ldots, v_{n-1}$ is bounded by

$$b_i^{\text{prop}} = t(i; b_0, b_1, \ldots, b_{n-1}) \qquad [75]$$

For example, equation 56 gives the function t corresponding to the function s given by equation 55.

If the operator being repeated is linear, then so is the operator made from k successive applications of it, for any k. Thus, in this case the overall error in v_i^*, $\varepsilon_i^{\text{overall}}$ = the error generated from the "initial errors," ε_0 through ε_{n-1}, + $\Sigma_{k=n}^i$ [the error propagated from the error generated in y_k, namely, $\varepsilon_k^{\text{gen}}$, and zero error in the $n - 1$ previous steps]. Since $s(i-j; \varepsilon_j, \varepsilon_{j+1}, \ldots, \varepsilon_{j+n-1})$ specifies the error propagation from ε_j through ε_{j+n-1} into ε_i, the term in the sum is $s(i-k+n-1; 0,0,\ldots,0,\varepsilon_k^{\text{gen}})$. Thus,

$$\varepsilon_i^{\text{overall}} = s(i; \varepsilon_0, \varepsilon_1, \ldots, \varepsilon_{n-1})$$
$$+ \sum_{k=n}^{i-1} s(i-k+n-1; 0,0,\ldots,0,\varepsilon_k^{\text{gen}}) + \varepsilon_i^{\text{gen}} \qquad [76]$$

It follows that

$$b_i^{\text{overall}} = t(i; b_0, b_1, \ldots, b_{n-1})$$
$$+ \sum_{k=n}^{i-1} t(i-k+n-1; 0,0,\ldots,0,b_k^{\text{gen}}) + b_i^{\text{gen}} \qquad [77]$$

The following example will illustrate the calculation of this important bound on the overall error in the result of a recurrence. Assume that the equation $y' = Ky$, $y(0) = \pi$ is being solved using the modified-Euler–Heun method with mop-up with $hK = 3/2$. According to the formula given after equation 74, the characteristic roots of the difference equation for propagated error are $\rho = 4$ and $\rho = 1/5$, that is,

$$\varepsilon_i^{\text{prop}} = c_0 4^i + c_1 (-1/5)^i \qquad [78]$$

The modified-Euler–Heun method requires two starting values y_0 and y_1 (that is, $n = 2$), where $y_0 = \pi$ is given and y_1 is computed from y_0 by some start-up method for differential equation solution. The constants c_0 and c_1 are determined using equation 78 from the errors in these two starting values:

$$\varepsilon_0 = c_0 + c_1$$
$$\varepsilon_1 = 4c_0 - c_1/5 \qquad [79]$$

producing

$$c_0 = (5\varepsilon_1 + \varepsilon_0)/21$$
$$c_1 = (20\varepsilon_0 - 5\varepsilon_1)/21 \qquad [80]$$

Thus

$$\varepsilon_i^{\text{prop from } y_0, y_1} = s(i; \varepsilon_0, \varepsilon_1)$$
$$= [4^i(5\varepsilon_1 + \varepsilon_0) + (-1/5)^i(20\varepsilon_0 - 5\varepsilon_1)]/21 \qquad [81]$$

From equation 81 it follows that

$$b_i^{\text{prop from } y_0, y_1} = t(i; b_0, b_1)$$
$$= [4^i(5b_1 + b_0) + (-1/5)^i(20b_0 - 5b_1)]/21 \qquad [82]$$

Equation 77 then gives the bound on the overall error in y_i:

$$b_i^{\text{overall}} = [4^i(5b_1 + b_0) + (-1/5)^i(20b_0 - 5b_1)]/21$$
$$+ \sum_{k=2}^{i-1} b_k^{\text{gen}}[(5)4^{i-k+1} - 5(-1/5)^{i-k+1}]/21 + b_i^{\text{gen}} \qquad [83]$$

We now need only the values of b_0, b_1, and b_k^{gen} for $k = 2,3,...,i$ to obtain the bound on the overall error in y_i. y_0 has only representation error, so $b_0 = R\pi$, where R depends on the parameters of the floating-point arithmetic used. b_1 is calculated from analysis of the error generated by arithmetic and approximation in the method used for start-up. b_k^{gen} for $2 \le k \le i$ is likewise the sum of an arithmetic error bound and an approximation error bound, but here on the error generated by one step of the modified-Euler–Heun method with mop-up. The calculation of these bounds will be left to the reader with two comments:

1. b_k^{arith} will depend on the computed values for y_{k-1} and y_{k-2}, that is, v_{k-1} and v_{k-2}. These values are derived from the solution of the difference equation for v_i, which we know to be parallel to the solution of the same difference equation for ε_i (equation 81):

$$v_i = [4^i(5y_1 + y_0) + (-1/5)^i(20y_0 - 5y_1)]/21 \qquad [84]$$

2. The value of b_k^{approx} involves $y^{(4)}$, and from $y' = Ky$ we see that

$$y^{(2)} = Ky' = K^2y, \; y^{(3)} = Ky^{(2)} = K^3y, \; y^{(4)} = Ky^{(3)} = K^4y \qquad [85]$$

We now return to our general discussion of the overall error in the recurrent application of an operator. This error is given by equation 76, and its bound is given by equation 77. If the operator in question is unstable, that is, the first term in equation 76 becomes infinite as $i \to \infty$, then the overall error will become infinite as $i \to \infty$. But note that even if the operator is stable the overall error given by equation 76 may still become infinite as $i \to \infty$. The individual terms may approach zero, but not fast enough to force the same behavior for the sum, which includes an infinite number of terms as $i \to \infty$.

If the operator being repeated is nonlinear, the linear recurrence relation for error propagation produced by the partial derivative approach is only an approximation to the actual recurrence relation for the error. The fact that the actual recurrence relation for the error is nonlinear means that the sum of the effects of two components of the error in v_k^* on v_i^* is not the same as the effect of the sum of the components of the error on v_i^*. Thus, the error generated in calculating v_k from v_{k-1}, v_{k-2}, ..., v_{k-n} cannot strictly be treated separately from the error propagated in the calculation of v_k. This problem strictly applies as well to the case of nonrecurrent operators treated in Chapter 7, but because the number of operators in sequence is low there the problem is negligible. With recurrent operators i can be quite large, and the second-order terms may not be negligible. However, if the linear recurrence solution given by the partial derivative approach is a reasonably good approximation, the bound given by equation 77 is acceptable. Operators for which the approximation is not good are very difficult to analyze, both with regard to the propagation of initial errors and to the overall error. They need to be handled step by step, rather than by using a recurrence relation.

8.5 Summary and Complements

The analysis of the error generated by recurrent methods is based on the solution of recurrence relations for the error. These recurrence relations are produced by the partial derivative approach applied to the operator that is recurrently repeated in the computation (see equation 11). The solution of this linear recurrence relation gives

$$\varepsilon_i = \sum_{k=1}^{n} \beta_k \rho_k^i \qquad [86]$$

From this it follows that absolute instability occurs when any ρ_k is greater than 1 in magnitude.

The recurrence that is applied in the computation, producing the variable v_i, is the same as that for the error ε_i if the recurrence relation for v_i is linear. If the recurrence relation for v_i is nonlinear, it is normally linearized before it is analyzed, for example, in the case of differential equations where the equation to be solved is assumed to be of the form $y' = Ky$. The linearization causes v_i and ε_i to satisfy the same difference equations here also. Thus the solutions for v_i and ε_i involve the same characteristic roots and differ only in the coefficients of the powers of the characteristic roots:

$$v_i = \sum_{k=1}^{n} \phi_k \rho_k^i \qquad [87]$$

Analysis of the relative error ε_i/v_i is thus straightforward.

We have noted, however, that the relative error of concern is ε_i/y_i, where y_i is the solution to the original problem rather than the solution v_i to the recurrence applied, which may be an approximation to the desired recurrence. Thus relative stability was defined as a description of the behavior of ε_i/y_i. We believe that this definition is most consistent with the rest of error analysis in numerical computing and applies to all recurrent numerical methods. However, for the analysis of methods for the solution of differential equations, many numerical analysts (like Ralston and Rabinowitz, 1978) have defined relative stability somewhat differently by choosing a different desired solution.

For the solution of differential equations of the form $y' = Ky$, one of the characteristic roots $\rho_k(hK)$ approaches e^{hK} as $h \to 0$. All the remaining characteristic roots can be called *parasitic*. Though some instability in the sense we have used earlier in this chapter can occur due to the nonparasitic root being greater than e^{hK}, the most dramatic instability occurs when some of the parasitic roots for the given value of hK are greater in magnitude than the nonparasitic root. Thus, considering the nonparasitic part of the solution to be the desired solution, many numerical analysts define a numerical method to be relatively unstable if for the given value of hK any of the parasitic roots is greater in magnitude than the nonparasitic root.

Returning to the approach recommended in this book, we use equation 86 to produce a bound on the error propagated into v_i from errors in the initial values $v_0, v_1, \ldots, v_{n-1}$. Note that we do not find a recurrence relation for the propagated error *bound*, but rather we find a recurrence relation for the propagated error itself and use the solution to produce a bound on the propagated error. The former approach would give an overly pessimistic bound for the error after many steps; the latter gives a more realistic bound.

A bound produced by the approach just described is not used by itself. Rather for a given number of applications of the repeated operator such a bound is produced (a) for the error due to the inputs to the first operator application and propagated through all applications of the operator, and (b) for each application of the operator, the error generated by that application and propagated through the remaining applications of the operator. For N applications of the operator these $N + 1$ bounds are summed to produce an overall bound (see equation 77).

The recurrence relations being solved involve parameters of the repeated operator being analyzed such as the step size. Therefore, the solutions to the recurrence relations and thus the overall bounds produced are functions of these parameters. These bounds are commonly used to choose the values of these parameters as well as to bound the error for given values of the parameters.

The error analysis together with the efficiency analysis of various methods for solving a problem can also be used to choose among these methods. Methods may differ in the work required per step, in the approximation error produced as a function of the step size, and in the degree to which this error (as well as the arithmetic error, which is often negligible) is increased or decreased by propagation. For a given accuracy requirement and class of problems, we can find for each method the approximate step size required and the work per step, and thus we can find the overall work required. For some methods this work will be large because poor approximation error properties require a small step; for other methods this work will be large because a limited region of adequate stability requires a small step; and for still other methods this work will be large because of a large amount of work per step.

The decision on which method to choose often depends on the particular problem, say the differential equation, being solved. A class of problems that have distinctive enough characteristics that a special group of methods has been developed for them is the so-called stiff differential equations (Gear, 1971; Ralston and Rabinowitz, 1978). These equations have the property that the stability of methods for their solution is strongly affected by an exponential component of the solution that has little or no contribution to the solution itself. In order to avoid the necessity of using a tiny step size to achieve satisfactory stability with standard methods, special methods have been developed whose better stability properties for this type of differential equation make them the methods of choice for solving such equations despite their somewhat poorer approximation error properties.

REFERENCES

Gear, C. W., *Numerical Initial Value Problems in Ordinary Differential Equations.* Prentice-Hall, Englewood Cliffs, N.J., 1971.

Ralston, A., and Rabinowitz, P., *A First Course in Numerical Analysis,* 2nd ed. McGraw-Hill, New York, 1978.

PROBLEMS

8.1 Assume one has a set of data points (x_i, y_i), for $i = 0, 1, 2, \ldots,$ n, where the x_i are equally spaced, and one wishes to fit the data by cubic splines, $p_i(x)$, such that $\hat{f}(x) = p_{i+1}(x)$ if $x \in [x_i, x_{i+1}]$, $p_{i+1}(x)$ agrees with $p_i(x)$ in value, slope, and curvature at x_i, and $p_{i+1}(x)$ exact-matches at x_i and x_{i+1} (this is the usual set of constraints for cubic splines). Assume that instead of using the boundary conditions $c_0 = c_n = 0$, where c_i is the second derivative at x_i, we are given measured values for c_0 and c_1. We can then use the recurrence relation given by equation 54 in Chapter 3 to compute $c_2, c_3, c_4, \ldots,$ and thus the splines $p_i(x)$.

(a) Show that equation 54 of Chapter 3 can be rewritten as

$$hc_{m+1} + 4hc_m + hc_{m-1} = \frac{6}{h}\delta^2 y_m$$

(b) Assume that there is a measurement error in our values for c_0 and c_1. Analyze the *absolute* stability of the operator defined by the above recurrence relation, assuming the y_i are error free. How does this stability depend on h?

8.2 Show that the recurrence $y_{i+1} = y_i + dy_i^3$, $y_0 = 1$ will be absolutely unstable for any value of $d > 0$ and will not be absolutely unstable if $-2 < d < 0$. (Note that the resulting difference equation does not have constant coefficients. You have to go back to first principles to solve this problem.)

8.3 For the predictor, corrector, and mop-up formulas determined in Problems 5.11 and 7.9, analyze the relative stability of repeating the step made from combining these operations. (*Hint:* In comparing $\rho_{max}(hK)$ to e^{hK}, you will find it simple to compare $\rho_{max}(hK) - \frac{2}{3}(hK)^2$ to $e^{hK} - \frac{2}{3}(hK)^2$ and expand each of these in Taylor series about $hK = 0$.)

8.4 For the predictor, corrector, and mop-up formulas determined in Problems 5.13 and 7.10, find the values of the parameters A_1 and C_1 for which repeating the step made from combining these operations is rel-atively stable.

8.5 Assume that a subroutine for computing e^z for $z > 0$ and $z = a$ multiple of 0.01 operates as follows. Solve the differential equation $y'(x) = y(x)$, $y(0) = 1$ at $x_{goal} = z$ using (simple) Euler's method with $h = 0.01$. Assume computation using hexadecimal floating-point numbers with a 14-digit fraction.

(a) Give a formula in terms of $\hat{y}_i$ for a bound on the magnitude of the error generated at the ith step. Do this determining the bound in the normal way and then substituting $\hat{y}_i$ for both y_i and e^{x_i} where they appear in the bound.

(b) In terms of z, find a bound on the error in $y(z)$ due to the error generated at the ith step. Use your answer to part a.

(c) In terms of z, give a bound on the overall error in $y(z)$.

(d) The correct answer to part c is

$$b_{y(z)} \approx (2.03 \times 16^{-13} + 5.05 \times 10^{-3})(1.01)^{100z-1}z$$

Compare the accuracy and efficiency of this method to a method which requires 100 arithmetic operations and produces an absolute error bounded in magnitude by $10^{-3}z$.

8.6 Assume that to solve the first-order ordinary differential equation $y' = f(x,y)$, with initial value $y(x_0) = y_0$, we wish to employ a recurrent method of the form

$$\hat{y}_{i+1} = A\hat{y}_i + Bhf(x_{i-1},\hat{y}_{i-1})$$

(a) Show that $A = 1$ and $B = 1$ if the method is to be appropriate for the case of small interval width h and decreasing $y(x)$ derivative magnitudes with derivative order.

(b) If $\partial f/\partial y = K$, show that the method is relatively stable for small $|hK|$. You may use the Taylor-series-based approximation $(1 + z)^{1/2} \approx 1 + \frac{1}{2}z - \frac{1}{8}z^2$.

(c) Assume $hK = -0.09$, $b_{y_0} = 8 \times 10^{-4}$, and $b_{y_1} = 8 \times 10^{-5}$. Give an approximate bound on the error in y_i due to these input errors, assuming i is large.

(d) Assume that $hK = -0.09$, $b_{y_0} = 0$, and $b_{y_1} = b$. Show that the error in y is bounded in magnitude by $1.25b[(0.9)^i - (0.1)^i]$.

(e) Assuming the same data as in part d, show that if an error bounded by b is generated at every step, for large i, $b_{y_i}^{\text{overall}} = 100b/9$.

(f) Assume that the method involves only one step. That is, assume $x_{\text{goal}} - x_0 = 2h$, $x_1 = x_0 + h$, and $\hat{y}_{\text{goal}} = y_1 + hf(x_0,y_0)$. If the relative error generated in calculating $f(x_0,y_0)$ is bounded by r_f and all arithmetic is done on a 24-digit binary floating-point computer, give a bound on the error in the computed value of $\hat{y}_{\text{goal}}$, given that x_0, y_0, y_1, and x_{goal} are provided as exact input.

(g) If the formula $\hat{y}_{i+1} = \hat{y}_i + hf(x_{i-1},\hat{y}_{i-1})$ were used as one estimator of y_{i+1} and Euler's method, which has approximation error $\varepsilon_{i+1}^{\text{approx}} = -(h^2/2)y_i'' + O(h^3)$, were used as another, what function of the two of these would produce a method with approximation error of $O(h^3)$?

8.7 Say whether the following statement is true or false and justify your choice: Decreasing the step size at step i of the application of a predictor-corrector method always decreases the error in y_k where $k > i$.

8.8 Determine the interval width h so that four decimal places of accuracy are obtained for $y(0.5)$ when applying Euler's method to the differential equation $y' = x^2 + y$, $y(0) = 1$.

CHAPTER NINE

Iterative Application of Operators: Convergence

When an operator T is repeated iteratively, the output of one step has no "correct value" but rather is an approximation to the limit that the sequence would achieve in the absence of generated error. Thus the error in x_i, the output of the ith step, is defined as

$$\varepsilon_i \triangleq x_i - \xi \qquad [1]$$

where

$$\xi = \lim_{i \to \infty} x_i \qquad [2]$$

Examples of iterative methods are those solving nonlinear equations, like Newton's method, those solving linear equations, like the Jacobi method and iterative improvement after Gaussian elimination using triangular factorization, and infinite series approximation methods, like the Newton divided-difference method. In analyzing such methods we normally have an analytical formula, implicit or explicit, for the desired value ξ and we wish to know the accuracy and efficiency of a method based on iterating a given operator T.

We will approach iterative methods much as we did recurrent methods. Only the definition of error has changed. We will first obtain an expression for ε_{i+1} in terms of ε_i in the absence of generated error, then determine ε_i as a function of i, and finally determine $\lim_{i \to \infty} \varepsilon_i$. Assuming convergence is obtained ($\lim_{i \to \infty} \varepsilon_i = 0$), we will be interested in the speed of convergence and the effect of generated error, producing from the latter analysis a bound on the error after infinitely many iterative steps. First, we will analyze single-step operators where the convergence is of the type called linear.

9.1 Single-Step Methods with Linear Convergence

A single-step method is one for which, in the absence of generated error, the transformation

$$x_{i+1} = T(x_i, x_{i-1}, x_{i-2}, ..., x_{i-n+1}) \qquad [3]$$

is given by

$$x_{i+1} = g(x_i) \qquad [4]$$

that is, the iterate is an explicit function only of the previous iterate. Remember that such an iteration is called a Picard iteration.

We want to analyze the error ε_i in x_i in the absence of generated error. To start, we wish to analyze convergence, that is, whether $\varepsilon_i \to 0$ as $i \to \infty$, but first we must determine whether such convergence, if it occurs, will be to a desired value, say the root of the nonlinear equation we wish to solve. Since convergence can occur only to a fixed point, if the desired solution is ξ, we check that $g(\xi) = \xi$.

Assuming $g(\xi) = \xi$, we must analyze the convergence of x_i to ξ, that is, of ε_i to 0. Analysis in the absence of generated error is of propagated error. The partial derivative approach to propagation analysis depends upon expanding g in a Taylor series about the point with no error, here ξ, giving

$$\varepsilon_{i+1} = \varepsilon_{g(x_i)} = g'(\eta_i)\,\varepsilon_i \qquad [5]$$

for some $\eta_i \in [x_i, \xi]$, assuming g is differentiable in $[x_i, \xi]$.

Convergence can be assured if for every i, $|\varepsilon_{i+1}| \le k|\varepsilon_i|$ for some positive constant $k < 1$. This condition can in turn be assured for a particular i if for all $\eta \in [x_i, \xi]$, $|g'(\eta)| \le k < 1$. If furthermore for all $\eta \in [x_i, \xi]$, $g'(\eta) \ge 0$, then ε_{i+1} will have the same sign as ε_i, so $x_{i+1} \in [x_i, \xi]$ and thus $[x_{i+1}, \xi] \subset [x_i, \xi]$ (see Figure 9–1a), so $|g'(\eta)| \le k < 1$ will also hold for $\eta \in [x_{i+1}, \xi]$. We can thus conclude that if for all $\eta \in [x_0, \xi]$,

$$0 \le g'(\eta) \le k < 1 \qquad [6]$$

then the sequence $\{x_i\}$ defined by $x_{i+1} = g(x_i)$ will converge. Note that this is a sufficient condition only; convergence may occur in the absence of this condition.

If $g'(\eta) < 0$ for some values of $\eta \in [x_i, \xi]$, then x_{i+1} may be on the opposite side of ξ from x_i, that is, $[x_i, \xi]$ may not include $[x_{i+1}, \xi]$ even though $|\varepsilon_{i+1}| < |\varepsilon_i|$ (see Figure 9–1b). In this case, to assure convergence we must have $|g'(\eta)| \le k < 1$ for $\eta \in [x_0, x']$, where x' is the farthest possible x_1 from x_0:

$$x' = \xi + \min_{\eta \in [x_0, \xi]} g'(\eta)(x_0 - \xi) \qquad [7]$$

FIGURE 9–1

Possible Relationships among x_i, x_{i+1}, ξ in a Convergent Iteration

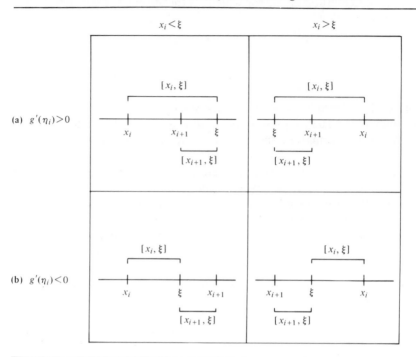

Since this condition is often difficult to check and $[x_0, x']$ is contained by $[x_0, 2\xi - x_0]$, the interval of width $2|x_0 - \xi|$ centered at ξ, we often assure convergence by checking that

$$|g'(\eta)| \leq k < 1 \qquad \text{for} \quad \eta \in [x_0, 2\xi - x_0] \qquad \text{[8]}$$

In summary, given an explicit formula for ξ and a value x_0, we check whether $0 \leq g'(\eta) \leq k < 1$ for $\eta \in [x_0, \xi]$. If so, convergence is assured. If $g'(\eta) \geq 1$ or $g'(\eta) \leq -1$ on this interval, we conclude that convergence cannot be assured from this x_0. If none of these conditions hold, that is, $-1 < g'(\eta) < 0$ for some $\eta \in [x_0, \xi]$, then we check whether $|g'(\eta)| \leq k < 1$ for $\eta \in [x_0, x']$, where x' is defined by equation 7, or more commonly for $\eta \in [x_0, 2\xi - x_0]$. If either of these conditions holds, convergence from x_0 is assured; if not, convergence cannot be assured.

If we have no explicit formula for ξ, estimates $\hat{\xi}$ for ξ and $\hat{x}'$ for x' must be made such that the intervals $[x_0, \hat{\xi}]$ and $[x_0, \hat{x}']$ certainly include $[x_0, \xi]$ and $[x_0, x']$, respectively, and the above process must be carried out for these intervals involving the estimates.

As an example, let us analyze whether the iteration, $x_{i+1} = x_i + (1/2)(x_i^2 - a)$, $x_0 = -a$, can be used to compute $-\sqrt{a}$, where $a > 0$. We first check that $-\sqrt{a}$ is a fixed point by substituting $-\sqrt{a}$ for both x_{i+1} and x_i in the formula for the iterative step: $-\sqrt{a} \overset{?}{=} -\sqrt{a} + (1/2)((-\sqrt{a})^2 - a)$. Since this checks, we investigate convergence by examining the derivative of $g(x) = x + (1/2)(x^2 - a)$, namely, $g'(x) = 1 + x$. For $0 < a < 1$, $0 < g'(x) < 1$ on $[x_0, \xi] = [-a, -\sqrt{a}]$, so convergence is assured. For $a > 2$, $|g'(-a)| > 1$ so convergence from $x_0 = -a$ can not be assured by this method. For $1 < a < 2$, $-1 < g'(a) < 0$, so we investigate $g'(x)$ on $[-a, -2\sqrt{a} + a]$. Since $|1 + x| < 1$ on this interval, convergence is assured for $1 < a < 2$.

Returning to the general analysis, consider equation 5 again. Assuming $\{x_i\}$ converges to ξ, $\{\eta_i\}$ also converges to ξ, since $\eta_i \in [x_i, \xi]$. So as $x_i \to \xi$,

$$\varepsilon_{i+1} \to g'(\xi) \varepsilon_i \qquad [9]$$

In the absence of generated error the error in the $(i+1)$th iterate is a constant times the error in the ith iterate. Such convergence is said to be *linear* with *convergence factor* $g'(\xi)$.

Note that convergence can occur only if $|g'(\xi)| \leq 1$ (in the limit the error gets smaller at each step). If $|g'(\xi)| < 1$ and g' exists in a neighborhood of ξ, there will be a neighborhood of ξ such that the iteration $x_{i+1} = g(x_i)$ converges to ξ in the absence of generated error if x_0 is in the neighborhood. For such an x_0 appropriately close to ξ,

$$\varepsilon_n \approx [g'(\xi)]^n \varepsilon_0 \qquad [10]$$

To generalize, an iteration converges (diverges) linearly to (from) ξ if as $x_i \to \xi$,

$$\varepsilon_{i+1} \to k_1 \varepsilon_i \qquad [11]$$

for some constant k_1. Convergence will be obtained for x_0 in some neighborhood of ξ if $|k_1| < 1$. Thus a small enough value of $|k_1|$ is desirable to achieve convergence. Furthermore, assuming $|k_1| < 1$, in the limit the error is diminished in n steps by $|k_1|^n$, so the speed of convergence to within a given error tolerance depends on $|k_1|$. More precisely, if $\varepsilon_{i+1} = k_1 \varepsilon_i$ for $i = 0, 1, 2, \ldots$, to achieve accuracy δ, $\log(\delta/|\varepsilon_0|)/\log|k_1|$ iterative steps are required (see Problem 9.1); if N_1 operations are required per iterative step and setup operations are ignored, the total number of operations required is

$$t_1 = N_1 \frac{\log(\delta/|\varepsilon_0|)}{\log|k_1|} \qquad [12]$$

Clearly the smaller the value of $|k_1|$, the fewer will be the number of steps required to achieve the desired accuracy, so small $|k_1|$ not only provides convergence but fast convergence.

All the preceding has been in the absence of generated error, that is, has been a discussion of propagated error. With no generated error an iterative method begins with an initial error ε_0 which is propagated into ε_1 by an application of g to x_0, then ε_1 is propagated into ε_2, etc. For convergence the propagation must be such that eventually the successive errors get smaller. That is, producing convergent iterations involves taking advantage of error propagation. Also note that with iterative methods, "stability" (error propagation to zero) has another name: "convergence."

But in real life each iterative step not only propagates errors — it also generates error. Let us assume the error generated at each step is bounded in magnitude by b^{gen}. The value of b^{gen} can be calculated by the techniques of Chapters 5–7, or perhaps also of Chapter 8 if the iterative step involves a recurrence, or of this chapter if the iterative step involves an iteration. Then ε_{i+1} = error propagated by step i + error generated by step i, so for linearly convergent iterations in the limit

$$\varepsilon_{i+1} = k_1\varepsilon_i + \varepsilon_i^{gen} \qquad [13]$$

Continuing to iterate will be advantageous as long as $|\varepsilon_{i+1}| < |\varepsilon_i|$. In the worst case

$$|\varepsilon_{i+1}| = |k_1|\,|\varepsilon_i| + b^{gen} \qquad [14]$$

so we can assure continued improvement in accuracy as long as

$$|k_1|\,|\varepsilon_i| + b^{gen} < |\varepsilon_i| \qquad [15]$$

that is, as long as

$$|\varepsilon_i| > \frac{b^{gen}}{1 - |k_1|} \qquad [16]$$

When

$$|\varepsilon_i| = \frac{b^{gen}}{1 - |k_1|} \qquad [17]$$

we can no longer ensure improvement in accuracy. That is, with linear convergence the error bound on x_i^*, assuming enough iterations are carried out, is

$$b_\infty = \frac{b^{gen}}{1 - |k_1|} \qquad [18]$$

Note that b_∞ increases as $|k_1|$ increases to 1. Thus (see Figure 9–2) for a convergence factor, k_1, of 0.9, $b_\infty = 10\,b^{gen}$ because far from the root (for $\varepsilon_i = 10\,b^{gen}$) the improvement in accuracy due to error prop-

FIGURE 9-2

Dependence of Improvement of Accuracy by One Iteration on Linear Convergence Factor

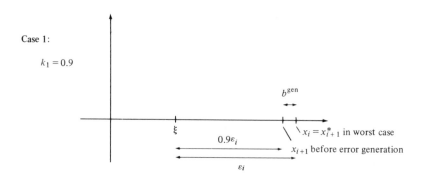

Case 1:

$k_1 = 0.9$

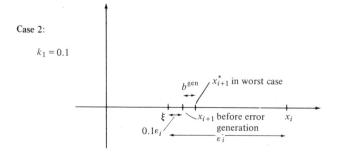

Case 2:

$k_1 = 0.1$

agation is small enough to be canceled out by error generation, whereas if $k_1 = 0.1$, $b_\infty = 1.1 b^{\text{gen}}$ because, with faster convergence, improvement far from the root is not canceled out by generated error. Thus, a small value of $|k_1|$ not only produces convergence and speed of convergence, it also produces accuracy.

9.2 Order of Convergence

We have seen three reasons to have as small a magnitude of the linear convergence factor as possible: existence of convergence, speed of convergence, and accuracy. Indeed we would like $k_1 = 0$. In Picard iteration $k_1 = g'(\xi)$, and if $g'(\xi) = 0$ we can analyze propagated error by carrying one step further in the Taylor series of equation 5:

$$g(x_i) = g(\xi) + g'(\xi)\varepsilon_i + \frac{g''(\zeta_i)}{2}\varepsilon_i^2 \qquad [19]$$

for some $\zeta_i \in [x_i, \xi]$, from which follows

$$\varepsilon_{i+1} = \varepsilon_{g(x_i)} = g(x_i) - g(\xi) = \frac{g''(\zeta_i)}{2}\varepsilon_i^2 \qquad [20]$$

so as $x_i \to \xi$,

$$\varepsilon_{i+1} \to \frac{g''(\xi)}{2}\varepsilon_i^2 \qquad [21]$$

To generalize as before, consider iterations with the property that, in the absence of generated error, as $x_i \to \xi$,

$$\varepsilon_{i+1} \to k_2\varepsilon_i^2 \qquad [22]$$

Such iterations are said to have *order of convergence* 2, or to have *quadratic* convergence, with convergence factor k_2. Newton's method is an example of an iteration with quadratic convergence, since there

$$g(x) = x - \frac{f(x)}{f'(x)} \qquad [23]$$

so

$$g'(x) = \frac{f(x)f''(x)}{[f'(x)]^2} \qquad [24]$$

from which we see that at a single root (where $f'(\xi) \neq 0$) $g'(\xi) = 0$.

Let us define more generally and precisely the order of convergence of a convergent sequence $\{x_i\}$. Intuitively and most commonly[†] it is the largest number α such that

$$\lim_{x_i \to \xi} [\varepsilon_{i+1}/\varepsilon_i^\alpha] = k_\alpha \neq 0 \qquad [25]$$

In the following we will assume this form, though the results adduced will also be able to be proved in a more complicated way using the more general form.

From the above we have seen that any single-step iteration for which g is appropriately differentiable has order of convergence 1 if $g'(\xi) \neq 0$ and 2 if $g'(\xi) = 0$ but $g''(\xi) \neq 0$. By the same argument, based on expanding $g(x_i)$ in a Taylor series about ξ, a single-step iteration with g appropriately differentiable has integer order of convergence n, where n is the smallest integer ≥ 1 such that $g^{(n)}(\xi) \neq 0$. Functions g which are

---

† To cover all peculiar cases the complete definition of the order of convergence is the least upper bound α of the numbers β such that the sequence $\{\varepsilon_{i+1}/\varepsilon_i^\beta\}$ has only finite limit points.

not appropriately differentiable near ξ (for instance, $g(x) = x + |x - \xi|^{3/2}$) and multistep iterations can produce noninteger orders of convergence.

If $\alpha > 1$ the convergence is said to be *superlinear*. Superlinear convergence corresponds to linear convergence with $|k_1| = 0$, as can be seen in the following. If in the limit as $x_i \to \xi$

$$\varepsilon_{i+1} = k_\alpha \varepsilon_i^\alpha, \qquad [26]$$

then this can be rewritten

$$\varepsilon_{i+1} = k_{1_i} \varepsilon_i \qquad [27]$$

where

$$k_{1_i} = k_\alpha \varepsilon_i^{\alpha-1}; \qquad [28]$$

the sequence can be thought to be converging linearly with a convergence factor that changes from step to step. (Of course, this is not true linear convergence, but the intuition is helpful.) As $\varepsilon_i \to 0$, $k_{1_i} \to 0$ for $\alpha > 1$, that is, the corresponding linear convergence factor is zero.

We have seen that a zero linear convergence factor is best with regard to existence of convergence, speed of convergence, and accuracy. Let us examine each of these properties for superlinearly convergent iterations. In examining the first two, we will assume no generated error.

We can show that if g has order of convergence $\alpha > 1$ at ξ, there exists a neighborhood of ξ such that if x_0 is in the neighborhood, the sequence will converge to ξ. Since

$$|x_{i+1} - \xi| \approx |k_\alpha||x_i - \xi|^\alpha = (|k_\alpha||x_i - \xi|^{\alpha-1})|x_i - \xi| \qquad [29]$$

we require only $|x_i - \xi| \leqslant |k_\alpha|^{1/(1-\alpha)}$ for convergence.

If equation 29 holds for x_i far from ξ (and it need not, even if $\alpha > 1$), we see that for $|x_i - \xi|$ large enough, $|x_{i+1} - \xi| > |x_i - \xi|$, so methods producing sequences with order of convergence greater than 1 may diverge if x_0 is not sufficiently close to ξ.

For $\alpha > 1$, assuming

$$\varepsilon_{i+1} = k_\alpha \varepsilon_i^\alpha \qquad [30]$$

the number of iterative steps required to achieve an error of magnitude δ starting from x_0 is obtained by solving $|\varepsilon_m| = \delta$ for m. Equation 30 produces

$$\varepsilon_m = k_\alpha \varepsilon_{m-1}^\alpha = k_\alpha (k_\alpha \varepsilon_{m-2}^\alpha)^\alpha = k_\alpha^{1+\alpha} \varepsilon_{m-2}^{\alpha^2}$$
$$= k_\alpha^{1+\alpha} (k_\alpha \varepsilon_{m-3}^\alpha)^{\alpha^2} = k_\alpha^{1+\alpha+\alpha^2} \varepsilon_{m-3}^{\alpha^3} \qquad [31]$$

Continuing this process we see

$$\varepsilon_m = k_\alpha^{(\alpha^m - 1)/(\alpha - 1)} \varepsilon_0^{\alpha^m} \qquad [32]$$

Solving

$$|k_\alpha^{(\alpha^m - 1)/(\alpha - 1)} \varepsilon_0^{\alpha^m}| = \delta \tag{33}$$

produces

$$m = \frac{1}{\log(\alpha)} \log\left[\frac{\log(|k_\alpha|)/(\alpha - 1) + \log(\delta)}{\log(|k_\alpha|)/(\alpha - 1) + \log(|\varepsilon_0|)}\right] \tag{34}$$

(see Problem 9.5a). If N_α operations are required per iterative step and setup operations are ignored, the total number of operations required is

$$t_\alpha = \frac{N_\alpha}{\log(\alpha)} \log\left[\frac{\log(|k_\alpha|)/(\alpha - 1) + \log(\delta)}{\log(|k_\alpha|)/(\alpha - 1) + \log|\varepsilon_0|}\right] \tag{35}$$

Note that the base of the logarithm is arbitrary.

If we are interested in limiting behavior, namely, for arbitrarily small $|\varepsilon_0|$ and δ, we can show (see Problem 9.5b) that if $\alpha > 1$, for any values of N_α, N_1, k_α, and k_1, then $t_\alpha < t_1$ (the superlinear method is more efficient), as predicted by superlinear methods corresponding to a zero linear convergence factor. Among superlinear methods if $|\varepsilon_0|$ and δ are arbitrarily small, the logarithms of these terms dominate the numerator and denominator of the argument of *log* in equation 35. Thus, in the limit the efficiency of a superlinearly converging iteration does not depend on the convergence factor. Rather the number of operations required is proportional to $N_\alpha/\log(\alpha)$.

Another way of arriving at this last result is as follows. Note from equation 32 that if a method has order of convergence $\alpha > 1$, a new method whose iterative step consists of m steps of the first method has order of convergence α^m. Thus any order of convergence can be obtained by combining an appropriate number of steps. So if we have two super-linear methods with orders of convergence α and β, respectively, where $\alpha < \beta$, we see that if $\alpha^m = \beta$, that is, $m = \log(\beta)/\log(\alpha)$, m steps of the first method will have the same order of convergence as the second method. Therefore, if one step of the second method requires more than $\log(\beta)/\log(\alpha)$ times the work of one step of the first method, the first will be preferred. That is, method 1 should be chosen if

$$\frac{\log(\beta)}{\log(\alpha)} N_\alpha < N_\beta$$

that is, if

$$\frac{N_\alpha}{\log(\alpha)} < \frac{N_\beta}{\log(\beta)}$$

the result obtained before.

Note that whereas in the above sense all superlinearly convergent methods are comparable, linearly convergent methods are not comparable to superlinearly convergent methods because for the former

$$\varepsilon_m = k_1^m \varepsilon_0 \tag{36}$$

Combining m steps changes the convergence factor but not the fact that the order of convergence is 1.

As an example, consider the secant method, which we will show in Section 9.4 to have order of convergence $(1 + \sqrt{5})/2 \approx 1.618$. Then, in terms of efficiency, the secant method will be preferred to Newton's method if one step of Newton's method requires more than $\log(2)/\log(1.618) \approx 1.44$ times as many operations as one step of the secant method. This is often the case, because Newton's method requires $f(x_i)$ and $f'(x_i)$ at each step whereas the secant method requires only $f(x_i)$.

As with linearly convergent methods, the analysis of convergence and speed of convergence was done in the absence of generated error. It remains to analyze the effect of generated error bounded by b^{gen} per step in superlinear methods. Consider quadratically convergent methods. In the worst case

$$|\varepsilon_{i+1}| = |k_2 \varepsilon_i^2| + b^{\text{gen}} \tag{37}$$

We can guarantee improvement, for $|\varepsilon_0|$ appropriately small, as long as $|\varepsilon_{i+1}| < |\varepsilon_i|$, that is,

$$|k_2| |\varepsilon_i|^2 + b^{\text{gen}} < |\varepsilon_i| \tag{38}$$

This condition no longer holds when

$$|k_2| |\varepsilon_i|^2 + b^{\text{gen}} = |\varepsilon_i| \tag{39}$$

that is, when

$$|\varepsilon_i| = \frac{1 - \sqrt{1 - 4b^{\text{gen}}|k_2|}}{2|k_2|} \tag{40}$$

(see Problem 9.8a,b to produce this result and to see what happened to the other root of the quadratic equation). Thus the error bound after many iterations is

$$b_\infty = \frac{1 - \sqrt{1 - 4b^{\text{gen}}|k_2|}}{2|k_2|} \tag{41}$$

If $4b^{\text{gen}}|k_2| \ll 1$, then

$$b_\infty \approx b^{\text{gen}} \tag{42}$$

(see Problem 9.8d), the result predicted by equation 18, which deals with b_∞ for linear convergence, with $k_1 = 0$.

For superlinearly convergent methods of order other than 2, we may solve the analog of equation 39, that is,

$$|k_\alpha| \, |\varepsilon_i|^\alpha + b^{\text{gen}} = |\varepsilon_i| \qquad [43]$$

for $|\varepsilon_i| = b_\infty$, presumably by an iterative method.

9.3 Obtaining Superlinear Convergence

Two methods are commonly used to obtain superlinear convergence. In the first we try to take an appropriate multiple of the change in the iterate produced by a linearly convergent method so as to produce quadratic convergence. In the second we try to use successive results of a linearly convergent method to compute a better value, such that the sequence of these better values is superlinearly convergent. Let us exemplify each of these in turn.

Consider the method

$$x_{i+1} = x_i - \lambda f(x_i) \qquad [44]$$

which is linearly convergent to the root of $f(x) = 0$ for an appropriate constant λ [this is called the method of constant slope (see Pizer, 1975, p. 215)]. At each step we change the iterate by

$$x_{i+1} - x_i = -\lambda f(x_i) \qquad [45]$$

But the step is not right for superlinear convergence. We guess that there is a constant β such that if we change x_i by β times the change given by our original method, superlinear convergence will be produced. That is, if the change produced is only half way to ξ, we would like it to be twice as much. If it is an overstep by a factor of 2 ($x_{i+1} - x_i$ is twice the distance to ξ from x_i), we would like it to be half as much. Thus we wish

$$x_{i+1} = x_i + \beta(x_{i+1}^{\text{old}} - x_i). \qquad [46]$$

$x_{i+1}^{\text{old}} - x_i$ is given by equation 45 in our case, so we wish to iterate using

$$x_{i+1} = x_i + \beta(-\lambda f(x_i)) \qquad [47]$$

choosing β so that superlinear convergence is produced.

Equation 47 is a single step iteration, $x_{i+1} = g(x_i)$, so we will produce quadratic convergence if $g'(\xi) = 0$, that is, if

$$1 - \beta \lambda f'(\xi) = 0 \qquad [48]$$

from which we arrive at

$$\beta = \frac{1}{\lambda f'(\xi)} \qquad [49]$$

But we do not know ξ so we approximate $f'(\xi)$ by $f'(x_i)$ at the ith step, so equation 47 becomes

$$x_{i+1} = x_i - \frac{f(x_i)}{f'(x_i)} \qquad [50]$$

This is Newton's method, which we have shown to converge quadratically.

We see that our first approach introduces a parameter into a linearly convergent method and then chooses the parameter to produce quadratic convergence. The linearly convergent method is never applied. It only serves as a basis upon which to compute analytically what the quadratically convergent method should be. In contrast is a method where successive iterates of the linearly convergent method are computed and then are combined to produce an improved iterate, such that the sequence of improved iterates is quadratically convergent.

In Section 7.4.4 we have already created a method for obtaining an improved iterate from three successive iterates of a linearly converging method: Aitken's δ^2-acceleration. It is left to Problem 9.9 to show that if the sequence $\{x_i\}$ produced by $x_{i+1} = g(x_i)$ is linearly convergent, then the sequence $\{z_i\}$ produced by

1. $u_{i+1} = g(z_i)$,
2. $v_{i+1} = g(u_{i+1})$, and
3. z_{i+1} = result of applying δ^2-acceleration to the sequence

$$\{z_i, u_{i+1}, v_{i+1}\}$$

[51]

is quadratically convergent. That is, one step of the quadratically convergent method consists of two steps of the linearly convergent method followed by δ^2-acceleration.

Since the method we have just described is quadratically convergent, there exists an interval about ξ such that if z_0 is in this interval the sequence converges to ξ. This is true even if the linearly converging (diverging) sequence produced by $x_{i+1} = g(x_i)$ is diverging, that is, has a convergence factor greater than 1 in magnitude. We have already seen this result, in Section 7.4.4. This approach of starting with a divergent method and accelerating is not normally recommended because the interval about ξ in which z_0 must fall to produce convergence may be very small and hard to locate, but the approach may in certain circumstances be useful.

9.4 Multistep Methods

A multistep method is one for which x_{i+1} is an explicit function of the n previous iterates for $n > 1$:

$$x_{i+1} = g(x_i, x_{i-1}, ..., x_{i-n+1}) \qquad [52]$$

As with recurrent methods we must express ε_{i+1} in terms of ε_i, ε_{i-1}, ..., ε_{i-n+1}, and solve this difference equation for ε_i as a function of i. Then we must check that $\lim_{i \to \infty} \varepsilon_i = 0$ (convergence occurs); and assuming this is true, we must use the expression for ε_i and thus ε_{i+1} to compute the order of convergence by finding the value α such that

$$\lim_{i \to \infty} \frac{\varepsilon_{i+1}}{\varepsilon_i^\alpha} = k_\alpha \qquad [53]$$

where $0 < |k_\alpha| < \infty$.

We can produce the required difference equation of propagated error by the standard partial derivative approach applied to equation 52:

$$\varepsilon_{i+1} = \sum_{k=i-n+1}^{i} \frac{\partial g}{\partial x_k} (\boldsymbol{\eta}^{(i)}) \, \varepsilon_k \qquad [54]$$

where $\eta_{k+1}^{(i)} \in [x_{i-k}, \xi]$ for $k = 0, 1, ..., n-1$. If there is convergence in the absence of generated error, equation 54 becomes in the limit

$$\varepsilon_{i+1} = \sum_{k=i-n+1}^{i} \frac{\partial g}{\partial x_k} (\xi, \xi, ..., \xi) \, \varepsilon_k \qquad [55]$$

It is often the case that each of the partial derivatives in equation 55 is zero, so higher-order terms must be carried in the Taylor series that produces equation 54:

$$\varepsilon_{i+1} = \sum_{k=i-n+1}^{i} \frac{\partial g}{\partial x_k} (\xi, \xi, ..., \xi) \, \varepsilon_k + \frac{1}{2} \sum_{k=i-n+1}^{i} \sum_{j=i-n+1}^{i} \frac{\partial^2 g}{\partial x_k \partial x_j} (\boldsymbol{\zeta}^{(i)}) \, \varepsilon_k \varepsilon_j$$

$$\qquad [56]$$

$$= \frac{1}{2} \sum_{k=i-n+1}^{i} \sum_{j=i-n+1}^{i} \frac{\partial^2 g}{\partial x_k \partial x_j} (\boldsymbol{\zeta}^{(i)}) \, \varepsilon_k \varepsilon_j$$

where $\zeta_{k+1}^{(i)} \in [x_{i-k}, \xi]$ for $k = 0, 1, ..., n-1$. Assuming convergence, in the limit equation 56 becomes

$$\varepsilon_{i+1} = \frac{1}{2} \sum_{k=i-n+1}^{i} \sum_{j=i-n+1}^{i} \frac{\partial^2 g}{\partial x_k \partial x_j} (\xi, \xi, ..., \xi) \, \varepsilon_k \varepsilon_j \qquad [57]$$

Let us take the secant method as our example. In this case $n = 2$ and

$$x_{i+1} = g(x_i, x_{i-1}) = \frac{x_{i-1} f(x_i) - x_i f(x_{i-1})}{f(x_i) - f(x_{i-1})} \qquad [58]$$

Thus

$$\frac{\partial g}{\partial x_i} = \frac{f(x_{i-1})/(x_i - x_{i-1})}{[f(x_i) - f(x_{i-1})]/(x_i - x_{i-1})} \left\{ \frac{f'(x_i)}{[f(x_i) - f(x_{i-1})]/(x_i - x_{i-1})} - 1 \right\} \qquad [59]$$

As $x_i \to \xi$ and $x_{i-1} \to \xi$,

$$f(x_{i-1}) \to f(\xi) + (x_{i-1} - \xi) f'(\xi) = (x_{i-1} - \xi) f'(\xi) \qquad [60]$$

and

$$\frac{f(x_i) - f(x_{i-1})}{x_i - x_{i-1}} \to f'(\xi) \qquad [61]$$

so

$$\frac{\partial g}{\partial x_i} \to \frac{[(x_{i-1} - \xi)/(x_i - x_{i-1})] f'(\xi)}{f'(\xi)} \left[\frac{f'(\xi)}{f'(\xi)} - 1 \right] = 0 \qquad [62]$$

By the symmetry of equation 58 $\partial g / \partial x_{i-1} \to 0$ also.

It can similarly be shown that in the limit

$$\frac{\partial^2 g}{\partial x_i^2} \to 0, \qquad \frac{\partial^2 g}{\partial x_{i-1}^2} \to 0, \quad \text{and} \quad \frac{\partial^2 g}{\partial x_i \partial x_{i-1}} \to \frac{f''(\xi)}{2f'(\xi)} \triangleq L \qquad [63]$$

(see Problem 9.10), which we assume to be nonzero. Thus from equation 57, in the limit

$$\varepsilon_{i+1} = \frac{f''(\xi)}{2f'(\xi)} \varepsilon_i \varepsilon_{i-1} = L \varepsilon_i \varepsilon_{i-1} \qquad [64]$$

Since equation 64 can be rewritten

$$\varepsilon_{i+1} = (L \varepsilon_{i-1}) \varepsilon_i \qquad [65]$$

it is clear that if $|\varepsilon_{i-1}|$ is appropriately small, $|\varepsilon_{i+1}| < |\varepsilon_i|$, so there will be convergence for some interval about ξ. To analyze the speed of this convergence we must solve the difference equation 64. The transformation

$$\phi_k \triangleq \ln(L \varepsilon_k) \qquad [66]$$

converts equation 64 into the linear difference equation

$$\phi_{i+1} - \phi_i - \phi_{i-1} = 0 \qquad [67]$$

which has the solution

$$\phi_i = c_1((1 + \sqrt{5})/2)^i + c_2((1 - \sqrt{5})/2)^i \qquad [68]$$

where c_1 and c_2 depend on the values of ϕ_0 and ϕ_1 (see Problem 9.11). As $i \to \infty$, the term corresponding to the smaller characteristic root in magnitude, $(1 - \sqrt{5})/2$, becomes negligible compared to the other term, so in the limit

$$\ln(L \varepsilon_i) = \phi_i = c_1((1 + \sqrt{5})/2)^i \qquad [69]$$

that is,

$$\varepsilon_i = \frac{1}{L} \exp[c_1((1 + \sqrt{5})/2)^i]$$

[70]

To determine the order of convergence of our multistep method, we must use this solution to the difference equation 64 to find that α for which

$$\lim_{i \to \infty} \frac{\varepsilon_{i+1}}{\varepsilon_i^\alpha} = k_\alpha$$

[71]

where $0 < |k_\alpha| < \infty$. From equation 70

$$\frac{\varepsilon_{i+1}}{\varepsilon_i^\alpha} = \frac{\dfrac{1}{L} \exp[c_1((1 + \sqrt{5})/2)^{i+1}]}{\left(\dfrac{1}{L}\right)^\alpha \exp[\alpha c_1((1 + \sqrt{5})/2)^i]}$$

[72]

which is constant in i (equal to $L^{\alpha-1}$) for

$$\alpha = (1 + \sqrt{5})/2 \approx 1.618$$

[73]

Thus the order of convergence of the secant method is 1.618 (the largest characteristic root of the linearized difference equation 67), and the convergence factor, k_α, is given by

$$k_\alpha = L^{\alpha-1} = \left[\frac{f''(\xi)}{2f'(\xi)}\right]^{(-1+\sqrt{5})/2}$$

[74]

Having shown that the secant method converges superlinearly, we can conclude using the results of Section 9.2 that (i) if f is twice differentiable in a neighborhood of ξ, there exists a neighborhood of ξ such that if x_0 and x_1 are in this neighborhood the iteration will converge in the absence of generated error; (ii) the bound on the error, b_∞, after an infinite number of iterative steps is approximately b^{gen}, the error generated in one step, and more precisely is the solution of

$$\frac{f''(\xi)}{2f'(\xi)} b_\infty^{(1+\sqrt{5})/2} + b^{gen} = b_\infty$$

[75]

(iii) on the basis of efficiency in the limit the secant method is to be preferred to any quadratically convergent method that takes more than $\log(2)/\log((1 + \sqrt{5})/2) = 1.44$ times as many operations per step. Since the secant method requires one evaluation of the function f per iterative step and a δ^2-accelerated linearly convergent method requires at least two evaluations of f per step of the quadratically convergent sequence, the secant method will be preferred on the basis of efficiency in the limit.

As we noted in Section 9.2, on this basis the secant method is often preferred to Newton's method as well.

9.5 Accumulation of Error

Until now we have assumed that the generated errors from various iterative steps were independent but all bounded in magnitude by the same value, b^{gen}. As convergence takes place, it may be closer to the truth that at each step the generated error is the same, ε^{gen}. How does this change the error bound after infinitely many iterations?

In the case of linear convergence, if we get close enough to ξ that we can assume

$$\varepsilon_{i+1}^{prop} = k_1 \varepsilon_i \qquad [76]$$

then

$$\varepsilon_{i+1} = k_1 \varepsilon_i + \varepsilon^{gen} \qquad [77]$$

Thus

$$\varepsilon_1 = k_1 \varepsilon_0 + \varepsilon^{gen}$$

$$\varepsilon_2 = k_1 \varepsilon_1 + \varepsilon^{gen} = k_1^2 \varepsilon_0 + (1 + k_1) \varepsilon^{gen}$$

$$\varepsilon_3 = k_1 \varepsilon_2 + \varepsilon^{gen} = k_1^3 \varepsilon_0 + (1 + k_1 + k_1^2) \varepsilon^{gen} \qquad [78]$$

$$\vdots$$

$$\varepsilon_n = k_1^n \varepsilon_0 + \sum_{i=0}^{n-1} k_1^i \varepsilon^{gen} = k_1^n \varepsilon_0 + \frac{1 - k_1^n}{1 - k_1} \varepsilon^{gen}$$

Assuming $|k_1| < 1$,

$$\lim_{n \to \infty} k_1^n = 0 \qquad \text{so} \qquad \varepsilon_\infty = \lim_{n \to \infty} \varepsilon_n = \frac{\varepsilon^{gen}}{1 - k_1} \qquad [79]$$

If we do not know ε^{gen} but only a bound on its magnitude, b^{gen}, then we see

$$b_\infty = \frac{b^{gen}}{1 - k_1} \qquad [80]$$

Comparing this result to our previous result on the assumption of independent errors, equation 18 [$b_\infty = b^{gen}/(1 - |k_1|)$], we see that for $k_1 > 0$ the result is the same. For $k_1 < 0$, however, the present assumption gives a smaller bound.

For quadratic convergence the bound does not change when one assumes that ε^{gen} is fixed (see Problem 9.12).

9.6 Summary and Complements

As with any computational method, an iterative algorithm must be analyzed in terms of efficiency and accuracy. We have seen that both of these depend on determining the order of convergence and convergence factor and that these are determined using the partial derivative approach to the analysis of error propagated by one step of the iterated operator. We have seen that the efficiency of an iterative method is basically determined by the product of the number of steps required for convergence and the amount of work per step. The number of steps required depends primarily on the order of convergence. For linearly convergent methods the number of steps required depends strongly on the convergence factor (see the expression before equation 12), but for superlinearly convergent methods the dependence of the number of steps required on the convergence factor is weak (see equation 34). Note that these expressions for the number of steps required are based on the assumption that at every step $\varepsilon_{i+1} = k_\alpha \varepsilon_i^\alpha$. Since this relation is only true if the error generated at each step is ignored and then only in the limit as $\varepsilon_i \to 0$, the numbers of steps given by these expressions are only approximate.

The accuracy of an iterative method carried until no further improvement can be obtained is determined by the order of convergence, the convergence factor, and the error generated at each step. For linearly convergent methods a bound b_∞ on the error in such a final result is given by equation 18, and for quadratically convergent methods it is given by equation 41. For other superlinearly convergent methods it is given by the solution to equation 43. For all superlinearly convergent methods if the convergence factor or the generated error per step is appropriately small, the bound on the error in the final result is approximately the bound on the error generated in the final step.

If the iteration is concluded with x_j before improvement has ceased but after $\varepsilon_{i+1}^{\text{prop}} \approx k_\alpha \varepsilon_i^\alpha$, then since

$$\varepsilon_j \approx k_\alpha \varepsilon_{j-1}^\alpha + \varepsilon_j^{\text{gen}},$$ [81]

$$\varepsilon_{j-1} \approx \left(\frac{\varepsilon_j - \varepsilon_j^{\text{gen}}}{k_\alpha}\right)^{1/\alpha}$$ [82]

that is,

$$x_{j-1} - \xi \approx \left(\frac{\varepsilon_j - \varepsilon_j^{\text{gen}}}{k_\alpha}\right)^{1/\alpha}$$ [83]

Subtracting approximation 83 from $x_j - \xi = \varepsilon_j$ produces

$$x_j - x_{j-1} \approx \varepsilon_j - \left(\frac{\varepsilon_j - \varepsilon_j^{\text{gen}}}{k_\alpha}\right)^{1/\alpha}$$ [84]

With values for x_j, x_{j-1}, α, and k_α and a bound b_j^{gen} on $|\varepsilon_j^{\text{gen}}|$, equation 84 can be used to give a bound b_j on $|\varepsilon_j|$. In particular, it can be shown that b_j is the larger in magnitude of the solution of equation 84 with b_j^{gen} in place of $\varepsilon_j^{\text{gen}}$ and the solution of equation 84 with $-b_j^{\text{gen}}$ in place of $\varepsilon_j^{\text{gen}}$ (see Problem 9.13).

Of course, all of the discussions of accuracy in this chapter have been of the error generated by the iterative method. Bounds on the errors due to any inputs other than x_0 must be analyzed separately by propagated error analysis techniques and be added to the generated error bounds discussed here.

All of this chapter has been presented in the context of the solution of nonlinear equations in one unknown. However, the same approach applies to any iterative method with a one-dimensional iterate. The iteration is of the form $x_{i+1} = g(x_i, x_{i-1}, ..., x_{i-n+1})$, which has a fixed point ξ, and we determine as above, first, whether convergence takes place; second, what is the order of convergence and convergence factor; using these, third, what is the time required; and, fourth, what is the bound on the final error.

For a multidimensional iteration, other problems arise, but the analysis is much the same.

REFERENCES

Pizer, S. M., *Numerical Computing and Mathematical Analysis*. Science Research Associates, Chicago, 1975.

PROBLEMS

9.1 **(a)** Devise an algorithm based on Newton's method for finding the nth root of a positive number for n an integer > 1.
(b) Under what conditions does the method converge?

9.2 Show that the iteration $x_{i+1} = x_i - f(x_i)/m$ (the method of constant slope) converges to a root ξ of $f(x)$ if both of the following conditions are satisfied:
(i) either
(1) $|f'(x)| < 2M|m|$ for some $0 \le M < 1$ in a symmetric interval about ξ,
or
(2) $|f'(x)| < |m|$ in a one-sided interval about ξ;
(ii) $f'(x)$ has the same sign as m in the interval in question.

9.3 Assume $f(x)$, $f'(x)$, and $f''(x)$ exist for $x \in [x_0^+, x_0^-]$. Further assume $f(x_0^+) > 0$ and $f(x_0^-) < 0$, so $f(x)$ must have a root $\xi \in [x_0^+, x_0^-]$.

(a) Show that the method of false position produces linear convergence to ξ.
(b) Assuming the result of part a, prove that if $f''(\xi) \ne 0$, one interval endpoint of the method of false position must eventually remain frozen (there exists an integer $N > 0$ such that either for $i > N$, $x_{i+1}^+ = x_i^+$, or for $i > N$, $x_{i+1}^- = x_i^-$, where the notation x_i^+ and x_i^- is defined in Section 3.4.1).

9.4 If $\varepsilon_{i+1} = k_1 \varepsilon_i$, for $i = 0, 1, 2, \ldots$, and $|\varepsilon_n| = \delta$, show $n = \log(\delta/|\varepsilon_0|)/\log(|k_1|)$.

9.5 **(a)** Show that the solution for the number of steps m to convergence if $\varepsilon_{i+1} = k_\alpha \varepsilon_i^\alpha$, for $i = 0, 1, \ldots$, shown to be the solution of equation 33, is given by equation 34.
(b) Given that the approximate time to achieve tolerance δ with a linearly convergent method is $t_1 = N_1 \log(\delta/|\varepsilon_0|)/\log(|k_1|)$ and that for a method with order of convergence α is

$$t_\alpha = \frac{N_\alpha}{\log(\alpha)} \log\left(\frac{\log(|k_\alpha|)/(\alpha - 1) + \log(\delta)}{\log(|k_\alpha|)/(\alpha - 1) + \log|\varepsilon_0|}\right)$$

show that if $\alpha > 1$, then for all N_α, N_1, k_α, and k_1 there exist positive values β, γ such that if $0 < \delta < \beta|\varepsilon_0| < \delta$, then $t_\alpha < t_1$.

9.6 **(a)** Show that Newton's method converges linearly to a root of multiplicity $m > 1$ with convergence factor $1 - 1/m$.
(b) Show that the iteration $x_{i+1} = x_i - mf(x_i)/f'(x_i)$ converges at least quadratically to the root.

9.7 A method that is quadratically convergent to roots of multiplicity greater than 1 can be devised by using the iteration $x_{i+1} = x_i - kf(x_i)/f'(x_i)$ and using the values of three successive iterates to choose the value of k to be used in the next two iterations. Devise such a method.

9.8 We have shown that improvement can be assured in a quadratically convergent iteration as long as $|k_2||\varepsilon_i|^2 - |\varepsilon_i| + b^{\text{gen}} < 0$, where k_2 is the convergence factor, b^{gen} is the bound on the error generated at each step, and ε_i is the error in x_i.

(a) Show that if $1 - 4 b^{\text{gen}} |k_2| > 0$ there is an interval of $|\varepsilon_i|$ values for which the condition assuring improvement holds. Give the endpoints of the interval.
(b) Describe the behavior of $|\varepsilon_i|$ for successive values of i if $|\varepsilon_0| >$ the larger endpoint of the interval; if $|\varepsilon_0|$ is in the interval; if $|\varepsilon_0| <$ the smaller endpoint of the interval.
(c) What is the behavior of $|\varepsilon_i|$ for successive values of i if the quantity $1 - 4 b^{\text{gen}} |k_2| < 0$, i.e., if $4 b^{\text{gen}} |k_2| > 1$.
(d) Using the first-degree Taylor approximation to $(1 - x)^{1/2}$, show that if $4 b^{\text{gen}} |k_2| << 1$, the smaller endpoint of the $|\varepsilon_i|$ convergence interval is approximately b^{gen}, i.e., $b_\infty \approx b^{\text{gen}}$.

9.9 Assume that (1) ξ is a fixed point of the function g, (2) g is twice differentiable in a neighborhood of ξ, and (3) $g'(\xi) \neq 1$. Consider the iteration defined by $z_{i+1} = $ the result of applying δ^2-acceleration to z_i, $g(z_i)$, and $g(g(z_i))$, that is,

$$z_{i+1} = g(g(z_i)) - \frac{[g(g(z_i)) - g(z_i)]^2}{g(g(z_i)) - 2g(z_i) + z_i}$$

By expanding $g(z_i)$ and $g(g(z_i))$ in appropriate Taylor series about ξ, show that

(a) ξ is a fixed point of this iteration.
(b) the iteration is quadratically convergent to ξ.

9.10 Assume that f is appropriately differentiable in a neighborhood of ξ, $f(\xi) = 0$, $f'(\xi) \neq 0$, and

$$g(x_i, x_{i-1}) = \frac{x_{i-1}f(x_i) - x_i f(x_{i-1})}{f(x_i) - f(x_{i-1})}$$

By taking derivatives and expanding $f(x_i)$ and $f(x_{i-1})$ in Taylor series about ξ, show that as $x_i \to \xi$ and $x_{i-1} \to \xi$,

$$\frac{\partial^2 g}{\partial x_i^2} \to 0, \quad \frac{\partial^2 g}{\partial x_{i-1}^2} \to 0, \quad \text{and} \quad \frac{\partial^2 g}{\partial x_i \, \partial x_{i-1}} \to \frac{f''(\xi)}{2f'(\xi)}$$

You may use the fact that as $x_i \to \xi$ and $x_{i-1} \to \xi$, $x_i - \xi \to k_\alpha(x_i - \xi)^\alpha$, where $\alpha > 1$, since this follows from

$$\frac{\partial g}{\partial x_i}(\xi,\xi) = 0 \quad \text{and} \quad \frac{\partial g}{\partial x_{i-1}}(\xi,\xi) = 0$$

9.11 Show that the linear, homogenous difference equation given by $\phi_{i+1} - \phi_i - \phi_{i-1} = 0$ has the solution $\phi_i = c_1((1 + \sqrt{5})/2)^i + c_2((1 - \sqrt{5})/2)^i$.

9.12 Show that if $\varepsilon_{i+1} = k_2\varepsilon_i^2 + \varepsilon^{\text{gen}}$ for $i = 0, 1, 2, \ldots$, $|\varepsilon^{\text{gen}}| < b^{\text{gen}}$, and b_∞ has the value given by equation 42, then b_∞ is the smallest value by which $|\varepsilon_i|$ can be bounded above as $i \to \infty$.

9.13 Given values Δ, k and b, consider the set of solutions for ε of $\Delta = \varepsilon - [(\varepsilon - \delta)/k]^{1/\alpha}$ over all $\delta \in [-b,b]$. Show that the maximum of these solutions in magnitude is either the solution for $\delta = b$ or that for $\delta = -b$.

9.14 Assume that you wish to compute $y = x^{1/2}$ for any input x such that $2 < x < 3$. Assume that you iterate using the formula $y_{i+1} = (y_i + x/y_i)/2$, $y_0 = 1.5$ on a decimal floating-point computer with a three-digit fraction. Evaluate in detail the quality of the method from all the aspects you think important.

9.15 Consider the following method of computing e^z. Iterate using $x_{i+1} = x_i - (\log(x_i) - z)$, $x_0 = 1$. Assume we have a subroutine for computing $\log(x)$ for any x.

(a) Show this method converges for $z > 0$ to $\xi = e^z$.
(b) Show the method converges linearly with convergence factor $1 - 1/\xi$.
(c) Bound the absolute error in x_∞ if each step generates an absolute error bounded in magnitude by 10^{-5}
(d) Using the fact that the above method converges linearly with convergence factor $1 - 1/\xi$, find a function s whose computation requires work well less than that of computing x_{i+1}^{old} by the formula below and such that the iteration with step defined by

$$x_{i+1}^{\text{old}} = x_i - (\log(x_i) - z)$$

$$x_{i+1} = s(x_{i+1}^{\text{old}}, x_i)$$

has improved speed of convergence over the original iteration.

9.16 Assume that a function $f(u)$ is defined by the result of the following iteration with parameter u:

$$v_{i+1} = \frac{1}{3}\left(2v_i + \left(\frac{u}{v_i}\right)^2\right), \qquad v_0 = u$$

(a) Assuming convergence, give $f(u)$ as a closed-form analytical expression.

(b) Show that the iteration converges to $f(u)$ if $u > 1$.

(c) If an error bounded in magnitude by 0.19 is made at each step when computing $f(8)$, give a bound on the relative error in the value computed.

(d) Give a bound on the error generated in v_{i+1}, assuming v_i is exact, if the computation is done on a decimal floating-point computer with a 6-digit fraction. Assume that the calculation is done as indicated in the formula above, that is, by dividing v_i into u, squaring the result, multiplying v_i by 2, adding this result to the previously computed square, and multiplying this sum by 0.333333.

CHAPTER TEN

Applications of the Analysis Techniques: Choice and Modification of Numerical Methods

10.1 Summary of Analysis Techniques

Techniques for the analysis of the time efficiency and accuracy of numerical methods have been the subject of this whole book. Independent of the numerical method used, the propagated error of an operator T computing output y from input x is bounded in magnitude by the techniques presented in Chapter 6. We have seen that there are two approaches for the computation of this bound, the partial derivative approach and the error magnification approach. These approaches are summarized in Figure 10–1. Both approaches give a bound on the magnitude of the error propagated by T. To bound the magnitude of the overall error, we must add this propagated error bound to a bound on the error generated by the numerical method used to approximate T.

Whereas the technique for propagated error analysis is independent of the numerical method used, the techniques for bounding generated error magnitude and for evaluating efficiency vary with the type of numerical method. We have seen that numerical methods can be categorized as nonrepetitive, recurrent, or iterative. Figure 10–2 summarizes the analysis techniques for generated error and efficiency for each of these cases. It should be noted that the analysis of the efficiency of recurrent and iterative methods requires the amount of work in the setup step and

FIGURE 10–1

Techniques for Bounding Propagated Error Magnitude of $y = T(x)$

Case	Technique
x is an n-vector and a bound on the magnitude of the error in each element is given.	Partial derivative approach given by equations 13 (absolute error) and 14 (relative error) in Chapter 6.
Only a bound on $\|\varepsilon_x\|$ is known.	Error magnification approach: $b_y = \|T\|_x\|b_x$ (absolute error); $r_y = S_x(T)r_x$ (relative error, fixed input x); $r_y = \text{cond}(T)r_x$ (relative error, range of inputs x).

FIGURE 10–2

Analysis Techniques for Numerical Methods

Method type	Efficiency	Generated error bound
Nonrepetitive methods	Count operations	Representation error bound + arithmetic error bound + approximation error bound. These may be determined by the methods of Chapters 5 and 7.
Recurrent methods	$t_{setup} +$ (No. of steps) $\times t_{step}$	Bound given in Chapter 8 by equation 77.
Iterative methods	$t_{setup} + t_\alpha$, with t_α given in Chapter 9 by equation 12 for linearly converging methods and by equation 35 for superlinearly converging methods.	Bound b_∞ given in Chapter 9 by equation 18 for linearly converging methods, by equation 41 for quadratically converging methods, and by solution of equation 43 for other superlinearly converging methods.

the amount of work in each repeated step. This efficiency analysis is in turn done by the methods given in Figure 10–2, applied to the step in question.

Similarly, the generated error bounds for recurrent and iterative methods involve the bounds on the magnitude of the generated error in the setup step and in each repeated step. These bounds are determined by the methods for analyzing generated error given in Figure 10–2.

10.2   Numerical Methods

In this section we will give a brief survey of the methods of choice for the various numerical problems we have been concerned with in this book, though there is no attempt to completely specify the appropriate method for all cases. Many numerical methods have been designed for special cases, and we will not attempt to survey these. Not all the methods surveyed have been discussed previously.

10.2.1   APPROXIMATION

Approximation with data having only representation error is in simple cases done by piecewise low-degree polynomial exact-matching, and by cubic spline exact-matching when smoothness is important. If the data are known to be cyclic, Fourier approximation using the Fast Fourier Transform (FFT) is used. Especially if the data have isolated areas of sharp rise or fall, it is useful to use rational approximation by functions of the form $(\Sigma_{i=1}^{n} a_i f_i(x))/(\Sigma_{i=1}^{m} b_i g_i(x))$, where the f_i and the g_i are given basis functions. A common example is where both the numerator and the denominator are polynomials: $f_i(x) = g_i(x) = x^{i-1}$.

In cases where the data error is dominated by representation error but for simplicity of evaluation one desires to fit the data with a function involving fewer parameters than the number of data points, approximation minimizing the Tchebycheff (ℓ_∞) norm is often used. This approach is applicable to any set of basis functions, including those mentioned above: polynomials, sinusoids, and rational functions.

When the data error is dominated by error from measurement or previous computation, least-squares approximation or some other norm-minimizing approximation should be chosen. Least-squares approximation is most common, though when data error is not normally distributed, other methods may give more accurate results. If the data are equally spaced, Fourier approximation after subtraction of linear tendency and odd reflection is especially appropriate. This approximation is normally done using the FFT algorithm. In cases where a simple approximation is desired, low-degree polynomial approximation can be chosen, but this should not be used for polynomials of degree higher than perhaps 4.

Least-squares splines are also possible where smooth approximation is desired.

Least-squares splines are an instance of *approximating splines,* that is, splines that need not exact-match the data, in contrast to *interpolating splines* like the exact-matching cubic splines that we have covered in some detail. A class of approximating splines of special interest is the *B*-splines. These are commonly used in specifying, as opposed to fitting, smooth curves, as in computer-aided design. They have the desirable property that a change in a data point affects only a small number of nearby spline pieces, unlike the interpolating cubic splines that we have emphasized. They also have the property of naturally following the data points while remaining smooth, though they do not minimize any norm of closeness or smoothness. They illustrate, for splines, the useful view of an approximating function as a linear combination of basis functions, where here each basis function is nonzero on only a few of the tabular argument intervals. The reader may find a discussion of these and other types of splines, on which much recent research has been carried out, in deBoor (1978).

10.2.2 DIFFERENTIATION

For numerical differentiation, the best advice is not to do it but rather to transform the problem into one requiring numerical integration, if possible. In cases where this is not possible, great care must be taken to differentiate a smooth approximation. Thus, numerical differentiation based on cubic spline approximation is preferable to the formulas based on simple exact-matching approximation in the case where the data error is dominated by representation error. Where the data error is dominated by measurement or computation error, numerical differentiation is especially inaccurate, and here least-squares approximation by a smooth function is critical to provide the function to be differentiated.

10.2.3 INTEGRATION

More often than not, the method of choice for numerical integration is Romberg integration. This is applicable whenever the function is available analytically or when the data are provided at 2^n equally spaced arguments for some n. Methods that sometimes require even fewer function evaluations when the function is analytically available are the class of methods collectively called Gaussian integration. These methods are based on the notion that one can choose the optimum points to evaluate the integrand. Thus, the methods specify the point at which the integrand must be evaluated on the integration interval. A method called Clenshaw-Curtis integration is based on Fourier approximation and thus uses the FFT as a subroutine. It has been reported by Gentleman (1972) that this method competes well with Romberg and Gaussian integration in terms of efficiency and approximation error and is better in terms of propagated error.

Both Romberg and Gaussian integrations are based on approximating the integrand by a polynomial and integrating this polynomial. If the integrand is very un-polynomial-like, the required interval width to obtain appropriate approximation error may be so small that the arithmetic error is undesirably large and the application of the method is inefficient. In such a case, one does better to fit the integrand by a function with properties more like the integrand itself. This is an instance of the general rule that basis functions for approximation should be chosen according to the shape of the function being approximated or its form as given by a model. For numerical integration it is also important that the approximating function be analytically integrable.

If the function being approximated is known to be oscillatory, a polynomial fit is not ideal, especially if there are a large number of oscillations over the integration interval. In this case, it is more profitable to approximate f by $p(x) \sin(\omega x)$, $p(x) \cos(\omega x)$, or a linear combination of these two, where $p(x)$ is a polynomial. Since for any polynomial the integral of either of these approximating functions is analytically computable, we can produce integration formulas just like those for pure polynomial approximating functions. The resulting method, which is produced in a way directly analogous to the Newton-Cotes rules, is called Filon integration if a quadratic polynomial is fitted (see Problem 10.1).

Similarly, if the integrand rises exponentially, we can approximate with $p(x)e^{ax}$, or if the integrand falls exponentially, with $p(x)e^{-ax}$. Both of these approximating functions are analytically integrable. In general we wish to approximate the integrand by

$$f(x) \approx w(x)p(x) \qquad [1]$$

where $w(x)p(x)$ is analytically integrable, $w(x)$ characterizes the predominant shape of $f(x)$, and $p(x)$ characterizes the slight variations of the function from $w(x)$ and thus is smooth.

The form of Gaussian integration with polynomial basis functions is called Gauss-Legendre integration. Gaussian integration is also applicable for integrands of the form $w(x)p(x)$. Common cases are $w(x) = e^{-x}$ with integration limits of 0 and ∞ (Laguerre-Gauss integration); $w(x) = e^{-x^2}$ with integration limits of $-\infty$ and ∞ (Hermite-Gauss integration); and $w(x) = 1/\sqrt{1 - x^2}$ with integration limits of -1 and 1 (Tchebycheff-Gauss integration).

10.2.4 SOLUTION OF DIFFERENTIAL EQUATIONS

For the solution of first-order ordinary differential equations, in many cases the most efficient method of adequate accuracy is the Adams predictor-corrector method, with a Runge-Kutta method used for start-up and interval halving. In cases where stability is such a problem that the decrease in efficiency is justified, a Runge-Kutta method is used by itself. Special methods are also available for the solution of stiff differential equations.

A class of predictor-corrector methods that have become quite popular are the variable-order, variable-step-size methods (see Gear, 1971). These methods, which are available for both nonstiff and stiff differential equations, involve the dynamic selection of not only the step size but also the order of the solution method, that is, the degree of solution polynomial for which the method is exact. The step size and order are chosen to obtain as efficient a solution as possible for a given accuracy. The methods include some criteria for deciding at each step whether to change the order and whether to change the step size. They change the order by increasing the number of previous values of the solution that are involved in the predictor and corrector. Since integration of $f(x)$ can be cast as the solution of the differential equation $y' = f(x)$, one can apply these ideas to numerical integration as well.

10.2.5 SOLUTION OF NONLINEAR EQUATIONS

Roots of nonlinear equations are most commonly found using the secant method or Newton's method. We have seen that unless the evaluation of $f'(x_i)$ involves very little additional calculation after $f(x_i)$ has been evaluated, the secant method is preferable. When there is some question about the convergence of the secant method, the hybrid secant method, in which the root is trapped and estimates not in the trapping interval are rejected and replaced by a false position estimate, should be used. Similarly, if Newton's method is chosen in a case where convergence is questionable, its hybrid form should be used.

Another quadratically convergent method is based on the combination of the method of false position with Aitken's δ^2-acceleration. A difficulty occurs with this combination without modification if one interval endpoint is frozen far from the root. In this situation many iterations of the simple method of false position are required before the convergence is linear enough for acceleration to be applied. However, Hamming (1971) has suggested a modification which avoids this behavior while maintaining the advantages of the method of false position. The modification involves, at the end of each iterative step, replacing the value of f at the interval endpoint not changed at that step, by γ times its value, where γ is a positive number less than 1. Hamming suggests using $\gamma = 1/2$. Ideally, γ should be even smaller for early iterations, increasing to 1 in the limit as the root is approached. When γ is held constant, this method can be shown to converge linearly, so Aitken's δ^2-acceleration is applicable. The combination of the modified false position method with Aitken's δ^2-acceleration converges quadratically and is thus competitive with the other superlinear methods described above. However, a quadratic step needs two iterations of the modified false position method, requiring two evaluations of f, whereas two steps of the secant method, which also require two evaluations of f, converge with order of convergence 2.6. Thus, except where the hybrid secant method chooses the false position option at almost every step because $f(x)$ has many near-zero derivatives near the root, the hybrid secant method is preferable to modified false position with acceleration.

10.2.6 SOLUTION OF LINEAR EQUATIONS

For the solution of systems of linear equations, the choice is between Gaussian elimination and iterative methods. Gaussian elimination is usually done by triangular factorization with pivoting based on scaled pivots. Choosing the pivot from the leftmost column of the submatrix remaining to be triangularized is usually adequate, but in situations where the matrix is ill conditioned, full pivoting will improve the accuracy. In full pivoting the pivot is chosen from all the elements of the submatrix, and both row and column permutations are required to place the pivot in the upper left of the submatrix. Column permutation is effectively a renaming of the unknowns, so when the solution is found the elements must be put back in the original order.

The result of Gaussian elimination is often improved using the iterative improvement scheme, which operates as follows. Let the solution produced by Gaussian elimination be x^0 and let the result of the ith iterative step be x^i. Then the iterative step consists of computing the vector $b - Ax^i$ in double precision, solving

$$A\Delta x^i = b - Ax^i \qquad [2]$$

and letting

$$x^{i+1} = x^i + \Delta x^i \qquad [3]$$

This step provides an improved solution without requiring retriangularization, since the already triangularized matrix A is used in each step. The improvement is obtained because significant digits that are lost when the right side of the equation is large can be kept when the right side of the equation is small, as it will be in iterations beyond the first. To make these low-order digits in the right side of the equation accurate, the multiplication of A by x^i and its subtraction from b must be done in double precision. The iteration usually converges to maximum accuracy in just two or three steps.

Iterative methods include not only the already discussed Jacobi and Gauss-Seidel methods, but Kaczmarz's projection method and the method of conjugate gradients. The Gauss-Seidel method is almost always used in preference to the Jacobi method. The Gauss-Seidel, projection, and conjugate gradients methods are all of the form

$$x^{(i+1,1)} = x^{(i,n)} + \alpha_1 p^1,$$

$$x^{(i+1,k+1)} = x^{(i+1,k)} + \alpha_{k+1} p^{k+1} \qquad (1 \leq k \leq n - 1) \qquad [4]$$

for a set of vectors p^k defined by the methods and values α_k computed from the coefficients of the equation and the present solution estimate $x^{(i+1,k)}$. The Gauss-Seidel method often does not converge but is especially applicable for sparse matrices. Kaczmarz's projection method always converges but may converge very slowly. The method of conjugate gradients is especially useful with symmetric positive definite matrices.

Sometimes the iterative methods have an advantage in accuracy over Gaussian elimination, and sometimes they have an advantage in efficiency as well. Since we are most familiar with the Gauss-Seidel method, let us compare Gaussian elimination with this iterative method, though the point is valid for other iterative methods as well.

The Gaussian elimination method requires approximately $n^3/3$ operations, where each operation consists of a multiplication or division and an addition or subtraction. An error generated early in this process is propagated by all the remaining $O(n^3)$ operations into the final x^i computed. Since the propagation tends to increase with the number of operations, this can be a bad propagation. Furthermore, though pivoting was designed to minimize bad propagation by any operation, the addition of numbers approximately equal in magnitude and opposite in sign can not be fully avoided, so the propagation by some of the operators can be quite severe. Therefore, with Gaussian elimination we can expect that the arithmetic error may be quite large.

The error generated by an iterative method, such as the Gauss-Seidel method, is approximately proportional to the error generated at any iterative step (see Chapter 9). But a step of the Gauss-Seidel method requires only n^2 of our operations. If the constant of proportionality between the final error and the generated error in a step is not too large, we can expect that the generated error due to the Gauss-Seidel method will be less than that due to Gaussian elimination. However, the Gauss-Seidel method may converge quite slowly, and we have seen that slow convergence in a linearly convergent method such as the Gauss-Seidel method corresponds to a large constant of proportionality. In cases where the iterative method produces less error despite a slow convergence rate, it may be advantageous to apply the method of Gaussian elimination and use its result as input to the Gauss-Seidel method, which in a few iterations can be made to produce results of better accuracy than the Gaussian elimination result.

10.2.7   MULTIDIMENSIONAL PROBLEMS

In this book we have restricted ourselves to one-dimensional problems, except in the case of linear equations. Multidimensional versions of most of these problems arise frequently. In almost all cases, multidimensional extensions of the methods of choice for one-dimensional problems are chosen to solve these multidimensional problems. Thus, in approximation there exist multidimensional polynomial approximation, multidimensional Fourier approximation, and multidimensional spline approximation. With systems of nonlinear equations there exist multidimensional forms of secant and Newton's methods.

Differential equations become multidimensional in a number of ways. First, there are systems of differential equations. These are solved by marching along a grid as with single differential equations, except that each step involves the prediction, correction, and mop-up or look-ahead calculation of many variables.

Second, there are differential equations of order higher than 1. For the solution of an nth-order equation to be specified fully, n conditions like the initial-value condition we have seen for $n = 1$ must be provided. If the n conditions are all at a fixed argument x_0, the problem is said to be an *initial-value problem*. In this case, the equation is transformed into many simultaneous first-order equations (see Pizer, 1975), which are solved by marching along a grid, as indicated above.

If the conditions provided with the differential equation are the solution values $y_0, y_1, \ldots, y_{n-1}$ at more than one argument $x_0, x_1, \ldots, x_{n-1}$, the problem is said to be a *boundary-value problem*. Such problems can be solved by iteratively solving an initial-value problem. We view the problem as finding the values of the derivatives of orders 1 through $n - 1$ at x_0 such that solution of the initial-value problem from the initial point x_0 gives the desired y_i value at each x_i, for $1 \leq i \leq n - 1$. Having found these derivative values, we evaluate the solution to that initial-value problem at any arguments of interest. Thus both the final evaluation of the solution at any argument as well as each iterative step involves the numerical solution of an initial-value problem, that is, recurrently marching along a grid (see Henrici, 1962).

In boundary-value problems an alternative to marching along a grid is to think of the solution at each grid point as an unknown, write a set of equations relating these unknowns, and solve this set simultaneously (see Henrici, 1962). Each equation in the set is based on locally approximating the solution by some function. Of special recent interest, among methods of this type, are ones where the approximations involved are splines. Such methods are called *finite-element methods* (see Strang and Fix, 1973). They allow repeating the solution with a grid that has smaller intervals in areas where analysis shows the accuracy to be inadequate.

A variant of this equation solution approach is to use the standard Picard technique of writing each equation to give the solution at one point in terms of the others and then using these transformed equations to calculate iteratively the values of these solutions. Thus, we have seen that boundary-value problems can be solved recurrently, directly, or iteratively.

Finally, full multidimensionality is achieved by partial differential equations. With these the grid at which the solution must be found is multidimensional. Nevertheless, these are basically boundary-value problems, and as with differential equations for functions of one variable the solution involves recurrently marching along the grid, directly solving a set of equations based on local approximation, or iteratively calculating the solution using versions of these equations that have been transformed for Picard iteration. Solutions of partial differential equations often have especially large problems of stability, and stiffness arises more frequently with them.

10.2.8 CONCLUSION

All of these methods are discussed in the literature. The list of references below, while not inclusive, contains useful books on numerical methods. With the background given by this book, we believe the student should be able to understand the methods and the strategies on which they are based and the analysis of their accuracy and efficiency. Using this understanding and analysis, he or she should be able to modify these methods to suit a particular problem. Furthermore, this should be true even for categories of problems that we have not discussed. These include optimization problems, eigenvector and eigenvalue problems, and multidimensional versions of the types of problems we have covered.

REFERENCES

Acton, F. S., *Numerical Methods That Work*. Harper & Row, New York, 1970.

Blum, E. K., *Numerical Analysis and Computation: Theory and Practice*. Addison-Wesley, Reading, Mass., 1972.

Conte, S. D., and deBoor, C., *Elementary Numerical Analysis*. 2nd ed. McGraw-Hill, New York, 1965.

deBoor, C., *A Practical Guide to Splines*. Springer-Verlag, New York, 1978.

Fike, C. T., *Computer Evaluation of Mathematical Functions,* Automatic Computation Series. Prentice-Hall, Englewood Cliffs, N.J., 1968.

Forsythe, G. E., Malcolm, M. A., and Moler, C. B., *Computer Methods for Mathematical Computations,* Automatic Computation Series. Prentice-Hall, Englewood Cliffs, N.J., 1977.

Franklin, J. N., *Matrix Theory,* Applied Mathematics Series. Prentice-Hall, Englewood Cliffs, N.J., 1968.

Gentleman, W. M., "Implementing Clenshaw-Curtis Quadrature." *Communications of the ACM,* **15**:337–346, 1972.

Gear, C. W., *Numerical Initial Value Problems in Ordinary Differential Equations.* Prentice-Hall, Englewood Cliffs, N.J., 1971.

Hamming, R. W., *Introduction to Applied Numerical Analysis.* McGraw-Hill, New York, 1971.

Hamming, R. W., *Numerical Methods for Scientists and Engineers,* 2nd ed. McGraw-Hill, New York, 1973.

Henrici, P., *Discrete Variable Methods for Ordinary Differential Equations.* Wiley, New York, 1962.

Hornbeck, R. W., *Numerical Methods.* Quantum Publs., New York, 1975.

Lanczos, C., *Applied Analysis.* Prentice-Hall, Englewood Cliffs, N.J., 1956.

Lapidus, L., and Seinfeld, J. H., *Numerical Solution of Ordinary Differential Equations.* Academic Press, New York, 1971.

Noble, B., *Applied Linear Algebra.* Prentice-Hall, Englewood Cliffs, N.J., 1969.

Ortega, J. M., *Numerical Analysis: A Second Course.* Academic Press, New York, 1972.

Pizer, S. M., *Numerical Computing and Mathematical Analysis.* Science Research Associates, Chicago, 1975.

Ralston, A., and Rabinowitz, P., *A First Course in Numerical Analysis,* 2nd ed. McGraw-Hill, New York, 1978.

Strang, G., and Fix, G. J., *An Analysis of the Finite Element Method.* Prentice-Hall, Englewood Cliffs, N.J., 1973.

PROBLEMS

10.1 We wish to approximate

$$I \triangleq \int_a^b f(x) \sin(\omega x)\, dx \text{ by } \hat{I} \triangleq \sum_{i=1}^m \int_{x_{2i-2}}^{x_{2i}} p_i(x) \sin(\omega x)\, dx,$$

where $x_k = a + kh$ with $h = (b - a)/(2m)$, and where $p_i(x)$ is a quadratic polynomial that matches $f(x)$ at x_{2i-2}, x_{2i-1}, and x_{2i}. Let $j = 2i - 1$. Then

$$\hat{I} = \sum_{i=1}^m \int_{x_{j-1}}^{x_{j+1}} p_i(x) \sin(\omega x)\, dx,$$

where $p_i(x)$ matches $f(x)$ at x_{j-1}, x_j, and x_{j+1}.
(a) Write $p_i(x)$ as a divided-difference formula where the points used are x_{j-1}, x_{j+1}, and x_j in the order specified.
(b) Integration by parts of $I_i \triangleq \int_{x_{j-1}}^{x_{j+1}} p_i(x) \sin(\omega x)\, dx$ produces

$$I_i = [-(1/\omega)p_i(x)\cos(\omega x) + (1/\omega^2)p_i'(x)\sin(\omega x)$$

$$\left. + (1/\omega^3)p_i''(x)\cos(\omega x)]\right|_{x_{j-1}}^{x_{j+1}},$$

since $p_i''(x)$ is a constant. Use this formula to show that if $\theta = \omega h$

$$\begin{aligned}
I_i/h = &-(1/\theta)[y_{j+1}\cos(\omega x_{j+1}) - y_{j-1}\cos(\omega x_{j-1})] \\
&+ (1/\theta^2)[\tfrac{3}{2}y_{j+1}\sin(\omega x_{j+1}) + \tfrac{3}{2}y_{j-1}\sin(\omega x_{j-1}) + \tfrac{1}{2}y_{j+1}\sin(\omega x_{j-1}) \\
&\quad + \tfrac{1}{2}y_{j-1}\sin(\omega x_{j+1}) - 2y_j(\sin(\omega x_{j+1}) + \sin(\omega x_{j-1}))] \\
&+ (1/\theta^3)[y_{j+1}\cos(\omega x_{j+1}) - y_{j-1}\cos(\omega x_{j-1}) - y_{j+1}\cos(\omega x_{j-1}) \\
&\quad + y_{j-1}\cos(\omega x_{j+1}) - 2y_j(\cos(\omega x_{j+1}) - \cos(\omega x_{j-1}))]
\end{aligned}$$

(c) Each of the terms above is of the form $Cy_k \sin(\omega x_n)$ or $Cy_k \cos(\omega x_n)$. For each term such that $k \neq n$, rewrite the term as the product of Cy_k and $\{\sin \text{ or } \cos\}(\omega(x_k + (n - k)h))$. Then use the formula for sines and cosines of sums of angles to show that

$$\begin{aligned}
I_i/h = &A(y_{j+1}\cos(\omega x_{j+1}) - y_{j-1}\cos(\omega x_{j-1})) \\
&+ B(y_{j+1}\sin(\omega x_{j+1}) + y_{j-1}\sin(\omega x_{j-1})) \\
&+ Cy_j \sin(\omega x_j),
\end{aligned}$$

where

$$A = (1/\theta^3)(1 - \cos(2\theta)) - (1/(2\theta^2)) \sin(2\theta) - 1/\theta,$$
$$B = (1/(2\theta^2))(3 + \cos(2\theta)) - (1/\theta^3) \sin(2\theta), \text{ and}$$
$$C = (4/\theta^3) \sin(\theta) - (4/\theta^2) \cos(\theta).$$

These three constants can be rewritten as

$$A = (-1/\theta^3)(\theta^2 + \theta \sin(\theta) \cos(\theta) - 2 \sin^2(\theta)),$$
$$B = (1/\theta^3)(\theta(1 + \cos^2(\theta)) - 2 \sin(\theta) \cos(\theta)), \text{ and}$$
$$C = (4/\theta^3)(\sin(\theta) - \theta \cos(\theta)).$$

(d)   Using $\hat{I} = \sum_{i=1}^{m} I_i$, show that

$$\hat{I} = h\left[A(f(b) \cos(\omega b) - f(a) \cos(\omega a) \right.$$
$$+ B\left(f(a) \sin(\omega a) + 2\sum_{i=1}^{m-1} f(a + 2ih) \sin(\omega(a + 2ih)) + f(b) \sin(\omega b) \right)$$
$$\left. + C\left(\sum_{i=1}^{m} f(a + (2i - 1)h) \sin(\omega(a + (2i - 1)h)) \right) \right]$$

10.2   For each of the following problems, assume the data arguments are equally spaced and say which of the following approximation methods you would use and why:

(1) Least-squares approximation with polynomials
(2) Exact-matching approximation with one polynomial
(3) Cubic-spline exact-matching approximation
(4) Fourier least-squares approximation
(5) Fourier exact-matching approximation

(a)   A physicist needs to find the position of a falling object at any time t, given 30 measured $(t_i, position_i)$ values. Assume the formula for a vacuum is correct:

$$position = \text{position at } t = 0$$
$$+ \text{ initial velocity} \times t$$
$$+ \text{ acceleration due to gravity} \times (t^2/2)$$

(b)   The cross-section of a sailboat hull is being designed. The designer has specified fifty (x,y) pairs, each of which represents a point on the hull cross-section. To compute drag, the positions of all points on the hull cross-section (not just of the fifty specified points) are needed.

(c)   A document retrieval system designer has a retrieval method with a parameter α, such that by varying α he can vary the trade-off between the value of the measure of relevance of the set of retrieved documents and the fraction of the relevant documents which are retrieved. He has measured exactly a set of 25 $(\alpha_i, relevance_i)$ pairs for a sample set of documents and wishes for any given α to find the corresponding value of the relevance measure for the whole population of documents from which he has a sample.

10.3 For each of the following situations, carefully specify which numerical method and parameters you would choose and why. If a standard method does not fit, say why, create a method, and justify your creation.

(a) To interpolate in a table telling on the hour for each hour between opening time and one hour after closing time, the average number of shoppers in line at the checkout counters at the supermarket on Tuesdays. That is, a job scheduler may wish to know the number expected to be in line at 1:20 P.M.

(b) To integrate over $[a,b]$ a function defined by data from a freshman physics experiment, where measurements are made at time $a \approx t_0$, t_1, $t_2, \ldots, t_n \approx b$, where each t_i is accurate but its value depends on the facility of the experimenter.

10.4 Assume $y(x)$ is the solution of the differential equation $y'(x) = f(x,y)$, $y(0) = y_0$, where $f(x,y)$ is given analytically and y_0 is a known constant. Further assume we cannot solve for y analytically.

Consider the problem of designing a library subroutine to give the solution to this differential equation at any value in $[-1,1]$ of the input parameter x. We would be willing to use much computer time in preparing the subroutine so that at each application of the subroutine the computer time used would be minimal. We would not want to apply a numerical method for the solution of differential equations for each execution of the subroutine because such methods are time consuming.

One possible method for resolving this problem would be to use a numerical method for the solution of differential equations to solve for y once at a set of points to be tabulated, and have the subroutine interpolate in the table. Ignoring roundoff, discuss in detail how you would go about implementing this method so that for every $x \in [-1,1]$, $y(x)$ could be found to within a constant given error ε. Justify your suggestions. Show your error analysis. Assume $|y^{(n)}(x)| < M_n$ for all n, where the values of M_n are given. Also assume the numerical method you choose for solving the differential equation has propagation factor 1, that is, the overall error at any step is the sum of the errors generated at all steps up to and including that step.

Be careful to address the question of the spacing of the tabular arguments and the interpolation method to be used with the table.

10.5 We wish to find numerically the zero of the solution of the differential equation

$$\frac{dy}{dx} = -x^2 y - x^3 - 1, \qquad y(0) = 1$$

with an error of magnitude less than 10^{-2}.

(a) What combination of standard numerical methods would you use to solve this problem? Justify your choice.
(b) Solve the problem.

10.6 Show that Hamming's modification of the method of false position converges linearly. What is the convergence factor? (*Hint:* First show that in the limit $x_i, x_{i+2}, x_{i+4}, \ldots$ are on one side of the root and $x_{i+1}, x_{i+3}, \ldots$ are on the other side of the root.)

APPENDIX A

Notation

1. Special Symbols and Notational Conventions

$\approx$	approximately equals
$\triangleq$	is defined as
$\overset{?}{=}$	must be tested for equality
$\equiv$	is identical to
$\not\equiv$	is not identical to
$:=$	is assigned the value
$\lesssim$	is less than or equal to a number approximately equal to
$\rightarrow$	approaches in the limit
lim sup	supremum of limit points
$\hat{f}$	estimate of f
f^{-1}	inverse of function f
f'	derivative of f
f''	second derivative of f
$y^{(n)}$	nth derivative of y
$\overline{x}$	complex conjugate of x, average of x
$\boldsymbol{x}$	n-vector x
x_i	element of n-vector x
x^j	x to the jth power, the jth n-vector x
$x^{(j)}$	the jth n-vector x
x_{max}	maximum value of an indexed set of variables x_i
$\{\,\}$	sequence
$(\cdot,\cdot)$	point in plane, row vector with two elements, inner product, open interval
$[\cdot]$	n-vector, matrix, closed interval
$[x_0, x_1, x_2, ..., x_n]$	smallest interval containing the x_i
∞	infinity
$\pm$	plus or minus
$\cdot$	placeholder, scalar multiplication, dot product
$!$	factorial

log	natural logarithm
e	base of natural logarithm
$\ni:$	such that
$\Sigma_{i=a}^{b}$	sum over i from a to b
$\Sigma_{\substack{i=a \\ i \neq c}}^{b}$	sum over i from a to b except $i \neq c$
$\Pi_{i=a}^{b}$	product over i from a to b
$\Pi_{\substack{i=a \\ i \neq c}}^{b}$	product over i from a to b except $i \neq c$
$\int_{a}^{b}$	integral over (a, b)
$\Sigma_{i \ni \,:\text{relation}}$	sum over i satisfying relation
$\psi(x)$	$\Pi_{i=0}^{n}(x - x_i)$
$f[x_0, x_1, \ldots, x_{lc}]$	k-1th divided difference of f at $x_0, x_1, \ldots,$ and x_k
δ, δ^k	central difference, kth central difference
Δ	forward difference
∇	backward difference
ℓ_p	norm in space of n-vectors
$\mathscr{L}_p$	norm in vector space of integrable functions
$[\ ]^T, (\ )^T$	vector or matrix transpose
$\det(\cdot)$	determinant
$\lvert x \rvert$	magnitude of x, ℓ_2 norm of n-vector x
$\lvert\lvert \cdot \rvert\rvert$	vector or operator norm
$\lvert\lvert \cdot \rvert\rvert_p$	ℓ_p or $\mathscr{L}_p$ norm
$\text{cond}(\cdot)$	condition number
$\text{cond}_p(\cdot)$	condition number using ℓ_p or $\mathscr{L}_p$ norm
$\text{cond}_{\text{range}}(\cdot)$	condition number for arguments in range
$S_x(\cdot)$	sensitivity at x
$T\mid_x$	T evaluated at x
$\in$	is a member of
$\subset$	is a subset of
$\cup$	union
$\cup_i$	union over i
$\max_i$	maximum over i
$\max_{\text{relation on } x}$	maximum over all x which satisfy the relation
$\max(\cdot, \cdot, \cdot)$	maximum of elements in list
x^*	computed result for x, with error
ε_x	error in x^*
b_x	bound on magnitude of absolute error in x^*
r_x	bound on magnitude of relative error in x^*
$._\text{gen}$	pertaining to generated error
$._\text{prop}$	pertaining to propagated error
$._\text{arith}$	pertaining to arithmetic error
$._\text{rep}$	pertaining to representation error
$._\text{approx}$	pertaining to approximation error

.comp	pertaining to computation error
.overall	pertaining to overall error
.composite	pertaining to composite integration error
.mult	pertaining to error in multiplication
.plus	pertaining to error in addition
.(c) prop from predictor	pertaining to error propagated from predictor to corrector
.acc	pertaining to δ^2-accelerated value
A/S	additions and subtractions
M/D	multiplications and divisions

2. Common Usage of Alphabetical Symbols

a	polynomial coefficient, interval limit as in integration, operator input, parameter
a_i	coefficient of polynomial or of vector
A	matrix, unknown differential equation coefficient
A_j	coefficient in barycentric Lagrange approximation, spline coefficient
$A^{(i)}$	matrix partially triangularized to i columns
A'_{ij}	constant-multiplied matrix element
A''_{ij}	scaled matrix element
b	polynomial coefficient, interval limit as in integration, n-vector, bound on magnitude of absolute error, operator input
$\{b^i\}$	basis of vector space
b_x	bound of magnitude of absolute error in x^*
b_i	bound on $\varepsilon_{T_i}^{\text{gen}}$
b_∞	absolute error magnitude bound on iterative result after ∞ steps
B	base of floating-point numbers, unknown differential equation coefficient
B_n	spline coefficient
c	polynomial coefficient, maximum error magnification factor, operator input, interval limit as in integration
c_n	second derivative value at x_n, estimated absolute error magnitude for $n+1$-point polynomial exact matching, constant in difference equation solution
$c(n)$	combination time for a problem of size n
C	unknown differential equation coefficient
d	vector in normal equation, interval limit as in integration, operator input

D	domain
exp	exponent of floating point number, exponential function
f	function, underlying function, fraction of floating point number
f_i, f^i	ith function
f_i^k	$f^k(x_i)$
$f(x,y)$	right side of differential equation
F	field, matrix in normal equation
g	function, cubic spline fit, iteration or recurrence function
G	vector space of functions
h	interval width
i	power, index
I	integral, interval $[x_0, x_1, ..., x_n, x]$
$\hat{I}_N$	numerical integration result with N panels
$I_{k,h}$	Romberg integration result in kth column with tabular interval h
j	power, index
J	range of evaluation arguments
k	index, multiple of fundamental frequency
k_α	convergence factor with order of convergence α
k_{1_i}	apparent linear convergence factor
K	$\partial f/\partial y$ value
$K(x)$	coefficient in polynomial exact-matching error
L	lower triangular matrix, stopping column in divided difference table, constant in secant equation difference equation
L_n	vector space of n-vectors
$L_j(x)$	jth Lagrange polynomial
m	number of approximating functions, degree of polynomial, extreme tabular argument index, number of iterative steps combined
m_i	number of paths from ith intermediate result to output
M	stopping column in divided difference table
n	number of digits in floating point fraction, number of subdivisions of an interval, size of n-vector, size of problem, degree of polynomial, number of subdivisions in divide and conquer, number of previous estimates used in iterative or recurrent step, multiplicity of root, number of steps in computation, extreme tabular argument index
n'	number of digits in floating-point fraction
$N, N+1$	number of data points

N	number of panels, number of recurrent operator applications
N_α	number of operations/step with order of convergence α
O	order
p	power in ℓ_p or $\mathscr{L}_p$ norm
$p(x)$	polynomial
$p_m(x)$	mth polynomial piece of spline
p^k	update vector in iterative linear equation solution
P	stopping column in divided difference table
q	polynomial
r_x	bound on magnitude of relative error in x^*
R	range, relative arithmetic error bound
s	$(x - x_0)/h$ — the number of intervals from x_0 to x, error calculation function
s_i	number of steps in a path from intermediate result to output
S	sum of squares, error calculation function operator
S_i	set of indices
t	time, polynomial argument, error magnitude bound calculation function
t_{setup}	computation time for setup step
t_{step}	computation time for repeated step
$t(n)$	time for a problem of size n
t_α	computation time with order of convergence α
T	operator
u	operator input, variable
u_i	result of i steps of approximate operator on accurate inputs
U	operator, upper triangular matrix
v	operator input
v_i	result of i steps of approximate operator on inaccurate inputs
$v^{(i)}$	vector of the last n values ending at the ith
V	vector space, operator
w	input, output
w_i	weight
x	argument, evaluation argument, n-vector, input
x'	convergence interval boundary
$x^{(i,k)}$	vector at kth stage of ith step
x^+, x^-	argument at which function is positive, negative
x_0	initial argument, particular argument
x_i	one of sequence of root estimates, one of sequence of arguments in recurrence, tabular arguments

x_{goal}	objective argument in solution of differential equation
x_{eval}	evaluation argument
x_{deriv}	argument for derivative
y	value of function at x; in particular, solution to differential equation; n-vector, output
y_i	$y(x_i)$, data values
$y_{i+1}^{(p)}$	predictor
$y_{i+1}^{(c)}$	corrector
$y_{i+1}^{(m)}$	result of mop-up
y_i'	$y'(x_i)$
z	argument
z_i	intermediate result
α	scalar, order of convergence
α_i	characteristic equation coefficient, coefficient of update vector in iterative linear equation solution
β	scalar, error tolerance, absolute error magnification factor, integer in weight
β_k	coefficient of characteristic equation solution
γ	integer in weight
γ_i	coefficient in solution of $y' = Ky$
δ	central difference, small value, error tolerance
δ^k	kth central difference
Δx^i	change between ith and $i+1$th solutions for x
ε	error, tolerance
ζ	argument value in an interval
$\zeta^{(i)}$	vector of values in intervals near the last n calculated values
ξ	root, argument value in an interval, fixed point
η	argument value in an interval
$\boldsymbol{\eta}^{(i)}$	vector of values in intervals near the last n calculated values
θ	exact order, argument value in an interval
θ_{xy}	angle between x and y
λ	scalar, slope
λ_j	coefficient of ε_j in recurrence error equation
μ	mean
υ	fundamental frequency
ρ	relative error magnification factor, characteristic root
σ_i	standard deviation of y_i
τ_i	relative error in v_i^*
ϕ	argument value in an interval
ϕ_j	coefficient of characteristic equation solution, solution of secant method difference equation

Specifications and Conventions of Programming Language Used

The algorithms presented in this book could be presented in many different languages. Indeed they have been expressed in two, Pascal in the text and FORTRAN in Appendix C. Pascal was chosen because of its readability, encouragement of good program structure, and popularity, and FORTRAN was chosen because of its popularity. We believe that most readers will find the Pascal programs more understandable and that little or no introduction to Pascal is necessary for this purpose. However, to support the reader unfamiliar with Pascal, this appendix presents the portion of the Pascal language that is used in the programs in this book. This portion is a subset of the version of Pascal supported by the Waterloo compiler. We believe that this subset is entirely standard Pascal, with the exception of the way in which formal FUNCTION parameters of subprograms are declared (see below).

All of the programs in the book are subprograms assumed to be internal to a main program or another subprogram. These subprograms are of two types, FUNCTION subprograms, used when only a single value is to be returned, and PROCEDURE subprograms, used when more than one value, e.g. an array of values, is to be returned. These subprograms consist of four parts:

1. A header comment describing the program, its input parameters, and the environment in which it works.

2. The subprogram header statement, giving the name and type of the subprogram, the names and types of the formal parameters of the subprogram, and, in the case of a FUNCTION subprogram, the type of the result.
3. A variable declaration section, giving the names and types of variables used in the program and an indication of the usage of each variable.
4. The executable portion of the program, consisting of a series of executable statements.

In all of these sections comments are used to communicate with the reader and are ignored by the computer. These comments are enclosed in curly braces, { and }.

```
{Normalize integral value}
```

The header comment consists only of comments. In the subprogram header statements, comments on each line describe the use of the parameter(s) specified on that line. Similarly, in the variable declaration section, comments on each line are used to describe the use of the variable(s) declared on that line. In the executable portion of the program, comments are used to describe the objective of the code that follows and that is indented from the comment.

The following will describe the structure and components of the last three subprogram parts. Throughout, capital letters have been used for Pascal keywords other than variable types, and lower case letters have been used for all other identifiers.

Subprogram Header Statement

The subprogram header statement consists of the keyword *FUNCTION* or *PROCEDURE*, specifying the subprogram type, followed by the name of the subprogram by which it is invoked, followed by a list of parameters, enclosed in parentheses.

```
PROCEDURE coeff_calc(
    y: real_array;            {data values}
    VAR a: sin_array;         {approximation coefficients}
    VAR line0,line1: real     {line coefficients in approximation}
                    );

    {real_array is defined as ARRAY[1..n] of real}
    {sin_array is defined as ARRAY[1..m] of real}
```

If the subprogram is of the FUNCTION type, the parenthesis closing the parameter list is followed by a colon and the type of the value returned.

```
FUNCTION trap_int(
    FUNCTION f(z:real) :real;    {function to be integrated}
    n: integer;                  {number of tabular intervals}
    xmin,xmax: real              {integration limits}
            ): real;
```

The parameter list gives names and types for the subprogram's formal parameters, that is, the dummy names used within the subprogram to refer to items passed by the invoking program. In this book four parameter types are used: *real*, or floating-point; *integer*, or fixed-point; *user-defined*, for an array parameter (see parameter *a* in the example subprogram header statement for *coeff_calc*), and *FUNCTION*, indicating a parameter that is the name of a FUNCTION subprogram.

User-defined types are necessary for us because array declarations are not allowed in parameter lists, although array parameters are allowed. Array parameters are obtained by defining a special array type in a block of code enclosing the subprogram. The definition of each such type is given in a comment in the subprogram being presented.

FUNCTION parameters are used to specify functions that are used by and invoked within the solution subprogram. For example, for the numerical integration subprogram *trap_int* the FUNCTION parameter *f* stands for the function calculating values of the integrand.

The parameter list consists of items separated by semicolons. Items corresponding to variables, namely, those of real, integer, or a user-defined type, consist of a list of one or more variables of the same type followed by a colon and the type identifier. FUNCTION parameters are specified by parameter list items of precisely the form of a FUNCTION subprogram header statement, with the formal parameters of this parameter subprogram specified by dummy names. This format is not standard Pascal, but is the form required by the Waterloo Pascal compiler.

In PROCEDURE subprogram header statements parameters whose values are to be returned to the calling program are preceded by the keyword *VAR*.

```
PROCEDURE Gauss_Seidel(
    A: real_matrix;              {coefficients}
    b: real_n_vector;            {right-hand-side of equations}
    VAR x: real_n_vector;        {solution vector}
    tol: real                    {used to stop iteration}
            );

    {real_matrix is defined as ARRAY[1..n,1..n] OF real}
    {real_n_vector is defined as ARRAY[1..n] OF real}
```

Variable Declaration Section

The types of all variables used in a subprogram are declared in a section headed by the keyword *VAR*. Following this is a list of items separated by semicolons. In each item there are one or more variable names, separated by commas, followed by a colon and the type of these variables. The type may be *real*, *integer*, a user-defined type, or an array of real or integer elements. As discussed under subprogram header statements, the definition of a user-defined type is given in an enclosing block of code but is specified in a comment in the subprogram being written.

```
VAR
    C: real_matrix;        {normalized "A" matrix}
    d: real_n_vector;      {normalized "b" vector}
    xcomp: real;           {new element estimate}
    maxnew,new,maxdif,diff: real;
                           {used for stopping criterion}
        j,k: integer;      {row and column indicators}
```

One-dimensional arrays are declared as having type *ARRAY* [*subscript-range*] of *element-type*, where the subscript-range field specifies the minimum and maximum subscript values for the array and the element-type field specifies, as *real* or *integer*, the type of the elements of the array. The format of the subscript-range field is the minimum subscript value, followed by two periods, followed by the maximum subscript value.

```
rombrow: ARRAY[0..100] OF real;  {last row of table}
```

For most of the arrays used in programs in this book, the size of the array is related to the size of the problem. Since Pascal does not strictly allow variable array size, this problem has been handled by declaring the array variable to be of a user-defined type, so that the size of the array is declared in a type definition in a block of code enclosing the subprogram being presented. This type definition is written in terms of an identifier *n* for the problem size, and this identifier in turn is defined as a constant in the block containing the definition. This allows us to make the type definition in terms of *n*, as in the example.

```
{real_n_vector is defined as ARRAY[1..n] OF real}
```

Two-dimensional arrays are declared just as one-dimensional arrays are, except that the subscript-range field consists of two "*minimum subscript value .. maximum subscript value*" subfields separated by commas. The first subfield gives the range for the first subscript, and the second subfield gives the range for the second subscript.

```
{real_matrix is defined as ARRAY[1..n,1..n] OF real}
```

Executable Program Portion

The executable portion of the program consists of a sequence of assignment statements and control groups. The elements of this sequence are separated by semicolons (but the sequence is not terminated by a semicolon), and they are preceded by the keyword *BEGIN* and followed by the keyword *END* and a closing semicolon.

Assignment statements consist of the target variable name followed by the symbol ": = " followed by an arithmetic expression. The arithmetic expression may consist of simple variable names, subscripted array variable names, arithmetic operators, function invocations, and parentheses.

```
xcomp := xcomp - C[j,k]*x[k];
```

Subscripted array variable names are specified by giving the array variable name followed by its subscript(s) enclosed in square brackets and separated by commas if there are more than one.

The arithmetic operators used are +, −, *, /, and *DIV*. Operator "*" denotes multiplication, "/" denotes division of reals to produce a real, and "DIV" denotes division of integers with truncation of the remainder to produce an integer.

Functions are invoked by mentioning, within an arithmetic expression, their name followed by actual parameter (argument) values enclosed in parentheses.

```
sintable[1] := sin(s);
```

The functions used in the subprogram presented are of two types, predefined (built-in) and user-defined. The former include the standard mathematical functions such as sin(x). User-defined functions are subprograms written by the user and enclosed by a block of code also enclosing the subprogram being presented.

```
IF sign(f(xnew)) = lftsign
```

User-defined functions are invoked either directly or by the invocation of a function that is a formal parameter, where the name of the user-defined function is used as the corresponding actual parameter (argument) in the invocation of the subprogram being presented.

```
func_sum := 0.5*(f(xmin) + f(xmax));
```

The effect of directly invoked functions required by a subprogram is indicated in its header comment. The purpose of parameter functions used by it is indicated in the parameter comment in the subprogram header statement.

Control groups are used either to select among statements or statement groups for execution or to repeat a statement or statement group. A statement group is defined as a sequence of statements separated by semicolons, preceded by the keyword *BEGIN*, and closed by the keyword *END*.

Selection is accomplished by the construct:

IF *condition*
      THEN *statement* or *statement group*
      ELSE *statement* or *statement group*

where the ELSE clause is optional. If the condition is true, the statement or statement group following the THEN keyword is executed. Otherwise the statement or statement group following the ELSE keyword is executed, with nothing being executed if no ELSE keyword is included.

```
IF index >0
   THEN lower_x := x[index-1]
   ELSE lower_x := -infinity;


IF sign(newval) = lftsign
   THEN
     BEGIN
       lftguess := new_root;
       lftval := newval
     END
   ELSE
     BEGIN
       rtguess := new_root;
       rtval := newval
     END
```

The condition used in an IF control group involves the comparison of values. Simple conditions consist of two arithmetic expressions separated by one of the following comparison operators: $=$ (is equal to), $<$ (is less than), $>$ (is greater than), $<=$ (is less than or equal to), $>=$ (is greater than or equal to). Conditions can be made by separating two parenthesized simple conditions by the keyword *OR* or *AND*.

```
IF ((xnew > rtguess) OR (xnew < lftguess))
   THEN BEGIN
     {Secant result no good; use false position}
     slope := (lftval-rtval)/(lftguess-rtguess);
     xnew := lftguess - lftval/slope
   END;
```

If OR is used, the condition is true if one or both of the simple conditions contained are true, and otherwise it is false. If AND is used, the condition is true if both of the simple conditions contained are true, and otherwise it is false.

Repetition is accomplished by any of three different control groups: the FOR group, the WHILE group, and the UNTIL group. The FOR group is used to repeat a statement or statement group for successive values of an index variable, with starting and ending values of the index variable specified in the FOR construct. The WHILE and UNTIL groups cause a statement or statement group to be repeated, with the repetition terminating when a condition in the construct is true (for UNTIL groups) or false (for WHILE groups).

The FOR group consists of a line of the form, *FOR index-variable = starting-value TO ending-value DO* followed by the body, which is the statement or statement group to be repeated.

```
{Compute sum of tabular values}
FOR i:=1 TO n-1 DO
   BEGIN
      x := x+h;
      func_sum := func_sum + f(x)
   END;
```

The effect is first to execute the body for each of successive (increasing) integer values of the index variable, with the starting value being used in the first execution of the body and the ending value being used in the final execution of the body. If the starting value is greater than the ending value, the body is not executed at all by this control group.

The FOR group can also be used when successive values of the index variable are to decrease by 1 rather than to increase by 1 as above. This behavior is obtained by replacing the keyword *TO* in the FOR line with the keyword *DOWNTO*.

```
{Back substitute}
FOR i:=n DOWNTO 1 DO
   BEGIN
      x[i] := b[i];
      {Subtract terms in already computed x[j]}
         FOR j:=i+1 TO n DO
            x[i] := x[i] - A[i,j]*x[j];
      {Solve for x[i]}
      x[i] := x[i]/A[i,i]
   END
```

In this situation the starting value must be greater than or equal to the ending value for the body to be executed once or more.

The WHILE group consists of a line of the form *WHILE condition DO* followed by the body, which is the statement or statement group to be repeated. The condition is of the same form as that for an IF control group.

```
{Iterate improvement until required accuracy is achieved}
WHILE (abs((xnew-xguess)/xnew) > tol) DO
   BEGIN
      {Compute improved estimate of root}
      xguess := xnew;
      xnew := xguess - f(xguess)/fprime(xguess);
   END;
```

Before each execution of the body the condition in the WHILE line is tested. If it is true, the body is executed and the cycle repeats. If the condition is false, the repetition is terminated. Since the condition is tested before each execution of the body, including the first, the body will not be executed at all if the condition is initially false.

The UNTIL group consists of the keyword *REPEAT* followed by the statement or statement group to be repeated followed by a line of the form *UNTIL condition*. The condition is of the same form as that for an IF control group.

```
REPEAT i:=i+1 UNTIL ((xeval<x[i]) OR (i=n)) ;
```

The effect is the same as for a WHILE control group with two differences. First, the repetition is terminated if the condition is true and continued if it is false. Second, the test is made after the execution of the body rather than before, so the body must be executed once.

APPENDIX C

Programs of Text in FORTRAN

PROGRAM 3–1
Barycentric Form of Lagrange Interpolation

```
C ***********************************************************************
C * BARYCENTRIC FORM OF LAGRANGE METHOD FOR POLYNOMIAL INTERPOLATION    *
C * GIVEN ARE                                                           *
C *     N,   THE DEGREE OF THE APPROXIMATING POLYNOMIAL;                *
C *     (X(J),Y(J)), 1<=J<=N+1,   THE DATA POINTS; AND                  *
C *     XEVAL,   THE EVALUATION ARGUMENT.                               *
C * THE APPROXIMATION AT XEVAL IS CALCULATED.                           *
C ***********************************************************************
C
C
      FUNCTION LGRNG(N,X,Y,XEVAL)
C
C   **PARAMETERS
      INTEGER N
      REAL X(100), Y(100)
      REAL XEVAL
C
C   **INTERNAL VARIABLES
      REAL A(100)
C                      **CONSTANTS IN SUMMED TERMS
      REAL NUM, DENOM
C                      **NUMERATOR, DENOMINATOR OF APPROXIMATION
      REAL TERM
C                    **TERM OF SUM
      INTEGER TERMNO
C                      **NUMBER OF TERMS (=N+1)
      INTEGER I,J
C                      **SUBSCRIPTS
C
C
C   **SET NUMBER OF TERMS
      TERMNO=N+1
C
C
C   **COMPUTE CONSTANTS IN SUMMED TERMS
      DO 20 J=1,TERMNO
         A(J)=1
         DO 10 I=1,TERMNO
            IF (I.NE.J) A(J)=A(J)/(X(J)-X(I))
```

339

```
10        CONTINUE
20     CONTINUE
C
C
C  **COMPUTE APPROXIMATION AT X=XEVAL
C  **COMPUTE NUMERATOR AND DENOMINATOR SUMS
       NUM=0
       DENOM=0
       DO 30 J=1,TERMNO
           TERM=A(J)/(XEVAL-X(J))
           DENOM=DENOM+TERM
           NUM=NUM+Y(J)*TERM
30     CONTINUE
C  **COMPUTE FINAL RESULT
       LGRNG=NUM/DENOM
C
C
       RETURN
       END
```

PROGRAM 3-2
Newton Divided-Difference Interpolation

```
C  ********************************************************************************
C  * NEWTON DIVIDED DIFFERENCE METHOD FOR POLYNOMIAL INTERPOLATION                *
C  * GIVEN ARE                                                                    *
C  *     N, THE NUMBER OF FIRST DIFFERENCES;                                      *
C  *     X(J), 1<=J<=N+1, SUCH THAT X(J)<X(J+1), THE N+1 DATA                     *
C  *         ARGUMENTS;                                                           *
C  *     P, THE LAST COLUMN IN THE DIFFERENCE TABLE WHERE SIGNS                   *
C  *         DO NOT FLUCTUATE NEAR THE PATH TO BE USED;                           *
C  *     XEVAL, THE EVALUATION ARGUMENT; AND                                      *
C  *     TOL, THE ERROR TOLERANCE.                                               *
C  * ALSO ASSUMED AVAILABLE IS A SUBROUTINE, F(I,DIFF), WHICH GIVES               *
C  * THE DIVIDED DIFFERENCE OF ORDER 'DIFF', WHERE 'I' IS THE INDEX               *
C  * OF THE LEAST DATA ARGUMENT TO BE USED TO FIND THAT DIFFERENCE.               *
C  ********************************************************************************
C
C
       FUNCTION NEWTON(N,X,P,XEVAL,TOL)
C
C  **PARAMETERS
       INTEGER N
       REAL X(100)
       INTEGER P
       REAL XEVAL,TOL
C
C  **INFIN IS DEFINED TO BE A VERY LARGE NUMBER
       COMMON INFIN
       REAL INFIN
C
C  **INTERNAL VARIABLES
       INTEGER INDEX
C                **INDEX OF LOWEST DATA ARGUMENT USED
       REAL DIFF
C                **DIFFERENCE BETWEEN ARGUMENT AND XEVAL
       INTEGER DIFORD
C                **ORDER OF HIGHEST DIVIDED DIFFERENCE USED
       REAL NEWTER
C                **NEW TERM IN APPROXIMATING SUM
       REAL PRVSIZ,PRVX
C                **PREVIOUS SUMMAND SIZE AND TABULAR ARGUMENT
       REAL FACTOR
C                **PRODUCT OF (X-X(J)) USED
```

```
      REAL LOWX,HIGHX
C                **NEXT LOWER AND HIGHER X THAT HAVE NOT YET BEEN
C                ** USED (OR +/- INFIN)
      REAL APPROX
C                **INTERMEDIATE ESTIMATION
C
C
C **FIND THE CLOSEST TABULAR ARGUMENT TO XEVAL
      DIFF=ABS(XEVAL-X(1))
      INDEX=1
      IF ((ABS(XEVAL-X(INDEX+1)).GE.DIFF).OR.(INDEX.GE.N)) GOTO 10
         INDEX=INDEX+1
         DIFF=ABS(XEVAL-X(INDEX))
10    CONTINUE
      IF ((INDEX.EQ.N).AND.(ABS(XEVAL-X(N+1)).LT.DIFF)) INDEX=N+1
C
C **INITIALIZE FOR APPROXIMATION COMPUTATION
      DIFORD=1
      NEWTER=F(INDEX,DIFORD)
      PRVX=X(INDEX)
      IF (INDEX.GT.1) GOTO 20
         LOWX=-INFIN
         GOTO 30
20    CONTINUE
         LOWX=X(INDEX-1)
30    CONTINUE
      IF (INDEX.LT.N+1) GOTO 40
         HIGHX=INFIN
         GOTO 50
40    CONTINUE
         HIGHX=X(INDEX+1)
50    CONTINUE
      FACTOR=1
      APPROX=0
C
C **ADD EACH NEW TERM AND FIND THE NEXT
60    CONTINUE
C        **ADD NEW TERM
         APPROX=APPROX+NEWTER
C        **SET UP FOR NEXT TERM
         PRVSIZ=ABS(NEWTER)
         FACTOR=FACTOR*(XEVAL-PRVX)
         DIFORD=DIFORD+1
C        **FIND NEXT CLOSEST ARGUMENT TO XEVAL
         IF (XEVAL-LOWX.LT.HIGHX-XEVAL) GOTO 90
            PRVX=X(INDEX+DIFORD)
            IF (INDEX+DIFORD.LT.N+1) GOTO 70
               HIGHX=INFIN
               GOTO 80
70          CONTINUE
               HIGHX=X(INDEX+DIFORD)
80          CONTINUE
            GOTO 120
90          CONTINUE
            INDEX=INDEX-1
            PRVX=X(INDEX)
            IF (INDEX.GT.1) GOTO 100
               LOWX=-INFIN
               GOTO 110
100         CONTINUE
               LOWX=X(INDEX-1)
110         CONTINUE
120      CONTINUE
C        **COMPUTE NEW TERM
         NEWTER=FACTOR*F(INDEX,DIFORD)
      IF ((ABS(NEWTER).LT.PRVSIZ).AND.(DIFORD.LT.N+1).AND.
     +           (ABS(NEWTER).GE.TOL)) GOTO 60
```

```
C
C
      IF (ABS(NEWTER).LT.PRVSIZ) GOTO 130
          NEWTON=APPROX
          GOTO 140
130   CONTINUE
          NEWTON=APPROX+NEWTER
140   CONTINUE
C
      RETURN
      END
```

PROGRAM 3–3
Cubic Spline Exact-Matching Approximation

```
C ******************************************************************************
C * CUBIC SPLINE EXACT-MATCHING APPROXIMATION                                  *
C * GIVEN ARE                                                                  *
C *     N, THE NUMBER OF INTERVALS;                                           *
C *     (X(I),Y(I)), 1<=I<=N+1, N+1 DATA POINTS SUCH THAT                     *
C *         X(I)<X(I+1) FOR 1<=I<=N;                                          *
C *     C(I), 1<=I<=N+1, N+1 SECOND DERIVATIVE VALUES AT THE                  *
C *         X(I); AND                                                         *
C *     XEVAL,  AN EVALUATION ARGUMENT.                                       *
C * MOST LIKELY, THE X(I), THE Y(I), AND TWO OF THE C(I) VALUES               *
C * WERE GIVEN TO A PREVIOUS PROGRAM, AND IT COMPUTED THE                     *
C * REMAINING C(I).                                                           *
C ******************************************************************************
C
C
      FUNCTION SPLINE(N,X,Y,C,XEVAL)
C
C  **PARAMETERS
      INTEGER N
      REAL X(100),Y(100),C(100)
      REAL XEVAL
C
C  **INTERNAL VARIABLES
      REAL H
C                          **WIDTH OF PARTICULAR INTERVAL
      REAL S,SCOMP
C                          **FRACTION OF INTERVAL AND ITS COMPLEMENT
      REAL A,B
C                          **COEFFICIENTS IN EVALUATION FORMULA
      REAL FACTOR
C                          **FACTOR IN COEFFICIENTS
      INTEGER I
C                          **SUBSCRIPT INDICATING TOP OF INTERVAL
C                            CONTAINING XEVAL
C
C
C  **FIND INTERVAL IN WHICH XEVAL FALLS
      I=2
10    IF ((XEVAL.LT.X(I)).OR.(I.EQ.N+1)) GOTO 20
      I=I+1
      GO TO 10
20    CONTINUE
C
C  **EVALUATE POLYNOMIAL FOR INTERVAL (X(I-1),X(I)) AT XEVAL
      H=X(I)-X(I-1)
      S=(XEVAL-X(I-1))/H
      SCOMP = 1.0-S
      FACTOR=H*H/6.0
      A=C(I-1)*FACTOR
      B=C(I)*FACTOR
```

```
C
C   **FINAL RESULT
      SPLINE=(A*SCOMP*SCOMP+Y(I-1)-A)*SCOMP+(B*S*S+Y(I)-B)*S
C
      RETURN
      END
```

PROGRAM 3–4a
Fourier Least-Squares Approximation with Subtraction of Linear Tendency and
Odd Reflection: Coefficient Calculation

```
C   ************************************************************************
C   * COEFFICIENT CALCULATION FOR FOURIER LEAST-SQUARES APPROXIMATION     *
C   *   WITH LINEAR TENDENCY SUBTRACTION                                  *
C   * GIVEN ARE                                                           *
C   *     N, THE NUMBER OF DATA POINTS;                                   *
C   *     M, THE NUMBER OF SINE TERMS IN THE APPROXIMATION;               *
C   *     SINVAL(I,J), 1<=I<=N, 1<=J<=M, TABLE OF SINUSOID VALUES WHERE   *
C   *             VALUE (I,J) IS SIN(J*PI*I/(N-1)); AND                   *
C   *     Y(I), 1<=I<=N, DATA VALUES AT EQUALLY SPACED ARGUMENTS.         *
C   * OUTPUT PARAMETERS ARE                                               *
C   *     A(J), 1<=J<=M, THE SINUSOID COEFFICIENTS IN THE                 *
C   *             APPROXIMATION; AND                                      *
C   *     LINE0 AND LINE1, LINE COEFFICIENTS.                            *
C   ************************************************************************
C
C
      SUBROUTINE COEFF(N,M,SINVAL,Y,A,LINE0,LINE1)
C
C   **PARAMETERS
      INTEGER N,M
      REAL SINVAL(100,100),Y(100),A(100)
      REAL LINE0,LINE1
C
C   **INTERNAL VARIABLES
      REAL FACTOR
C                     **FACTOR IN COEFFICIENT CALCULATION
      INTEGER I,J
C                     **SUBSCRIPTS
      INTEGER NLESS1,NLESS2
C                     **LOOP INDEX LIMITS
C
C
C
C   **INITIALIZE LOOP LIMITS
      NLESS1=N-1
      NLESS2=N-2
C
C   **COMPUTE LINE COEFFICIENTS
      LINE0=Y(1)
      LINE1=(Y(N)-Y(1))/(N-1)
C
C   **SUBTRACT LINEAR TENDENCY FROM DATA VALUES
      DO 10 I=1,NLESS1
          Y(I)=Y(I)-(LINE0+LINE1*(I-1))
10    CONTINUE
C
C   **COMPUTE SINUSOID COEFFICIENTS
      FACTOR=2.0/(N-1)
      DO 30 J=1,M
         A(J)=0
         DO 20 I=1,NLESS2
             A(J)=A(J)+Y(I+1)*SINVAL(I,J)
20       CONTINUE
         A(J)=FACTOR*A(J)
```

```
30      CONTINUE
C
        RETURN
        END
```

PROGRAM 3–4b

Fourier Least-Squares Approximation with Subtraction of Linear Tendency and
Odd Reflection: Evaluation of Approximation at Argument

```
C  *********************************************************************************
C  * FOURIER LEAST SQUARES APPROXIMATION WITH LINEAR TENDENCY                      *
C  *     SUBTRACTION - EVALUATION OF APPROXIMATION AT AN ARGUMENT                  *
C  * GIVEN ARE                                                                     *
C  *     N, THE NUMBER OF DATA POINTS;                                             *
C  *     M, THE NUMBER OF SINE TERMS IN THE APPROXIMATION;                         *
C  *     A(J), 1<=J<=M, COEFFICIENTS OF THE SINE TERMS OF THE                      *
C  *           APPROXIMATION;                                                      *
C  *     LINE0 AND LINE1, LINE COEFFICIENTS;                                       *
C  *     X1, THE FIRST TABULAR ARGUMENT;                                           *
C  *     X, THE EVALUATION ARGUMENT; AND                                           *
C  *     H, THE TABULAR ARGUMENT INTERVAL WIDTH.                                   *
C  * THE APPROXIMATION AT X IS COMPUTED.                                           *
C  *********************************************************************************
C
C
       FUNCTION FOURI(N,M,A,LINE0,LINE1,X1,X,H)
C
C  **PARAMETERS
       INTEGER N,M
       REAL A(100)
       REAL LINE0,LINE1,X1,X,H
C
C  **INTERNAL VARIABLES
       REAL S
C                        **ARGUMENT IN UNITS OF H
       REAL SINTAB(100),COSTAB(100)
C                        **TABLES OF SINE AND COSINE OF
C                          (J*PI*(X-X1)/((N-1)*H)
       REAL APPROX
C                        **TEMPORARY APPROXIMATION RESULT
       INTEGER I
C                        **SUBSCRIPT
C
C
C  **INITIALIZE
       PI=3.141592
C
C  **COMPUTE LINE TERMS
       S=(X-X1)/H
       APPROX= LINE1*S + LINE0
C
C  **COMPUTE TABLE OF SINES AT S
       S=PI*S/(N-1)
       SINTAB(1)=SIN(S)
       COSTAB(1)=COS(S)
       DO 10 J=2,M
          SINTAB(J)=SINTAB(J-1)*COSTAB(1)+COSTAB(J-1)*SINTAB(1)
          COSTAB(J)=COSTAB(J-1)*COSTAB(1)-SINTAB(J-1)*SINTAB(1)
10     CONTINUE
C
C  **COMPUTE SINE TERMS
```

```
      DO 20 J=1,M
         APPROX=APPROX+A(J)*SINTAB(J)
20    CONTINUE
C
C  **FINAL RESULT
      FOURI=APPROX
C
      RETURN
      END
```

PROGRAM 3-5
Trapezoidal Rule for Numerical Integration

```
C  *******************************************************************************
C  * TRAPEZOIDAL RULE FOR NUMERICAL INTEGRATION OF F(X) OVER AN                  *
C  *      INTERVAL, XMIN TO XMAX                                                 *
C  * GIVEN ARE                                                                   *
C  *      F, THE FUNCTION TO BE INTEGRATED;                                      *
C  *      N, THE NUMBER OF TABULAR ARGUMENTS; AND                               *
C  *      XMIN AND XMAX, THE LIMITS OF INTEGRATION.                              *
C  * AN APPROXIMATION TO THE INTEGRAL IS COMPUTED.                              *
C  *******************************************************************************
C
C
      FUNCTION TRAP(F,N,XMIN, XMAX)
C
C  **PARAMETERS
      REAL F
      INTEGER N
      REAL XMIN,XMAX
C
C  ** INTERNAL VARIABLES
      REAL H
C                        **TABULAR ARGUMENT WIDTH
      REAL X
C                        **TABULAR ARGUMENT
      REAL FSUM
C                        **WEIGHTED SUM OF F(X) VALUES
      INTEGER NLESS1
C                        **LOOP LIMIT (=N-1)
      INTEGER I
C                        **LOOP COUNTER
C
C
C  **INITIALIZE
      NLESS1=N-1
      H=(XMAX-XMIN)/N
      FSUM=.5*(F(XMIN) + F(XMAX))
      X=XMIN
C
C  **COMPUTE WEIGHTED SUM OF TABULAR VALUES
      DO 10 I=1,NLESS1
         X=X+H
         FSUM=FSUM + F(X)
10    CONTINUE
C
C  **NORMALIZE INTEGRAL VALUE
      TRAP=H*FSUM
C
      RETURN
      END
```

PROGRAM 3–6
Simpson's Rule for Numerical Integration

```
C ********************************************************************************
C * SIMPSON'S RULE FOR NUMERICAL INTEGRATION OF F(X) OVER AN              *
C *       INTERVAL, XMIN TO XMAX                                          *
C * GIVEN ARE                                                            *
C *     F, THE FUNCTION TO BE INTEGRATED;                                *
C *     N, THE NUMBER OF TABULAR ARGUMENTS (N IS ASSUMED EVEN AND AT     *
C *         LEAST 4); AND                                                *
C *     XMIN AND XMAX, THE LIMITS OF INTEGRATION.                        *
C * AN APPROXIMATION TO THE INTEGRAL IS COMPUTED.                        *
C ********************************************************************************
C
C
      FUNCTION SIMP(F,N,XMIN, XMAX)
C
C **PARAMETERS
      REAL F
      INTEGER N
      REAL XMIN,XMAX
C
C ** INTERNAL VARIABLES
      INTEGER M
C                          **NUMBER OF PANELS
      INTEGER MLESS1
C                          **LOOP LIMIT (=M-1)
      REAL H
C                          **TABULAR ARGUMENT WIDTH
      REAL X
C                          **TABULAR ARGUMENT
      REAL EVNSUM,ODDSUM
C                          **SUMS OF EVEN AND ODD TERMS
      INTEGER I
C                          **LOOP COUNTER
C
C
C **INITIALIZE
      M=N/2
      MLESS1=M-1
      H=(XMAX-XMIN)/N
      ODDSUM=0
      EVNSUM=0
      X=XMIN
C
C **COMPUTE SUMS OF EVEN AND ODD TERMS
      DO 10 I=1,MLESS1
         X=X+2*H
         ODDSUM = ODDSUM + F(X-H)
         EVNSUM = EVNSUM + F(X)
10    CONTINUE
      ODDSUM = ODDSUM + F(X+H)
C
C **WEIGHT THE SUMS AND NORMALIZE INTEGRAL VALUE
      SIMP=(H/3)*(F(XMIN) + 4*ODDSUM + 2*EVNSUM +F(XMAX))
C
      RETURN
      END
```

PROGRAM 3–7
Euler's Method for Solving First-Order
Ordinary Differential Equation

```
C ****************************************************************************
C * EULER'S METHOD FOR SOLVING A FIRST-ORDER ORDINARY DIFFERENTIAL           *
C *     EQUATION: Y'(X)=YPR1ME(X,Y) AND Y(XO)=YO                             *
C * GIVEN ARE                                                                *
C *     YPRIME(X,Y), THE DERIVATIVE FUNCTION OF Y(X);                        *
C *     XO AND YO, THE INITIAL X AND Y VALUES;                              *
C *     H, STEP SIZE; AND                                                    *
C *     N, THE NUMBER OF STEPS.                                             *
C * RETURNED ARE                                                            *
C *     Y(I), 1<=I<=N+1, THE SOLUTION VALUES.                               *
C ****************************************************************************
C
C
      SUBROUTINE EULER(YPRIME,XO,YO,H,N,Y)
C
C **PARAMETERS
      REAL YPRIME
      REAL XO,YO,H
      INTEGER N
      REAL Y(100)
C
C  **INTERNAL VARIABLES
      REAL XI
C                        **TABULAR ARGUMENT
      INTEGER I
C                        **SUBSCRIPT USED WITH SOLUTION ARRAY
C
C  **INITIALIZE
      Y(1)=YO
      XI=XO
C
C  **RECURRENTLY COMPUTE Y(I) VALUES
      DO 10 I=1,N
         Y(I+1)=Y(I)+H*YPRIME(XI,Y(I))
         XI=XI+H
10    CONTINUE
C
      RETURN
      END
```

PROGRAM 3–8
Modified Euler-Heun Predictor-Corrector Method for Solving First-Order
Ordinary Differential Equation

```
C ****************************************************************************
C * MODIFIED EULER-HEUN METHOD FOR SOLVING A FIRST-ORDER ORDINARY            *
C *     DIFFERENTIAL EQUATION: Y'(X)=YPRIME(X,Y) AND Y(XO)=YO                *
C * GIVEN ARE                                                                *
C *     YPRIME(X,Y), THE DERIVATIVE OF Y(X);                                *
C *     XO AND YO, THE INITIAL X AND Y VALUES;                              *
C *     H, THE STEP SIZE; AND                                               *
C *     N, THE NUMBER OF STEPS.                                             *
C * RETURNED ARE                                                            *
C *     Y(I), 1<=I<=N+1, THE SOLUTION VALUES.                               *
C ****************************************************************************
C
C
      SUBROUTINE EULHN(YPRIME,XO,YO,H,N,Y)
```

```
C
C **PARAMETERS
      REAL YPRIME
      REAL X0,Y0,H
      INTEGER N
      REAL Y(100)
C
C  **INTERNAL VARIABLES
      REAL XI
C                      **TABULAR ARGUMENT
      REAL LFDER,RTDER
C                      **DERIVATIVES AT LEFT AND RIGHT ENDS OF INTERVAL
      REAL PRED
C                      **APPROXIMATE VALUE AT RIGHT END
      INTEGER I
C                      **SUBSCRIPT USED WITH SOLUTION ARRAY
C
C  **INITIALIZE
      Y(1)=Y0
C
C **COMPUTE FIRST STEP USING SIMPLE EULER PREDICTION
      LFDER=YPRIME(X0,Y0)
      PRED=Y0+H*LFDER
      RTDER=YPRIME(X0+H,PRED)
      Y(2)=Y0+H*(LFDER+RTDER)/2
      XI=X0+H
C
C  **RECURRENTLY COMPUTE Y(I) VALUES
      DO 10 I=2,N
         LFDER=YPRIME(XI,Y(I))
         PRED=Y(I-1)+2*H*LFDER
         RTDER=YPRIME(XI+H,PRED)
         Y(I+1)=Y(I)+H*(LFDER+RTDER)/2
         XI=XI+H
10    CONTINUE
C
      RETURN
      END
```

PROGRAM 3–9
Bisection Method for Solving Nonlinear Equations

```
C ************************************************************************************
C * BISECTION METHOD FOR FINDING ROOTS OF EQUATIONS                                  *
C * GIVEN ARE                                                                        *
C *     F, THE FUNCTION FOR WHICH THE ROOT IS DESIRED;                               *
C *     LGUESS AND RGUESS, ESTIMATES OF THE ROOT SUCH THAT                           *
C *           LGUESS < ROOT AND RGUESS > ROOT; AND                                   *
C *     TOL, A THRESHOLD ON THE ERROR BOUND ON THE ROOT                              *
C *           ESTIMATE.                                                              *
C ************************************************************************************
C
C
      FUNCTION BISECT(F,LGUESS,RGUESS,TOL)
C
C **PARAMETERS
      REAL F
      REAL LGUESS,RGUESS,TOL
C
C **INTERNAL VARIABLES
      REAL LSIGN
C             **SIGN OF F(LGUESS)
      REAL XNEW
C             **NEW ROOT ESTIMATE
```

```
C
C **INITIALIZE
      LSIGN=SIGN(1.0,F(LGUESS))
C
C **ITERATE IMPROVEMENT UNTIL DESIRED ACCURACY IS REACHED
10    CONTINUE
C     **COMPUTE IMPROVED ESTIMATE OF ROOT
         XNEW=(LGUESS+RGUESS)/2
C     **REPLACE THE GUESS AT WHICH F AGREES IN SIGN WITH THE
C     **  SIGN OF XNEW WITH THE VALUE OF XNEW
         IF (SIGN(1.0,F(XNEW)).EQ.LSIGN) GOTO 20
            RGUESS=XNEW
            GOTO30
20          LGUESS=XNEW
30    IF(((RGUESS-LGUESS)/2).GT.TOL) GOTO 10
C
C **RETURN FINAL RESULT
      BISECT=XNEW
      RETURN
      END
```

PROGRAM 3–10
Method of False Position for Solving Nonlinear Equations

```
C ******************************************************************************
C * METHOD OF FALSE POSITION FOR FINDING ROOTS OF EQUATIONS                    *
C * GIVEN ARE                                                                  *
C *    F, THE FUNCTION FOR WHICH THE ROOT IS DESIRED;                          *
C *    LGUESS AND RGUESS, ESTIMATES OF THE ROOT SUCH THAT                      *
C *          LGUESS < ROOT AND RGUESS > ROOT; AND                             *
C *    TOL, A THRESHOLD ON THE RELATIVE DIFFERENCE BETWEEN SUCCESSIVE          *
C *          ROOT ESTIMATES.                                                   *
C ******************************************************************************
C
C
      FUNCTION FALSE(F,LGUESS,RGUESS,TOL)
C
C **PARAMETERS
      REAL F
      REAL LGUESS,RGUESS,TOL
C
C **INTERNAL VARIABLES
      REAL LSIGN
C              **SIGN OF F(LGUESS)
      REAL NEWRT,OLDRT
C              **MOST RECENT AND NEXT MOST RECENT ROOT ESTIMATES
      REAL LVAL,RVAL,NEWVAL
C              **F(LGUESS),F(RGUESS),F(NEWRT)
      REAL SLOPE
C              **SLOPE OF LINE THROUGH F AT LGUESS AND RGUESS
C
C **INITIALIZE
      OLDRT=LGUESS
      NEWRT=RGUESS
      LVAL=F(LGUESS)
      RVAL=F(RGUESS)
      LSIGN=SIGN(1.0,LVAL)
C
C **ITERATE IMPROVEMENT UNTIL DESIRED ACCURACY IS REACHED
10    IF (ABS((OLDRT-NEWRT)/OLDRT).LE.TOL) GOTO 40
C     **COMPUTE IMPROVED ESTIMATE OF ROOT
         OLDRT=NEWRT
         SLOPE=(LVAL-RVAL)/(LGUESS-RGUESS)
         NEWRT=LGUESS-LVAL/SLOPE
```

```
C     **REPLACE THE GUESS AT WHICH F AGREES IN SIGN WITH THE
C      *SIGN OF XNEW WITH THE VALUE OF XNEW
       NEWVAL=F(NEWRT)
       IF (SIGN(1.0,NEWVAL).EQ.LSIGN) GOTO 20
          RGUESS=NEWRT
          RVAL=NEWVAL
          GOTO30
20        LGUESS=NEWRT
          LVAL=NEWVAL
30     GOTO 10
40     CONTINUE
C
C **RETURN FINAL RESULT
       FALSE=NEWRT
       RETURN
       END
```

PROGRAM 3–11

Newton's Method for Solving Nonlinear Equations

```
C *********************************************************************************
C * NEWTON'S METHOD FOR FINDING ROOTS OF EQUATIONS FOR ONE                        *
C *      EQUATION IN ONE UNKNOWN                                                  *
C * GIVEN ARE                                                                     *
C *    F, THE FUNCTION FOR WHICH THE ROOT IS DESIRED;                             *
C *    FPRIME, THE FUNCTION GIVING THE DERIVATIVE OF F;                           *
C *    XGUESS, A GUESS AT THE ROOT; AND                                           *
C *    TOL, A THRESHOLD ON THE RELATIVE DIFFERENCE BETWEEN                        *
C *         SUCCESSIVE ROOT ESTIMATES.                                            *
C *********************************************************************************
C
C
       FUNCTION NEWTON(F,FPRIME,XGUESS,TOL)
C
C **PARAMETERS
       REAL F,FPRIME
       REAL XGUESS,TOL
C
C **INTERNAL VARIABLES
       REAL XNEW
C                              **IMPROVED ESTIMATE OF ROOT
C
C
C **INITIALIZE
       XNEW=XGUESS-F(XGUESS)/FPRIME(XGUESS)
C
C **ITERATE IMPROVEMENT UNTIL DESIRED ACCURACY IS ACHIEVED
10     IF (ABS((XNEW-XGUESS)/XNEW).LE.TOL) GOTO 20
C         **COMPUTE IMPROVED ESTIMATE OF ROOT
          XGUESS=XNEW
          XNEW=XGUESS-F(XGUESS)/FPRIME(XGUESS)
       GOTO 10
20     CONTINUE
C
C **RETURN THE FINAL RESULT
       NEWTON=XNEW
       RETURN
       END
```

PROGRAM 3–12
Secant Method for Solving Nonlinear Equations

```
C  ***************************************************************************
C  * SECANT METHOD FOR FINDING ROOTS OF EQUATIONS FOR ONE                   *
C  *     EQUATION IN ONE UNKNOWN                                            *
C  * GIVEN ARE                                                             *
C  *     F, THE FUNCTION FOR WHICH THE ROOT IS DESIRED;                    *
C  *     NEWG AND OLDG, TWO GUESSES AT THE ROOT; AND                       *
C  *     TOL, A THRESHOLD ON THE RELATIVE DIFFERENCE BETWEEN               *
C  *            SUCCESSIVE ROOT ESTIMATES.                                 *
C  ***************************************************************************
C
C
       FUNCTION SECANT(F,NEWG,OLDG,TOL)
C
C  **PARAMETERS
       REAL F
       REAL NEWG,OLDG
       REAL TOL
C
C  **INTERNAL VARIABLES
       REAL NEWVAL,OLDVAL
C                         **VALUE OF F AT NEWG AND OLDG
       REAL XNEW
C                         **IMPROVED ROOT ESTIMATE
       REAL SLOPE
C                         **SLOPE OF SECANT
C
C
C  **INITIALIZE
       OLDVAL=F(OLDG)
C
C  **ITERATE IMPROVEMENT UNTIL DESIRED ACCURACY IS ACHIEVED
10     CONTINUE
         NEWVAL=F(NEWG)
C        **COMPUTE SLOPE OF SECANT THROUGH NEWG AND OLDG
           SLOPE=(NEWVAL-OLDVAL)/(NEWG-OLDG)
C        **COMPUTE IMPROVED ROOT ESTIMATE
           XNEW=NEWG-NEWVAL/SLOPE
C        **MAKE NEWG AND OLDG THE TWO MOST RECENT ROOT ESTIMATES
           OLDG=NEWG
           NEWG=XNEW
           OLDVAL=NEWVAL
       IF (ABS((NEWG-OLDG)/OLDG).GT.TOL) GOTO 10
C
C  **RETURN THE FINAL RESULT
       SECANT=XNEW
       RETURN
       END
```

PROGRAM 3–13
Hybrid Secant-False Position Method

```
C  ********************************************************************************
C  * HYBRID SECANT-FALSE POSITION METHOD FOR FINDING ROOTS OF ONE                *
C  *     EQUATION IN ONE UNKNOWN: F(X)=0                                          *
C  * GIVEN ARE                                                                    *
C  *     F, THE FUNCTION FOR WHICH THE ROOT IS DESIRED;                           *
C  *     LGUESS AND RGUESS, ROOT ESTIMATES WHERE LGUESS < ROOT                    *
C  *           AND RGUESS > ROOT; AND                                             *
C  *     TOL, A THRESHOLD ON THE RELATIVE DIFFERENCE BETWEEN                      *
C  *           ROOT ESTIMATES.                                                    *
C  ********************************************************************************
C
C
       FUNCTION HYBRID(F,LGUESS,RGUESS,TOL)
C
C  **PARAMETERS
       REAL F
       REAL LGUESS,RGUESS
       REAL TOL
C
C  **INTERNAL VARIABLES
       REAL XNEW
C                          **NEW ROOT ESTIMATE
       REAL XOLD,XPREV
C                          **NEXT TWO MOST RECENT ROOT ESTIMATES
       REAL LFTVAL,RTVAL
C                          **F(LGUESS),F(RGUESS)
       REAL NEWVAL,OLDVAL
C                          **F(XNEW),F(XOLD)
       REAL PREVAL
C                          **F(XPREV)
       REAL SLOPE
C                          **SLOPE OF LINE THROUGH F
       INTEGER LSIGN
C                          **SIGN OF F(LGUESS)
C
C
C
C  **INITIALIZE
       XOLD=LGUESS
       XPREV=RGUESS
       LFTVAL=F(LGUESS)
       RTVAL=F(RGUESS)
       LSIGN=SIGN(1.0,LFTVAL)
       OLDVAL=LFTVAL
       PREVAL=RTVAL
C
C  **ITERATE IMPROVEMENT UNTIL DESIRED ACCURACY IS ACHIEVED
10     CONTINUE
C          **COMPUTE IMPROVED ESTIMATE OF ROOT BY SECANT METHOD
           SLOPE=(OLDVAL-PREVAL)/(XOLD-XPREV)
           XNEW=XOLD-OLDVAL/SLOPE
         IF ((XNEW.LE.RGUESS).AND.(XNEW.GE.LGUESS)) GOTO 20
C          **SECANT RESULT NO GOOD; USE FALSE POSITION
           SLOPE=(LFTVAL-RTVAL)/(LGUESS-RGUESS)
           XNEW=LGUESS-LFTVAL/SLOPE
20         CONTINUE
C          **UPDATE RECENT ROOT ESTIMATES AND FUNCTION VALUES
           XPREV=XOLD
           XOLD=XNEW
           PREVAL=OLDVAL
           NEWVAL=F(XNEW)
           OLDVAL=NEWVAL
```

```
C        **TAKE XNEW IN PLACE OF THE ONE OF LGUESS AND RGUESS
C        **  FOR WHICH F AGREES IN SIGN WITH F AT XNEW
         IF (SIGN(1.0,NEWVAL).EQ.LSIGN) GOTO 30
             RGUESS=XNEW
             RTVAL=NEWVAL
             GOTO 40
30           CONTINUE
             LGUESS=XNEW
             LFTVAL=NEWVAL
40           CONTINUE
         IF (ABS((XOLD-XPREV)/XOLD).GT.TOL) GOTO 10
C
C **RETURN THE FINAL RESULT
         HYBRID=XNEW
         RETURN
         END
```

PROGRAM 3–14
Elementary Algorithm for Gaussian Elimination

```
C ********************************************************************************
C * SOLVE AX=B; SET OF N LINEAR EQUATIONS IN N UNKNOWNS                          *
C * GIVEN ARE                                                                    *
C *    N, THE NUMBER OF EQUATIONS AND UNKNOWNS (<=100);                          *
C *    A, THE MATRIX OF COEFFICIENTS; AND                                        *
C *    B, THE N-VECTOR CONTAINING THE RIGHT-HAND-SIDE VALUES OF                  *
C *          THE EQUATIONS.                                                       *
C * OUTPUT PARAMETER IS                                                          *
C *    X, THE SOLUTION VECTOR.                                                   *
C ********************************************************************************
C
C
      SUBROUTINE GAUSS(N,A,B,X)
C
C **PARAMETERS
      INTEGER N
      REAL A(100,100)
      REAL B(100),X(100)
C
C **INTERNAL VARIABLES
      REAL AMULT
C                    **ROW MULTIPLIER
      INTEGER I,J,K
C                    **ROW AND COLUMN INDICATORS
      INTEGER IPLUS1,NLESS1
C                    **LOOP DELIMITERS (=I+1 AND N-1)
      INTEGER ICOMP
C                    **LOOP COUNTER (I+ICOMP=N)
C
C
C **INITIALIZE
      NLESS1=N-1
C
C **TRIANGULARIZE
C **ELIMINATE THE I-TH COLUMN BELOW THE I-TH ROW
      DO 30 I=1,NLESS1
         IPLUS1= I+1
         DO 20 J=IPLUS1,N
C            **COMPUTE MULTIPLIER FOR J-TH ROW
             AMULT=A(J,I)/A(I,I)
```

```
C                    **COMPUTE NONZERO ELEMENTS OF J-TH ROW
                     DO 10 K=IPLUS1,N
                        A(J,K)=A(J,K)-AMULT*A(I,K)
10                   CONTINUE
C                    **COMPUTE NEW B(J)
                     B(J)=B(J)-AMULT*B(I)
20            CONTINUE
30      CONTINUE
C
C **BACK SUBSTITUTE
        DO 50 ICOMP=1,N
          I=N-ICOMP+1
          X(I)=B(I)
C             **SUBTRACT TERMS IN ALREADY COMPUTED X(J)
              IPLUS1=I+1
              DO 40 J=IPLUS1,N
                 X(I)=X(I)-A(I,J)*X(J)
40            CONTINUE
C             **SOLVE FOR X(I)
              X(I)=X(I)/A(I,I)
50      CONTINUE
C
        RETURN
        END
```

PROGRAM 3-15
Jacobi Method for Solution of Linear Equations

```
C  ********************************************************************************
C  * SOLVE AX=B; SET OF N LINEAR EQUATIONS IN N UNKNOWNS BY THE JACOBI METHOD   *
C  *                                                                            *
C  * GIVEN ARE                                                                  *
C  *    N, THE NUMBER OF EQUATIONS AND UNKNOWNS (<=100);                        *
C  *    A, THE N X N MATRIX OF COEFFICIENTS;                                    *
C  *    B, THE N-VECTOR CONTAINING THE RIGHT-HAND-SIDE VALUES                   *
C  *        OF THE EQUATIONS;                                                   *
C  *    XOLD, A GUESS AT THE SOLUTION VECTOR; AND                              *
C  *    TOL, A TOLERANCE FOR STOPPING ITERATION. THE TEST FOR                  *
C  *        CONVERGENCE IS ONE OF MANY POSSIBILITIES:  IS THE SIZE             *
C  *        OF THE VECTOR OF DIFFERENCES BETWEEN THE OLD AND NEW               *
C  *        VECTORS SMALL ENOUGH RELATIVE TO THE SIZE OF THE NEW               *
C  *        VECTOR?  THE SIZE OF A VECTOR IS TAKEN TO BE THE                   *
C  *        MAXIMUM ELEMENT MAGNITUDE.                                          *
C  * OUTPUT PARAMETER IS                                                        *
C  *    XNEW, THE SOLUTION VECTOR.                                             *
C  ********************************************************************************
C
C
        SUBROUTINE JACOBI(N,A,B,XOLD,XNEW,TOL)
C
C **PARAMETERS
        INTEGER N
        REAL A(100,100)
        REAL B(100),XOLD(100),XNEW(100)
        REAL TOL
C
C **INTERNAL VARIABLES
        REAL C(100,100)
C                      **NORMALIZED "A" MATRIX
        REAL D(100)
C                      **NORMALIZED "D" MATRIX
        REAL MAXDIF,DIF,MAXNEW,NEW
C                      **USED WITH ITERATION STOPPING CRITERION
        INTEGER J,K
C                      **ROW AND COLUMN INDICATORS
```

```
C
C
C **NORMALIZE MATRIX
      DO 20 J=1,N
         DO 10 K=1,N
            IF (K.NE.J) C(J,K)=A(J,K)/A(J,J)
10       CONTINUE
         D(J)=B(J)/A(J,J)
20    CONTINUE
C
C **ITERATE IMPROVEMENT UNTIL REQUIRED ACCURACY IS ACHIEVED
C **COMPUTE NEW ESTIMATE
30    CONTINUE
         MAXNEW=0
         MAXDIF=0
         DO 50 J=1,N
            XNEW(J)=D(J)
            DO 40 K=1,N
               IF (K.NE.J) XNEW(J)=XNEW(J)-C(J,K)*XOLD(K)
40          CONTINUE
C           **FIND MAXIMUM ABSOLUTE DIFFERENCE BETWEEN OLD AND
C           **  NEW ELEMENTS
            DIFF=ABS(XNEW(J)-XOLD(J))
            IF (DIFF.GT.MAXDIF) MAXDIF=DIF
C           **FIND MAXIMUM NEW ELEMENT VALUE
            NEW=ABS(XNEW(J))
            IF (NEW.GT.MAXNEW) MAXNEW=NEW
50       CONTINUE
C        **LET PRESENT ESTIMATE BE IMPROVED ESTIMATE
         DO 60 J=1,N
            XOLD(J)=XNEW(J)
60       CONTINUE
      IF (MAXDIF/MAXNEW.GT.TOL) GOTO 30
C
      RETURN
      END
```

PROGRAM 3–16
Gauss-Seidel Method for Solution of Linear Equations

```
C ********************************************************************************
C * SOLVE AX=B; SET OF N LINEAR EQUATIONS IN N UNKNOWNS BY THE          *
C *      GAUSS-SEIDEL METHOD                                            *
C * GIVEN ARE                                                           *
C *    N, THE NUMBER OF EQUATIONS AND UNKNOWNS (<=100);                 *
C *    A, THE MATRIX OF COEFFICIENTS;                                   *
C *    B, A VECTOR CONTAINING THE RIGHT-HAND-SIDE VALUES OF THE         *
C *         EQUATIONS;                                                  *
C *    X, A GUESS AT THE SOLUTION VECTOR WHICH CHANGES AS THE           *
C *         GUESS IMPROVES AND HOLDS THE SOLUTION AT THE TIME OF        *
C *         RETURN; AND                                                 *
C *    TOL, A TOLERANCE FOR STOPPING ITERATION. THE TEST FOR           *
C *         CONVERGENCE IS ONE OF MANY POSSIBILITIES:  IS THE SIZE     *
C *         OF THE VECTOR OF DIFFERENCES BETWEEN THE OLD AND NEW       *
C *         VECTORS SMALL ENOUGH RELATIVE TO THE SIZE OF THE NEW       *
C *         VECTOR?  THE SIZE OF A VECTOR IS TAKEN TO BE THE           *
C *         MAXIMUM ELEMENT MAGNITUDE.                                  *
C ********************************************************************************
C
C
      SUBROUTINE GAUSEI(N,A,B,X,TOL)
C
C **PARAMETERS
      INTEGER N
      REAL A(100,100)
```

```
      REAL B(100),X(100)
      REAL TOL
C
C **INTERNAL VARIABLES
      REAL C(100,100)
C                    **NORMALIZED "A" MATRIX
      REAL D(100)
C                    **NORMALIZED "D" MATRIX
      REAL XCOMP
C                    **NEW ELEMENT ESTIMATE
      REAL MAXDIF,DIF,MAXNEW,NEW
C                    **USED WITH ITERATION STOPPING CRITERION
      INTEGER J,K
C                    **ROW AND COLUMN INDICATORS
C
C
C **NORMALIZE MATRIX
      DO 20 J=1,N
         DO 10 K=1,N
            IF (K.NE.J) C(J,K)=A(J,K)/A(J,J)
10       CONTINUE
         D(J)=B(J)/A(J,J)
20    CONTINUE
C
C **ITERATE IMPROVEMENT UNTIL REQUIRED ACCURACY IS ACHIEVED
30    CONTINUE
C        **COMPUTE NEW ESTIMATE
         MAXNEW=0
         MAXDIF=0
         DO 50 J=1,N
            XCOMP=D(J)
            DO 40 K=1,N
               IF (K.NE.J) XCOMP=XCOMP-C(J,K)*X(K)
40          CONTINUE
C           **FIND MAXIMUM ABSOLUTE DIFFERENCE BETWEEN OLD AND
C           **   NEW ELEMENTS
            DIFF=ABS(XCOMP-X(J))
            IF (DIFF.GT.MAXDIF) MAXDIF=DIF
C           **USE NEW VALUE IMMEDIATELY
            X(J)=XCOMP
C           **FIND MAXIMUM NEW ELEMENT VALUE
            NEW=ABS(XCOMP)
            IF (NEW.GT.MAXNEW) MAXNEW=NEW
50       CONTINUE
      IF (MAXDIF/MAXNEW.GT.TOL) GOTO 30
C
      RETURN
      END
```

PROGRAM 6–1
Gaussian Elimination with Scaling and Pivoting

```
C *********************************************************************************
C * SOLVE AX=B; SET OF N LINEAR EQUATIONS IN N UNKNOWNS BY                        *
C *     GAUSSIAN ELIMINATION WITH SCALING AND PIVOTING                           *
C * GIVEN ARE                                                                     *
C *   N, THE NUMBER OF EQUATIONS AND UNKNOWNS (<=100);                            *
C *   A, THE N X N MATRIX OF COEFFICIENTS; AND                                    *
C *   B, THE N-VECTOR CONTAINING THE RIGHT-HAND-SIDE VALUES                       *
C *       OF THE EQUATIONS.                                                       *
C * OUTPUT PARAMETER IS                                                           *
C *   X, THE SOLUTION VECTOR.                                                     *
C *********************************************************************************
C
C
      SUBROUTINE GAUSSP(N,A,B,X)
```

```
C
C **PARAMETERS
      INTEGER N
      REAL A(100,100)
      REAL B(100),X(100)
C
C **INTERNAL VARIABLES
      REAL SCALE(100)
C                     **SCALING VECTOR
      INTEGER P(100)
C                     **PERMUTATION VECTOR
      REAL MAX
C                     **LARGEST SCALED PIVOT MAGNITUDE
      REAL MAXROW
C                     **ROW CONTAINING MAX
      REAL SMAG
C                     **MAGNITUDE OF SCALED PIVOT CANDIDATE
      REAL AMULT
C                     **ROW MULTIPLIER
      INTEGER TEMP
C                     **PERMUTATION VECTOR ELEMENT BEING SWITCHED
      INTEGER ICOMP
C                     **LOOP INDEX (I+ICOMP=N)
      INTEGER IPLUS1,NLESS1
C                     **LOOP LIMITS (=I+1 AND N-1)
      INTEGER I,J,K
C                     **ROW AND COLUMN INDICATORS
C
C
C **INITIALIZE
      NLESS1=N-1
      DO 20 I=1,N
C         **INITIALIZE PERMUTATION VECTOR
          P(I)=I
C         **COMPUTE SUM OF ROW MAGNITUDES
          SCALE(I)=0
          DO 10 J=1,N
              SCALE(I)=SCALE(I) + ABS(A(I,J))
10        CONTINUE
20    CONTINUE
C
C **TRIANGULARIZE WITH PIVOTING
      DO 70 I=1,NLESS1
C         **FIND PIVOT
          MAX=0
          DO 40 J=I,N
              SMAG=ABS(A(P(J),I)/SCALE(P(J)))
              IF (SMAG.LE.MAX) GOTO 30
                  MAX=SMAG
                  MAXROW=J
30            CONTINUE
40        CONTINUE
C         **SWITCH ROWS
          TEMP=P(MAXROW)
          P(MAXROW)=P(I)
          P(I)=TEMP
C         **ELIMINATE SUBDIAGONAL ELEMENTS IN I-TH COLUMN
          IPLUS1=I+1
          DO 60 J=IPLUS1,N
C             **COMPUTE NONZERO ELEMENT OF J-TH ROW
              AMULT=A(P(J),I)/A(P(I),I)
              DO 50 K=IPLUS1,N
                  A(P(J),K)=A(P(J),K) - AMULT*A(P(I),K)
50            CONTINUE
              B(P(J))=B(P(J)) - AMULT*B(P(I))
60        CONTINUE
70    CONTINUE
```

```
C
C **BACK SUBSTITUTE
      DO 100 ICOMP=1,N
         I=N-ICOMP+1
         X(I)=B(P(I))
C        **SUBTRACT TERMS IN ALREADY COMPUTED X(J)
            IPLUS1 = I+1
            IF (IPLUS1.EQ.N) GOTO 90
              DO 80 J=IPLUS1,N
                 X(I)=X(I) - A(P(I),J)*X(J)
80            CONTINUE
90            CONTINUE
C        **SOLVE FOR X(I)
            X(I)=X(I)/A(P(I),I)
100   CONTINUE
C
      RETURN
      END
```

PROGRAM 7–1
Romberg Integration

```
C ********************************************************************************
C * ROMBERG INTEGRATION TO INTEGRATE F(X) FROM XMIN TO XMAX                      *
C * GIVEN ARE                                                                    *
C *     F, THE FUNCTION TO BE INTEGRATED;                                        *
C *     XMIN AND XMAX, THE LIMITS OF INTEGRATION;                                *
C *     MAXIT, THE MAXIMUM NUMBER OF ITERATIONS (ASSUMED <=100); AND             *
C *     TOL, AN ERROR TOLERANCE.                                                 *
C ********************************************************************************
C
C
      FUNCTION ROMB(F,XMIN,XMAX,TOL)
C
C **PARAMETERS
      REAL F
      REAL XMIN,XMAX
      INTEGER MAXIT
      REAL TOL
C
C **INTERNAL VARIABLES
      REAL H
C                 **INTERVAL WIDTH
      REAL ARG
C                 **EVALUATION ARGUMENT
      REAL ROMROW(100)
C                 **LAST ROW OF TABLE
      REAL NEWTAB, OLD1, OLD2
C                 **TEMPORARY TABLE ENTRIES
      REAL DENOM
C                 **4, 3, 2, OR 1
      INTEGER N
C                 **(XMAX-XMIN)/H
      INTEGER K
C                 **LAST TABLE ROW COMPUTED
      INTEGER I,J
C                 **LOOP COUNTERS
C
C
C **INITIALIZE
      H=XMAX-XMIN
      N=1
C     **SIMPLE TRAPEZOIDAL RULE
      ROMROW(1)=(H/2)*(F(XMAX)+F(XMIN))
      K=1
```

```
C
C **COMPUTE AS MANY TABLE ROWS AS NECESSARY
10    CONTINUE
          OLD1=ROMROW(1)
C         **COMPUTE NEW TRAPEZOIDAL RULE FOR INTERVAL H/2
              NEWTAB=0
              ARG=XMIN+H/2
              DO 20 J=1,N
                  NEWTAB=NEWTAB+F(ARG)
                  ARG=ARG+H
20            CONTINUE
              ROMROW(1)=(OLD1+H*NEWTAB)/2
C         **COMPUTE REMAINDER OF K-TH ROW
              DENOM=0
              IF (K.LT.2) GOTO 40
              DO 30 I=2,K
C                 **SAVE VALUE IN PREVIOUS ROW
                      OLD2=ROMROW(I)
                  ROMROW(I)=ROMROW(I-1)+(ROMROW(I-1)-OLD1)/(DENOM-1)
                  OLD1=OLD2
                  DENOM=DENOM+4
30            CONTINUE
40            ROMROW(K+1)=ROMROW(K)+(ROMROW(K)-OLD1)/(DENOM-1)
C         **SET UP NEXT ROW COMPUTATION
              K=K+1
              H=H/2
              N=2*N
      IF((ABS(ROMROW(K)-ROMROW(K-1)).GE.TOL).OR.(K.LT.MAXIT)) GOTO 10
C
C **RETURN FINAL RESULT
      ROMB=ROMROW(K)
C
      RETURN
      END
```

Index

Problem numbers appear in **boldface.**